CLOSED
TELEV
FOR TECH

Volu

Industrial and Co
(Principles and Circuits)

K. J. BOHLMAN
T.Eng(C.E.I.), F.S.E.R.T., A.M.Inst.E

Senior Lecturer in Television
Lincoln College of Technology

LONDON
NORMAN PRICE (PUBLISHERS) LTD

NORMAN PRICE (PUBLISHERS) LTD
17 TOTTENHAM COURT ROAD, LONDON, W1P 9DP

© NORMAN PRICE (PUBLISHERS) LTD., 1978

ISBN 0 85380 100 2

Printed in Great Britain by
A. BROWN & SONS LTD., *Hull*

AUTHOR'S PREFACE

This book is based on course material given to technicians in industry and was prompted by the lack of published literature on the subject.

The text is aimed at the technician who is required to adjust and maintain CCTV equipment in industrial and commercial applications. Essentially, this volume is non-mathematical but a few simple calculations have been included to assist the reader in grasping the basic principles of the important topics.

In this first volume the emphasis is towards monochrome television equipment which is in more general use in industry. A later volume will deal with the implications of colour.

It is assumed that the reader has a basic knowledge of semiconductor devices and electronic circuits. However, the elementary principles of the circuits peculiar to CCTV cameras and monitors have been given special attention. The needs of the domestic television engineer who is considering entering the CCTV field have been catered for as well as the requirements of allied technicians in industry who are often called upon to maintain this specialized equipment.

CCTV is a vast and complicated subject and it is not possible in a book of this nature fully to explore all of the topics. Anyone studying the subject for the first time must first obtain a good understanding of an elementary system and this is the main aim of this book.

I am most grateful for the valuable discussions and assistance given by colleagues at the college and for the kind assistance of manufacturers who have supplied me with photographs and circuit diagrams of their products.

CONTENTS

1	Introduction and Basic Elements of Light	*page* 1
2	Lenses	23
3	CCTV Signals and Principles	53
4	Television Camera Tubes	76
5	Camera Circuit Operation	95
6	The Monitor Display Tube	141
7	The Video Monitor	159
8	Basic Adjustments to Camera and Monitor	190
9	Lighting	200
10	The Signal Cable	204
11	Fault-Finding Charts	212

Appendices

A	EFFECT OF SYNC. SEPARATION INTEGRATING NETWORK ON COMPOSITE SYNC. PULSE TRAIN	*page* 225
B	REACTANCE TRANSISTOR STAGE	237
C	MAXIMUM VIEWING DISTANCE	241
D	THE LENS EQUATION	243

Bibliography	245
Index	247

CHAPTER 1
INTRODUCTION AND BASIC ELEMENTS OF LIGHT

CLOSED circuit television is a modern tool of industry, enabling many industrial problems to be tackled with an efficiency which cannot be bettered by other means. CCTV allows complete automated industrial processes to be controlled by one man in a central position, using cameras mounted in strategic positions around the plant. By viewing the scene from any camera on his monitor, the operator can quickly direct maintenance staff to deal with breakdowns or blockages as they arise thus eliminating expensive delays. CCTV enables a security man to keep constant vigilance on the vital areas of an art gallery or factory site by day and night thereby guarding against theft and vandalism. The smooth flow of traffic through modern motorway interchange points or across a major bridge road can be achieved with the minimum of manpower by using cameras placed at strategic points. From the monitors in the traffic control centre the controller is instantly aware of congestions or breakdowns and can quickly put into operation diversion or lane closure plans or summon assistance if needed. During a training course at a factory or educational establishment, CCTV may be employed to eliminate frequent visits to view a production line process or a piece of equipment where limited space or travel distance may be a problem. Video tape recordings of the process or equipment can be made under ideal conditions and replayed to the class on a large screen monitor. Particular items can be replayed many times to emphasize an important point and small detail can be viewed simultaneously by the whole class. The ability of a CCTV system to view and display documents or written information is used extensively in industry and at airports for information transmission, stock cards, sales or flight information etc. being readily viewed at remote distances.

This versatile and expanding art has many applications, a few of which have been briefly mentioned. In its simplest form a CCTV system may consist of a camera, picture monitor and coaxial connecting lead. At the other end of the scale a system may incorporate full studio facilities providing monochrome and colour pictures of broadcast quality. In this introductory book we shall be dealing with the most common type of monochrome equipment which is used for non-critical applications in industry and commerce.

Figure 1.1 shows the basic elements of a simple CCTV system. Light from the subject or scene to be televised is focused by the camera lens on to the face of the camera

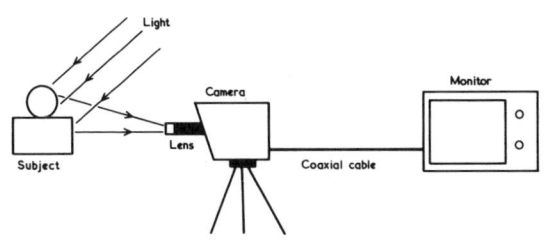

FIG. 1.1 A SIMPLE CCTV SYSTEM

tube. In the camera tube, the optical image formed is converted into an electrical charge image. An internal electron beam moving at line and field rate sequentially scans the charge image causing an electrical voltage to appear at the camera tube output. After amplification in the camera amplifiers, the electrical (video) signal voltage is fed to the monitor *via* the coaxial cable. This transmission line must possess as small an attenuation as is practicable and must propagate all frequency components of the signal equally. At the monitor the video signal is used to produce an image on the screen of the original scene. In addition to the video signal, the camera must supply sync. pulses to the monitor to synchronize the display tube's electron beam with that of the camera tube. CCTV simply means that the video signals are confined to the transmission cable between the camera and monitor thus making the system 'private' rather than being broadcast by a radio frequency transmitter which makes general reception possible.

The scene may be illuminated by sunlight but when necessary artificial light sources may have to be used solely or in conjunction with natural light. During the course of his duties, the CCTV technician may be required to set up a new viewing situation or modify an existing one to improve the displayed picture. It is clear, therefore, that in addition to being familiar with camera, monitor and ancillary equipment electronics the technician should have a basic understanding of the properties of light and elementary optics. In this chapter we shall consider the important properties of light and their significance in relation to CCTV.

ELECTROMAGNETIC WAVES

Radiant heat from an electric fire travels through space in the form of electromagnetic waves. In contrast to the convection and conduction of heat no medium is required to propagate radiant heat, which may pass through a vacuum. An electromagnetic wave consists of an oscillating electric field together with an associated magnetic field, co-existing at right angles to one another, figure 1.2. The movement of these two fields through space constitutes an electromagnetic wave. The plane containing the electric and magnetic fields is referred to as the **wavefront**.

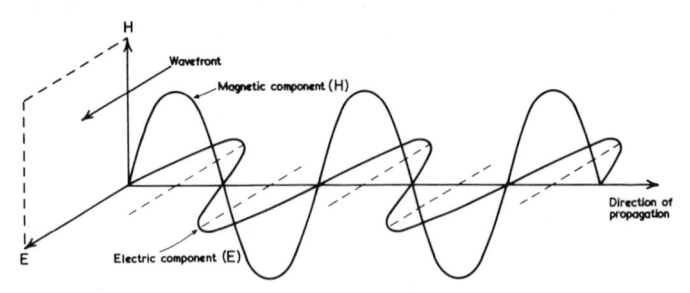

FIG. 1.2   REPRESENTATION OF AN ELECTROMAGNETIC WAVE

Light rays, radio waves, X-rays and cosmic rays like radiant heat travel through space in the form of electromagnetic waves and with high velocity (3×10^8 metres per second in a vacuum). It is evident, therefore, that these different forms of radiation are all essentially of the same form. What then distinguishes one form of energy from another? The properties of the different types of electromagnetic energy are determined by their frequency (or wavelength), see figure 1.3. For convenience in this diagram the spectrum is divided into sections but the sections have no precise boundaries since the behaviour of e.m. waves does not change sharply at given frequencies. The wavelength (λ) of an e.m. wave may be found from

$$\lambda = \frac{v}{f} \text{ metres}$$

[where f is the frequency (Hz) and v is the velocity (metres/sec)].

Visible light occupies a very small section of the e.m. spectrum with wavelengths in the approximate range of 380–780 nano metres (1 nano metre = 10^{-9} metre). When all

INTRODUCTION AND BASIC ELEMENTS OF LIGHT

of the frequencies present in the visible spectrum reach the eye simultaneously we see 'white' light.

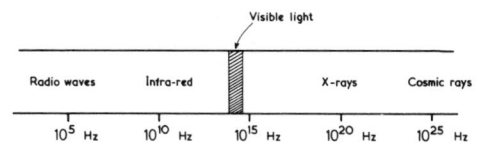

FIG. 1.3   ELECTROMAGNETIC SPECTRUM

IMPORTANT PROPERTIES OF LIGHT
(a) Rectilinear Propagation

By considering the shadow cast by an object placed in the path of a beam of light as in figure 1.4, it is evident that the light must travel in straight lines and not bend round the object. 'Straight line travel' or 'rectilinear propagation' as it is sometimes called still

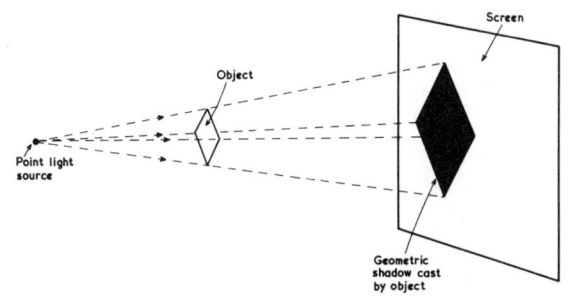

FIG. 1.4   SHADOW CAST BY AN OBJECT ILLUSTRATING RECTILINEAR PROPAGATION

holds good over great distances as is evident by the shadow cast by the moon on the earth during an eclipse of the sun, for example.

It should be appreciated that although figure 1.4 shows the light rays travelling in straight lines, this is only approximately true. On closer examination it is found that some of the light passes into the geometric shadow of the object. This is the result of the diffraction of light which occurs whenever light passes the edge of an object.

(b) Reflection

When we see an object, light passes directly from the object to our eyes and sets up the sensation of vision. Objects are only visible if they are themselves sources of light or they reflect light from a source into the eye. Nearly all of what we see about us is the result of the reflection of light. Most bodies of everyday life, *e.g.* flowers, paper, cloth, people, etc. reflect light in an irregular way as illustrated in figure 1.5(a). This is because these types of objects have surfaces that are rough on a microscopic scale and cause the

(a)   Irregular reflection
      – most bodies

(b)   Regular reflection
      – mirrors

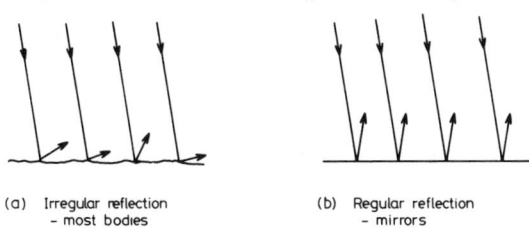

FIG. 1.5   REFLECTION

light to be reflected in an irregular manner. At each point on the object the laws of reflection are obeyed but the angle of the reflected ray varies.

On the other hand, smooth surfaces like polished metal or glass with a metallic coating on the back or front surface reflect light in a regular way, and are used as mirrors. The angle of the reflected light is now the same at every point, figure 1.5(b).

A ray of light such as CO (figure 1.6) falling on the surface of a plane (flat, non-curved) mirror makes an angle i with the *normal* ON to the mirror called the 'angle of

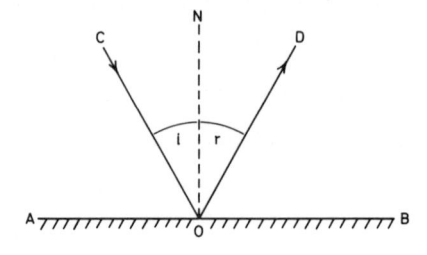

FIG. 1.6   REFLECTION AT A PLANE MIRROR SURFACE

incidence'. The angle r made with the normal by the reflected ray is called the 'angle of reflection'. Experimentally, it may be established that
(i) The angle of incidence equals the angle of reflection.
(ii) The incident ray, the reflected ray and the normal at the point of incidence all lie in the same plane.

These are the two laws of reflection.

Figure 1.7(a) shows how an image is formed in a mirror from a small point object A which is assumed to be emitting light rays in all directions. At each incident point on

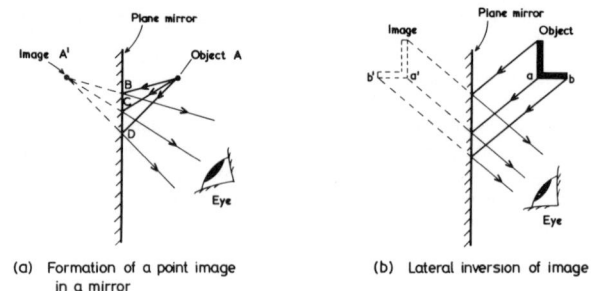

(a) Formation of a point image in a mirror

(b) Lateral inversion of image

FIG. 1.7   MIRROR IMAGES

the mirror, B, C, D etc., the laws of reflection are obeyed. To an observer, the rays appear to come from A' (at the same distance as A to the mirror but on the other side of it) where there appears to be an image of the object A. Image A' is a **virtual image,** since light does not actually pass through the image point but only **appears** to do so.

If a man is looking in a mirror and combing his hair with his left hand, he appears to be using his right hand. This can be explained by reference to figure 1.7(b) which shows the formation of an image from an L-shaped object. The image of point a on the object is at a' at an equal distance behind the mirror. Point b on the object which is to the right of a produces an image at b' on the left of a' i.e. the right-hand side of the object becomes the left-hand side of the image and *vice versa*. Thus the object is **laterally inverted** to an observer. Any technician who has fitted up a mirror for a hospital patient to view a television screen will have met the problem of caption reversal—the simple remedy being to reverse the line scan coil connections.

INTRODUCTION AND BASIC ELEMENTS OF LIGHT

Mirrors in everyday use are usually made by depositing silver on the rear surface of a section of glass sheet. This affords protection to the reflecting surface but is a disadvantage for certain applications. When light falls on a rear-silvered mirror, some of the incident light is reflected by the front surface of the glass. The amount of reflected light is small (about 5%) when compared with the reflection from the silvered surface, but causes a faint second image to appear as illustrated in figure 1.8.

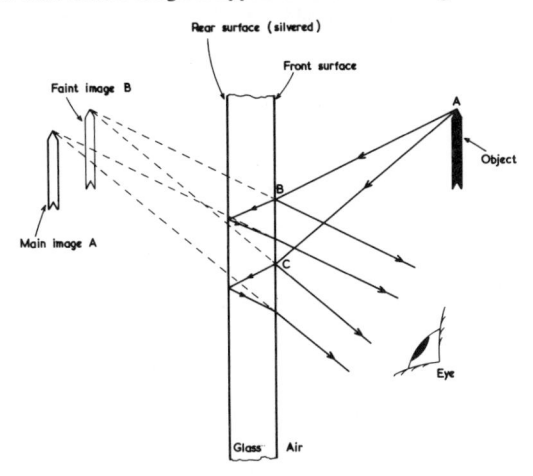

FIG. 1.8 TWIN IMAGES FORMED BY A REAR-SILVERED MIRROR

When a mirror is used in t.v. camera operations to view a subject indirectly (which may be necessary when the subject is in a difficult position to take a direct shot or when an unusual view is required) the presence of a second image on the sensitive surface of the camera tube would degrade the contrast and possibly reduce the definition. Twin images can be avoided by using a **front-silvered** mirror which forms only one image.

If, in figure 1.8, the object is moved farther away from the mirror, the rays AB and AC approach one another until finally the rays coincide when the object is at infinity and the light from it is parallel; there will then be only one image. Thus, a rear-silvered mirror is satisfactory when the object distance is quite large. However, when the light falling on the mirror is unavoidably convergent or divergent the use of a front-silvered mirror will avoid double images.

Before proceeding, it may be mentioned here that the lines shown so far in the diagrams representing rays, give the direction that light energy is travelling. It should be appreciated that a light ray has a finite width. A collection of light rays is known as a 'beam'. Rays which are being received from a point on a most distant object, e.g. the sun are substantially parallel, figure 1.9(a). The light coming from a lamp forms a divergent beam as in figure 1.9(b) whereas a light source and a lens can provide a convergent beam, figure (1.9 c).

(a) Parallel light beam   (b) Divergent light beam   (c) Convergent light beam

FIG. 1.9 LIGHT BEAMS

(c) Refraction

When light passes from one medium to another, say, from air to glass or from air to water, the rays are bent unless the rays strike the boundary of the two mediums at right angles. This causes the rays to travel in a different direction. On account of the change in direction, the rays are said to be **refracted**.

In figure 1.10 the incident ray AO is refracted at the air-to-glass boundary and travels in the new direction OB through the glass medium. As some reflection always

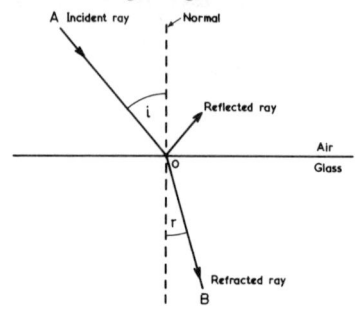

FIG. 1.10   REFRACTION OF LIGHT AT A PLANE SURFACE

accompanies refraction, a reflected ray has been shown in the diagram. It is important to note that *if a light ray is reversed it travels along its original path, i.e.* a light ray travelling in the direction BO through the glass will take the direction OA on entering the air medium.

The laws of refraction are:
(i) At the point of incidence, the incident and refracted rays and the normal all lie in the same plane.
(ii) For the two media concerned

$$\frac{\sin i}{\sin r}$$

is a constant where i is the angle of incidence and r is the angle of refraction.

This constant is called the **refractive index** (n) for the two media. As the magnitude of the constant depends upon the colour of the light, it is usually specified as that obtained with yellow light. The refractive index for air-to-glass is about 1·5 and for air-to-water about 1·33.

Refraction is a consequence of the fact that light travels more slowly through water or glass than through air which is less dense. It may be shown that

$$v = \frac{c}{n},$$

where n is the refractive index, c is the velocity of light in a vacuum and v is the velocity of light in the medium.

Figure 1.11 illustrates the useful points to remember concerning refraction. When a ray of light passes from air to a denser medium of glass the ray is *bent towards the normal*, diagram (a). If a ray is passing from glass to a less dense medium of air the ray is *bent away from the normal*. These properties are helpful in understanding the basic action of lenses which are considered in Chapter 2.

When a ray of light passes through a glass block with parallel sides, the ray that emerges is parallel with the ray entering the block, figure 1.12. Thus, the glass block has not changed the direction of the ray but has displaced it sideways; this is sometimes called 'lateral displacement'. Therefore, if an object is viewed *via* a glass section which is held obliquely to the object, it will appear to be displaced sideways. If the incident ray strikes the face of the glass block at right angles there will be no lateral displacement.

INTRODUCTION AND BASIC ELEMENTS OF LIGHT 7

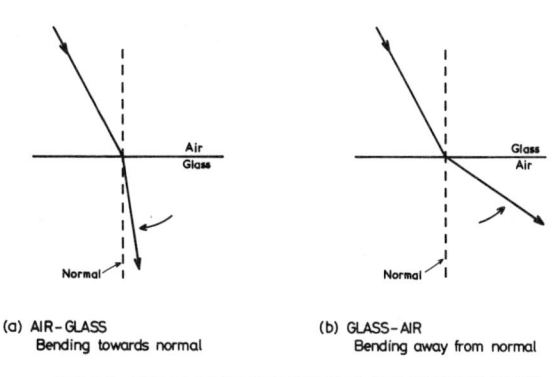

(a) AIR-GLASS
Bending towards normal

(b) GLASS-AIR
Bending away from normal

FIG. 1.11 REFRACTION ON ENTERING AND LEAVING GLASS

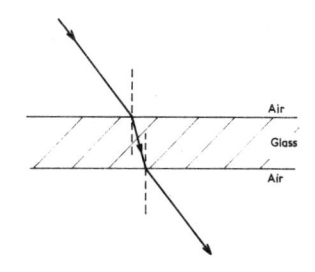

FIG. 1.12 REFRACTION THROUGH A GLASS BLOCK

Clearly, the thinner the glass the less will be the lateral displacement.

A further aspect of refraction is considered in figure 1.13. When a ray of light travelling in glass makes a small angle x with the normal at the glass-to-air boundary as

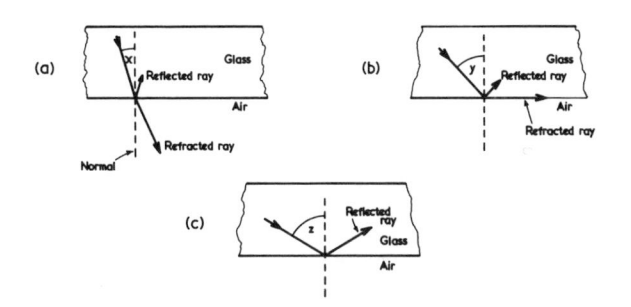

FIG. 1.13 TOTAL INTERNAL REFLECTION

in diagram (a), only a small portion of the light is reflected. Most of the light is transmitted into the air medium and is bent away from the normal. As the incident angle is increased a situation is reached where the refracted ray travels along the glass-to-air boundary but the portion of light reflected is still quite small, diagram (b). If the angle of incidence is now increased above the *critical angle y*, there is no refraction at the boundary and all of the incident light is reflected, diagram (c). This phenomenon is called 'total internal reflection'. It should be noted that this phenomenon cannot occur when light travels from a less dense to a more dense medium, *e.g.* air-to-glass or water-to-glass.

The critical angle *(y)* for light rays travelling from a more dense to a less dense medium may be found from

$$y = \arcsin\left(\frac{n_1}{n_2}\right),$$

where n_1, is the refractive index of the less dense medium and n_2 is the refractive index of the more dense medium. For a glass-to-air boundary the critical angle is about 42°. Thus, if the angle of incidence inside the glass exceeds 42°, total internal reflection will occur. With a glass-to-water boundary the critical angle is about 63°.

Transparent prisms of various shapes are used extensively in optical instruments and also have applications in some t.v. cameras. Prisms are easily manufactured and the effects they produce are readily calculated. Consider a ray of single wavelength falling on the face *AB* of a triangular prism *ABC*, figure 1.14. At *Y* the ray is refracted towards the normal making angle *b* less than angle *a*. On arrival at the face *AC* the ray

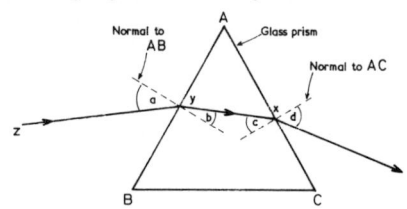

FIG. 1.14 REFRACTION THROUGH A TRIANGULAR GLASS PRISM

is again refracted but with the bending at *X* being away from the normal, *i.e.* angle *d* is greater than angle *c*. Any other ray parallel to *ZY* which is refracted by the prism will meet the faces *AB* and *AC* at the same angles as the ray *ZY* and will thus suffer the same degree of bending. Therefore, if a parallel beam of light falls on a prism of this type, it will emerge after refraction as a parallel beam.

Figure 1.15 shows how a glass prism may be used to rotate the path of a beam of light. Mirrors, of course, could be used for this purpose but there is the difficulty of

(a) Beam turned through 90°   (b) Beam turned through 180°

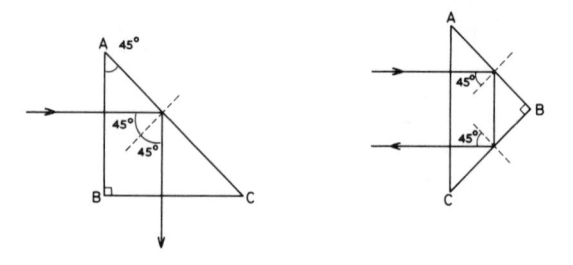

FIG. 1.15 ROTATING A LIGHT BEAM USING A GLASS PRISM

mounting a mirror so rigid that it does not move, hence prisms are more frequently used. In diagrams (a) and (b) right-angled prisms are used and in both cases it will be noted that the ray meets the glass-to-air surface at an angle of 45° which is a little greater than the critical angle (the exact value of the critical angle depends upon the kind of glass used). Thus, total internal reflection takes place, *i.e.* the glass-to-air

INTRODUCTION AND BASIC ELEMENTS OF LIGHT 9

surface acts as a perfect mirror. There will be some loss of light due to reflection back into the air as the light enters the prism and back into the glass as the beam leaves the prism, but this is quite small.

(d) Diffraction

It is more than four hundred years since experimenters first noticed that over small regions colours were present at the edges of shadows cast by objects. At that time there was no explanation for the observations, but it is now known that the colour fringes are due to *diffraction* which is the property of light to spread round corners. Diffraction can be successfuly explained only by using the 'wave theory' of light which cannot be considered in detail here. However, consider a point source of light placed at O and sending out rays in all directions, figure 1.16.

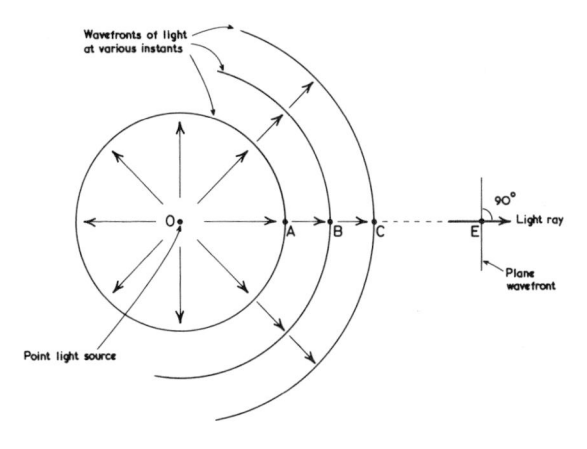

FIG. 1.16 WAVEFRONTS

Suppose that after a short interval the light energy emitted to the right of O has reached a position A. The surface of a sphere of which OA is the radius is the wavefront (see figure 1.2) at this particular instant. If the light source is of one colour, then at every point on this surface the light energy will have the same frequency and be of the same phase, *i.e.* the light is *coherent* at every point. As time elapses the wavefront moves outward and it may take up surfaces of spheres of radii OB and later OC. At a point a long way from O, such as E, the wavefront is the surface of a sphere of very large radius and the wavefront is substantially plane. Light from the sun is in the form of plane waves when it reaches earth as it is a long way off.

Figure 1.17 shows a wavefront which has reached the surface AD from a source O. According to Huygens, every point on the surface of the wavefront behaves as a new centre of wave disturbance, *i.e.* it can be considered that every point produces a secondary wavelet. It is possible, using these secondary wavelets, to construct a new wavefront at a later instant. As can be seen the new wavefront touches all secondary wavelets. Using this idea, consider now figure 1.18 which shows a light wave advancing on a narrow slit. From a few points on the wavefront at the slit exit, secondary wavelets have been constructed to show the advance of the wavefront as it leaves the slit. It is evident that the secondary wavelets direct energy into the geometric shadow: this is diffraction. The variations of brightness at the edges of the shadow are due to 'interference' between the secondary wavelets. Investigation shows that diffraction occurs when light passes the edge of any obstacle whether its edges are straight or round.

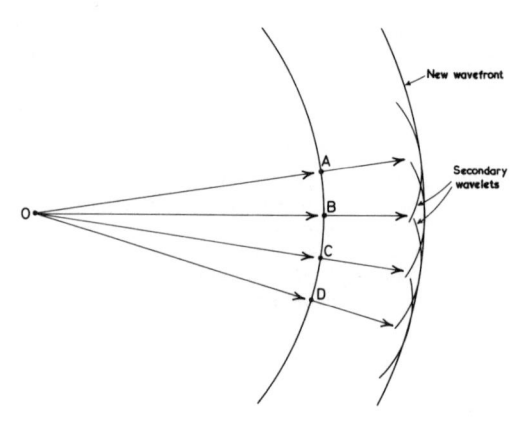

FIG. 1.17   SECONDARY WAVELETS

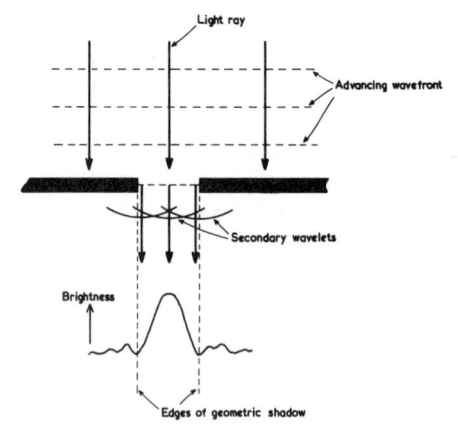

FIG. 1.18   DIFFRACTION AT A NARROW SLIT

COLOUR

In a CCTV system, the camera will in most circumstances be 'looking' at a coloured scene or subject. If the system is a monochrome one, what particular feature of the colours making up the scene is relayed from the camera to the monitor? When a colour CCTV system is used instead, what additional information must the camera convey to the monitor? To answer these questions some knowledge of coloured light is required.

Spectral Colours

When a beam of white light as in figure 1.19 is allowed to fall on a glass prism, the ray that emerges is no longer a beam of white light but a divergent beam containing all the colours of the rainbow and their intermediate tints. This experiment, due to Newton, showed that white light is not the purest kind of light but on the contrary is a mixture of a vast range of colours. Each colour has a specific wavelength but the difference in wavelength between adjacent colours of the spectrum is so small that the vast range of colours are not distinguishable by the human senses. Instead, we see a gradual blending of the numerous colour radiations which produces a graduation of

INTRODUCTION AND BASIC ELEMENTS OF LIGHT

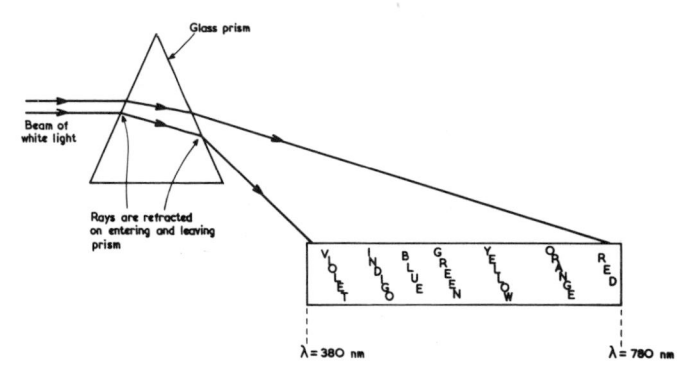

FIG. 1.19   SPECTRAL COLOURS

distinct hues called SPECTRAL COLOURS. These are Red, Orange, Yellow, Green, Blue, Indigo and Violet. (Mnemonic: ROY. G. BIV).

Note that the amount of refraction or bending that occurs is greater for the shorter wavelengths, *i.e.* more at the 'blue' end of the spectrum than the 'red' end.

Characteristics of a Colour

Any colour may be defined by three characteristics.

(1) Hue—Hue is the quality of a colour that is most noticeable to the human senses. It is the preferred term for colour, *e.g.* when white light falls on to a red object the light reflected has a red hue. Similarly, if the object is green the light reflected has a green hue, figure 1.20 (a) and (b). Thus, hue depends upon the *dominant* wavelength of the light energy.

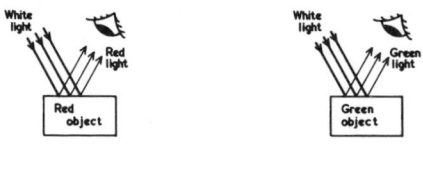

(a) Object has a red hue   (b) Object has a green hue

FIG. 1.20   HUE

(2) Saturation—Saturation is the quality of a colour that defines its depth. The saturation of a colour depends upon its dilution with white light. For example, when white light is added to a colour of pure red hue a paler shade of red is produced which we commonly call pink. As illustrated in figure 1.21, the dominant wavelength of pink is in the red part of the spectrum thus pink has the *same hue* as red. The greater the amount of white light added, the paler or more *desaturated* the hue becomes. Desaturated colours are usually referred to as pale shades or pastel colours. Most of the colours that we see about us are desaturated to various extents.

(3) Luminance—This term describes the brightness of a colour as assessed by the human eye. Consider figure 1.22 which shows three lamps throwing out separate red, blue and green beams all having the *same* energy. Now, to an observer the colours will not appear to be of the same brightness even though the beams are projected with equal energy. It will be found that the green light will appear the brightest of the three, with the red the next brightest and the blue producing the smallest sensation of brightness.

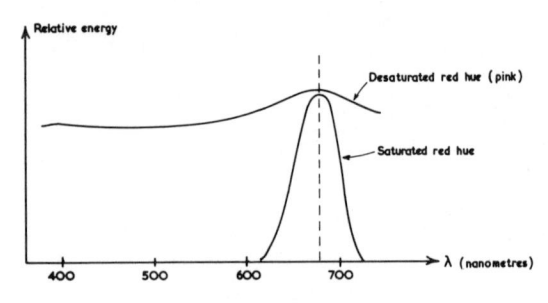

FIG. 1.21 SATURATION

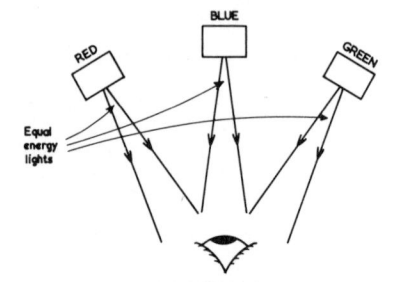

FIG. 1.22 LUMINANCE

The reason for this is because of the non-linear response of the eye to lights of different colours, see figure 1.23. It will be noted that the response of the eye is greatest (most sensitive) at about 550 nm which corresponds to a yellow-green hue. The response drops off at the red end of the spectrum and even more at the blue end.

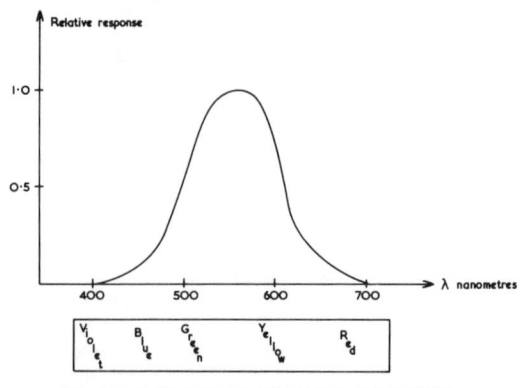

FIG. 1.23 RESPONSE OF STANDARD HUMAN EYE

INFORMATION TO BE CONVEYED
Mono CCTV

In a monochrome CCTV system it is the *luminance* information of the scene presented to the camera that is conveyed to the monitor. Since the camera is attempting to replace the human seeing senses in the studio, it would appear that the response of the camera tube to various colours should closely follow the response of the human eye (figure 1.23). However, the interpretation of a coloured scene in black and white is an

INTRODUCTION AND BASIC ELEMENTS OF LIGHT 13

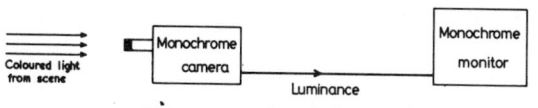

FIG. 1.24 MONOCHROME CCTV

unnatural process and is therefore subject to standards which are a matter of opinion. As a result of statistical tests carried out by the BBC research department into viewers' preferences, it was concluded that the response of the camera tube need not closely follow the curve of figure 1.23. In fact, a wider frequency response was found to be the best with the peak shifted from 550 nm to 460 nm. The colour response of a camera tube is determined by the chemical composition and processing of the light sensitive target. Clearly, the response given to the camera tube will affect the ability of a viewer to guess what the actual colours might be. Not that the viewer will necessarily wish to know what the actual colours are, but would soon be critical if unable to distinguish the differences between the flesh tones of white, yellow or browned skinned people on a monochrome display tube.

If threads of different coloured cottons, held against a white background, are viewed from a distance of about 10 metres it is difficult to say what the actual colours of the threads are. The threads themselves are quite distinguishable, *i.e.* the eye can perceive the fine detail but the colours show up as variations of light and dark. We are thus only conscious of the differing luminance levels of the threads. To preserve fine detail in a t.v. system a comparatively large bandwidth is required to handle the luminance information or signal. With a 625-line system a video bandwidth extending up to 5·5 MHz is used by the broadcasting authorities. For non-critical mono CCTV applications, an upper video frequency of 3 MHz may be quite acceptable but with modern cameras 10 MHz is not uncommon as the bandwidth restrictions of off-air transmissions do not necessarily apply to CCTV.

A monochrome system reproduces a colour scene in shades of white going from peak white, through shades of grey to black. In a CCTV industrial situation it may so happen that the contrast of the important scene detail is insufficient, *e.g.* when, say, a camera is monitoring the level of water against a background of similar luminance level. When presented with such a situation, one is inclined to think that the matter is outside one's control; after all the camera is faithfully (we hope) reproducing the scene presented to the camera lens. However, bearing in mind that some colours show a greater luminance level than others it may be possible using some coloured paint to vary the luminance level of the scene background. In this way the contrast of the relevant scene details may be improved.

On the other hand there may be a coloured object (or shiny one) which is showing up on the monitor as a peak white due to its high luminance content (the object may not of course be an important scene detail). If the scene is static and is being viewed continuously over long periods (not uncommon in industry), any objects of high luminance can cause burns on the light sensitive target of the camera tube (see page 80). This situation may be remedied by applying some coloured paint to the offending object (if at all possible) to reduce the luminance level. Figure 1.25 shows the decreasing order of luminance (from top to bottom) for some common colours. It is possible to tackle both the difficulties mentioned by other means *e.g.* by careful

WHITE	← Maximum luminance
YELLOW	
CYAN	
GREEN	
MAGENTA	
RED	
BLUE	
BLACK	← Zero luminance

FIG. 1.25 LUMINANCE LEVELS FOR DIFFERENT COLOURS

positioning and choice of the type of lighting source used or by fitting colour filters to the lens system.

Colour CCTV
For colour television the information concerning the luminance, hue and saturation of the scene colours has to be conveyed to the monitor as illustrated in figure 1.26.

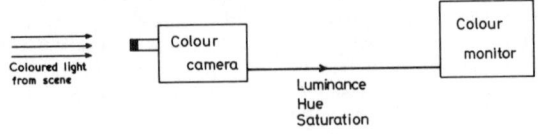

FIG. 1.26 COLOUR CCTV

We have noted that the eye is not conscious of colour changes when they appear in fine detail. Thus, there is no need to transmit the hue and saturation information at wide bandwidth. A bandwidth of 0—1 MHz is found to be satisfactory for these signals which, together with the luminance signal (0—5.5 MHz), results in a high definition colour t.v. system.

COLOUR MIXING
The principle of all colour printing, colour photography and colour television is based on the mixing of usually three colours to obtain all of the colours present in the original. In colour television, the three colours (called primary colours) are red (615 nm), green (532 nm) and blue (470 nm). By suitable mixing of the three primaries it is possible to produce nearly all the colours that occur in nature.

There are two ways that we can mix colours: by mixing coloured lights (the additive system); or by mixing coloured pigments, *e.g.* inks, dyes and paints (the subtractive system).

Additive Colour Mixing
If two of the t.v. primary colours are projected from lamps on to a white screen so that they partly overlap, the colour that we see in the overlap area is the result of the additive mixture of the original primaries. Figure 1.27 shows the result when various coloured lights are additively mixed. Note that when all the primaries are additively mixed (in suitable proportions) the result is white.

In addition, white is produced when the following lights are additively mixed:

$$\text{Blue} + \text{Yellow} = \text{White}$$
$$\text{Green} + \text{Magenta} = \text{White}$$
$$\text{Red} + \text{Cyan} = \text{White}$$

Subtractive Colour Mixing
When white light falls on a coloured object, the object appears that colour because the pigment reflects only its own colour whilst absorbing or subtracting the other radiations of the white light. We have shown that white light is composed of red, green and blue lights. Thus, if white light falls on a red object, the red light is reflected but the green and blue components are absorbed or subtracted from the incident white light, figure 1.28 (a). If the object is yellow, the pigment will reflect the red and green components but absorb the blue component, figure 1.28 (b). When, say, magenta light falls on a blue object as in diagram (c), the pigment reflects the blue component of the magenta light but absorbs the red component.

It is well known that coloured objects appear to change their colour when seen under different colour light sources, *e.g.* under street lighting (sodium lamps). The yellow object of diagram (d) appears red when viewed under a red light source. No light

INTRODUCTION AND BASIC ELEMENTS OF LIGHT

RED + GREEN = YELLOW

GREEN + BLUE = CYAN

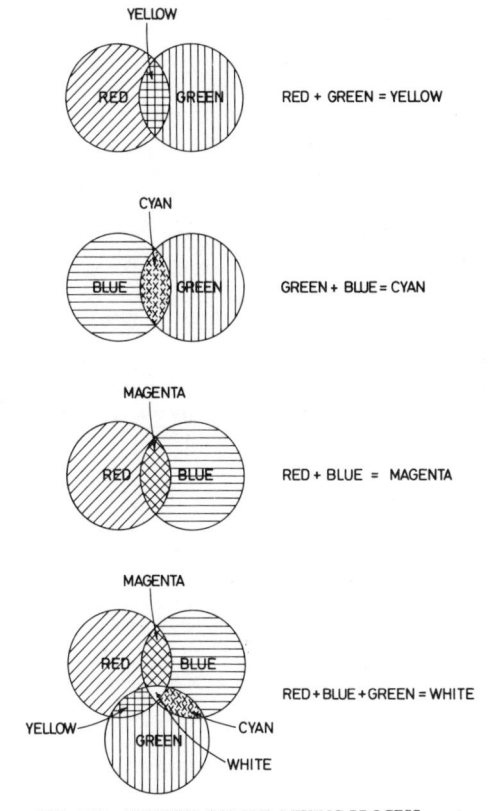

RED + BLUE = MAGENTA

RED + BLUE + GREEN = WHITE

FIG. 1.27   ADDITIVE COLOUR MIXING PROCESS

(a) Red light reflected (blue and green absorbed).

(b) Red and green light reflected (blue absorbed).

(c) Blue light reflected (red absorbed).

(d) Yellow pigment reflects the red light, thus object appears red.

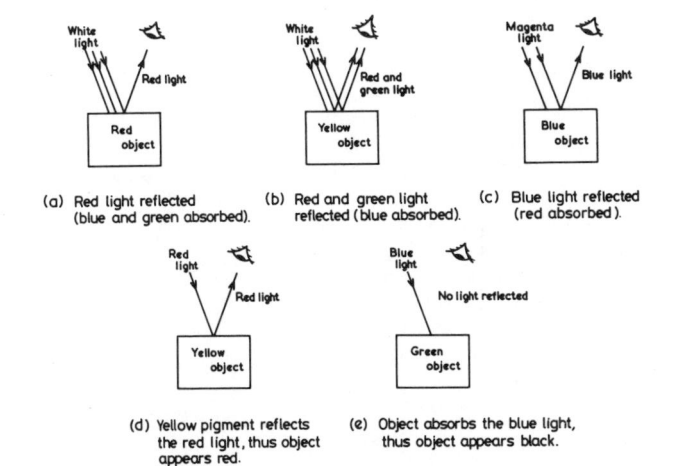

(e) Object absorbs the blue light, thus object appears black.

FIG. 1.28   REFLECTION AND ABSORPTION BY COLOUR PIGMENTS

would be reflected from the object in diagram (e) assuming a pure blue light source and a pure green pigment.

When pigments are mixed the resultant colour is that which is *common* to the original colours of the mixture. For example, a mixture of yellow and magenta paints produces red.

$$\text{YELLOW} + \text{MAGENTA} = (red + \text{green}) + (red + \text{blue}).$$

A mixture of cyan and magenta inks will result in blue since this is the common colour

$$\text{CYAN} + \text{MAGENTA} = (\text{green} + blue) + (blue + \text{red}).$$

When cyan, magenta and yellow pigments are mixed the result is black since there is no common colour. This method of mixing is used, of course, by artists and the colours yellow, cyan and magenta are referred to as the artists' primary colours but are commonly called yellow, blue and red.

The same process applies to the mixing of coloured filters. The yellow filter of figure 1.29 will pass red and green light whereas the magenta filter will pass red and blue light. Thus when the two filters are 'mixed' the light passing through the combination is that which is common, *i.e.* red light.

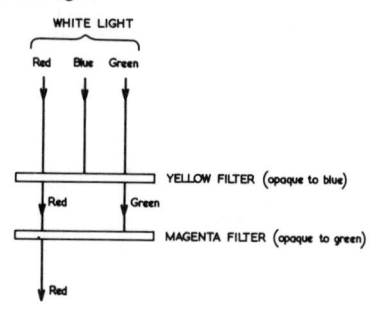

FIG. 1.29 MIXING OF COLOUR FILTERS

The Human Eye

The eye is divided into two parts by a crystalline lens made up of layers of transparent tissue. The shape of the lens can be altered to some extent by the ciliary muscles. Immediately in front of the lens is the iris. This is the coloured portion of the eye that can be seen through the transparent bulge of the cornea. The iris is an opaque

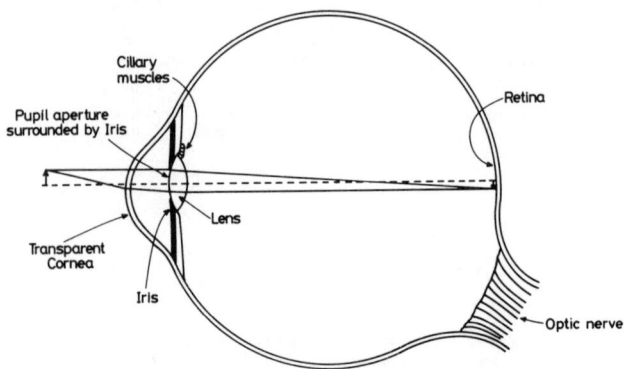

FIG. 1.30 THE HUMAN EYE

INTRODUCTION AND BASIC ELEMENTS OF LIGHT 17

screen with a small hole in the centre called the pupil. The size of the pupil is controlled involuntarily by the amount of light falling on the eye. At the back of the eye is the retina. In order to see an object clearly it must be focused on the centre of this surface. The retina is composed of a number of layers and at the back are the sensitive 'seeing' elements. There are two types of elements called rods and cones (on account of their shape when viewed under a microscope). When light falls on these receptors, electrical impulses are fed to the brain *via* the optic nerve.

In normal vision, in good light, the image we are studying is brought to a focus on a very small region at the centre of the retina. This small area is confined to cones only, thus these receptors must be responsible for detailed vision and colour sensation. Away from the centre of the retina, the cones get less and less in proportion to the rods. Whilst the rods give no detailed vision and little colour sensation, they allow us to see in poor light, *e.g.* moonlight but the detail is poor. They also allow a general perception over an arc of about 190°.

It has long been believed that there are three different kinds of cone pigments which are responsive to the primary colours red, green and blue. The behaviour of the eye is consistent with this idea which is illustrated in figure 1.31. The 'red' cones show a maximum sensitivity at about 590 nm, the 'green' around 550 nm and the 'blue' at 460

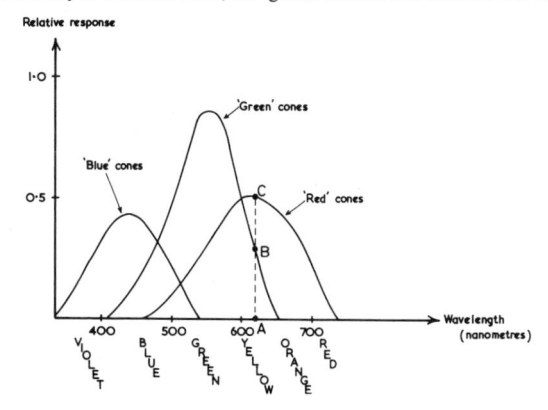

FIG. 1.31 RELATIVE RESPONSE OF INDIVIDUAL CONES

nm. If a particular monochromatic yellow light denoted by *A* on the diagram is focused on to the retina it will stimulate the 'red' and 'green' cones with responses *AC* and *AB* respectively. The 'red' and 'green' cones will transmit the information to the brain *via* the optic nerve where the sensation of yellow will be registered. It is thus logical to deduce that if red and green light simultaneously arrive at the retina, the stimulation of the 'red' and 'green' cones will give the impression of a yellow which is not distinguishable from the monochromatic yellow. Thus yellow can be matched by a suitable mixture of the red and green primaries. It is, of course, the tricolour nature of colour vision which allows colour television in its three-colour form to succeed. Integration of the three curves produces the overall response of the human eye shown in figure 1.23.

PROPERTIES OF HUMAN VISION WHICH HAVE AN IMPORTANT BEARING ON TELEVISON

(1). The eye cannot distinguish between light that is coloured because it is of a particular wavelength and a mixture of lights of different wavelengths. Thus only three primary colours are required to produce most natural colours.

(2). Certain colours appear brighter than others, even when they are all projected with equal energy; any colour or mono t.v. system must take this property into account.

(3). The eye is very sensitive to luminance changes in a scene and a comparatively wide video bandwidth is needed to preserve the fine detail. On the other hand, the eye has difficulty in differentiating between colours when they appear in fine detail. Thus a saving in video bandwidth can be made when relaying or transmitting colour information.

(4). When they eye is presented with a scene it will retain a lasting impression of that scene for a period equal to about 1/25th of a second, *i.e.* the eye has a time lag. This effect is called **persistence of vision** and enables us to see a continuous picture on a television screen using a rapid sequence of scanning.

White Light

There are many different light sources that we would subjectively classify as 'white light' but these have quite different spectral energy distributions as illustrated in figure 1.32. For example, the light from a gas-filled tungsten lamp has a preponderance of energy at the red end of the spectrum and is classified as a 'warm white'. On the other

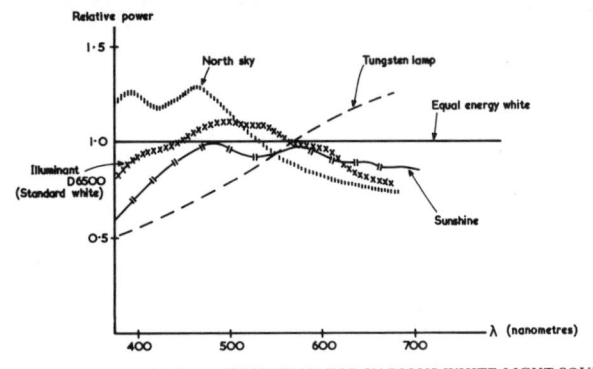

FIG. 1.32 SPECTRAL ENERGY DISTRIBUTION FOR VARIOUS WHITE LIGHT SOURCES

hand the light emanating from a north sky has more energy at the blue end of the spectrum and is referred to as a 'cold white'. As the type of white light source used will have a marked effect on the actual colours seen by a television camera, it is sometimes necessary to specify what white light source is to be used. For colour television, the standard white Illuminant D_{6500} was chosen as it matches standard daylight more closely than other standard illuminants. The type of white light emitted from the screen of a monochrome television display tube depends upon the chemical composition of the screen coating. Some monochrome display tubes emit light which matches the standard illuminant to meet the requirements of television studios.

Equal energy white is a hypothetical white which would be produced if all the visible radiations of the e.m. spectrum were of equal energy. This type of white is a useful concept for simplifying theoretical studies in calorimetry work.

Colour Temperature

A bar of steel gradually heated begins to emit light when its temperature is about 500°C. It then emits a dull red light. If the heating is continued, the colour of the light changes through red, orange, yellow until, when the temperature is over 1000°C, the light emitted appears to be white. Thus we can say that when the temperature of a body is low it emits radiations of long wavelength. As the temperature is increased, radiations of shorter wavelength are released.

The heat radiated from a body also depends upon the nature of its surface; a dull black body radiating better at any particular temperature than a shiny white one. For purposes of comparison we can imagine a perfectly black body being heated. Such a radiator is said to be a **black body radiator** or **full radiator**.

INTRODUCTION AND BASIC ELEMENTS OF LIGHT

Some sources of light have an energy distribution matching that of a black body radiator when the full radiator is held at a particular temperature. Thus, for some purposes these light sources can be conveniently classified by quoting their colour temperature. This is defined as the temperature (Kelvin scale)* to which a full radiator would have to be raised to emit light of the same energy distribution. It should be realised that the colour temperature is **not** the actual temperature of the light source itself.

Some light sources cannot be precisely matched by a full radiator and in this case the **correlated colour temperature** may be quoted. This is the temperature to which a full radiator would need to be raised to emit light that will nearly match the given light source.

A tungsten filament lamp operating at a temperature of 2800 K emits a white light corresponding to a colour temperature of 2854 K, *i.e.* the temperature to which a full radiator would have to be raised to emit a white light of the same character.

Light Units (S.I.)

In judging the brightness of a light source, the eye takes into account factors other than that of the energy behind the light source. We have noted that the eye is more sensitive to green than to red or blue light when they are all projected with equal energy. Thus for such reasons, the intensity of a particular light source has to be estimated by **visual comparison** with a standard source.

Luminous Intensity

The luminous intensity (I) expresses the visual sensation of power from a light source. The unit used is the **candela** (cd) and is defined as 1/60th of the light emitted from 1 cm^2 of the surface of a 'full radiator' when held at the freezing point of platinum. It is possible for standard lamps to be calibrated in terms of this unit, to which other light sources may be compared.

Luminous Flux

Light energy from a light source is emitted as a continuous stream of energy. We speak of the light energy as 'flux' and the name luminous flux (Φ) is the quantity of light energy emitted per second. The unit of luminous flux is the **lumen** which is defined as the quantity of light energy per second which is emitted in unit solid angle by a uniform point light source of 1 candela.

Consider figure 1.33 where a point light source of 1 candela is placed at the centre of a transparent sphere of radius r. 'Flux' from the source will be continuously streaming across the surface of the sphere. One lumen is the quantity of light streaming across an area r^2 on the surface of the sphere in one second. Unit solid angle is the solid angle subtended by the area r^2 at the centre of the sphere.

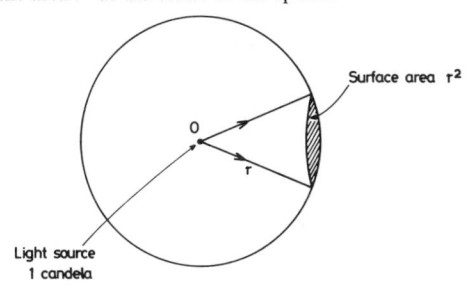

FIG. 1.33 LIGHT ENERGY EMITTED IN UNIT SOLID ANGLE

*Temperature (Kelvin scale) = Degrees Centigrade + 273 *e.g.* 0°C = 273 K and 100°C = 373 K.

Illumination

Illumination (E) is defined as the luminous flux per unit area falling on the part of the surface under consideration. The modern unit of illumination is the **lux** which is equal to 1 lumen per square metre. The original English unit was the foot-candle which equals 1 lumen per square foot. Since 1 square metre = 10·76 square feet, 1 foot-candle (or 1 foot lambert) = 10·76 lux.

If Φ is the luminous flux incident on a particular area A the illumination E is given by

$$E = \frac{\Phi}{A}$$

Typical *minimum* illumination levels for CCTV cameras range from 10—250 lux (at the lens surface).

The illumination of a surface decreases as the distance from the light source is increased. Provided the dimensions of the light source are small compared with the distance, the intensity of illumination is inversely proportional to the square of the distance.

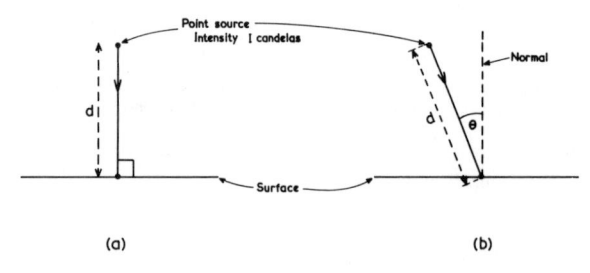

FIG. 1.34   INVERSE SQUARE LAW OF ILLUMINATION

When the light is incident at right angles to a surface as in figure 1.34 (a), the illumination

$$E = \frac{I}{d^2} \text{ lux.}$$

If the light strikes the surface at an angle θ to the normal, the illumination is given by

$$E = \frac{I \cos \theta}{d^2} \text{ lux, diagram (b).}$$

To show the magnitude of practical values consider figure 1.35. It is required to determine the intensity of illumination at a point A on a surface 4 metres from a small lamp of luminous intensity 200 candelas, also at some other point B on the surface 2·5 metres from A.

Intensity of illumination at A;

$$E = \frac{200}{4^2} \text{ lux} = \underline{12\cdot 5 \text{ lux.}}$$

INTRODUCTION AND BASIC ELEMENTS OF LIGHT 21

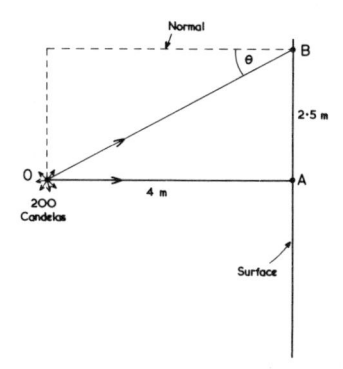

FIG. 1.35 CALCULATING THE ILLUMINATION AT A SURFACE (see text)

Intensity of illumination at B:

Now the distance $OB = \sqrt{4^2 + 2 \cdot 5^2}$

$ = 4 \cdot 72$

Thus $E = \dfrac{200 \cos \theta}{(4 \cdot 72)^2}$

$ = \dfrac{200 \times 4/4 \cdot 72}{(4 \cdot 72)^2}$ lux

$ = 7 \cdot 6$ lux.

The luminous intensity of a lamp depends upon the direction from which the lamp is viewed. For example, a vertically mounted coil filament provides a greater intensity in a direction at right angles to the filament coil than from its ends. Also, the use of a reflector increases the luminous flux in a particular direction. Thus, in determining the illumination which will be given by a particular lamp it is necessary to refer to polar diagrams of the lamp in question. This is not particularly convenient for CCTV purposes, therefore measurement is preferred to calculation. A suitable instrument for measuring illumination levels consists of a photovoltaic cell connected to a microammeter. The scale of the meter is calibrated by placing the instrument at a number of distances from a standard lamp of known luminous intensity. The scale of the meter may be calibrated to read lux directly or the current readings may be compared with a prepared graph to obtain the illumination level. The 'lux meter' may have a basic scale reading from 0—50 lux which may be increased by a factor of 10 or 100 by using specially designed opaque covers into which a number of small holes are made to control the amount of light falling on the light-sensitive cell.

Luminance

The luminance of a surface in a particular direction may be defined as the luminous flux per unit area *coming* from the surface. This is to be distinguished from the illumination of a surface which is the luminous flux per unit area which is *falling* on the surface. For example, the illumination of this page of print is fairly uniform, but the luminance varies considerably. The luminance of the white paper is high since the white surface scatters and reflects a large amount of the incident light, whereas the luminance of the black print is low since little light is reflected by the black pigment.

Luminance (L) may be measured in terms of light intensity units (candelas per square metre) or in light flux units (lumens per square metre).

Reflection Factor

The difference in the luminance values between the black print of this page and the white background is due to the difference in reflection properties of the black and white surfaces. It is thus necessary to be able to compare the reflection properties of different materials and surfaces. This is fulfilled by the introduction of a reflection factor.

$$\text{Reflection factor } (\rho) = \frac{\text{lumens per square metre reflected from a surface}}{\text{lumens per square metre incident upon the surface}}$$

$$= \frac{L}{E}$$

Transmission Factor

Some materials pass light more readily than others, *i.e.* they possess different transmission properties.

$$\text{Transmission factor } (\tau) = \frac{\text{lumens per square metre transmitted by the material}}{\text{lumens per square metre incident upon the material}}$$

CHAPTER 2
LENSES

For CCTV or photography a lens unit normally consists of a combination of thin glass lenses. There are two main types of lenses which act upon a narrow parallel beam of light falling on the lens in the following manner.

(a) A converging or **convex** lens causing the light to converge to a point as in figure 2.1 (a).

(b) A diverging or **concave** lens causing the light to diverge on the far side of the lens as if it were originating from an apparent point on the side of the light source, figure 2.1 (b).

(a) CONVERGING LENS   (b) DIVERGING LENS

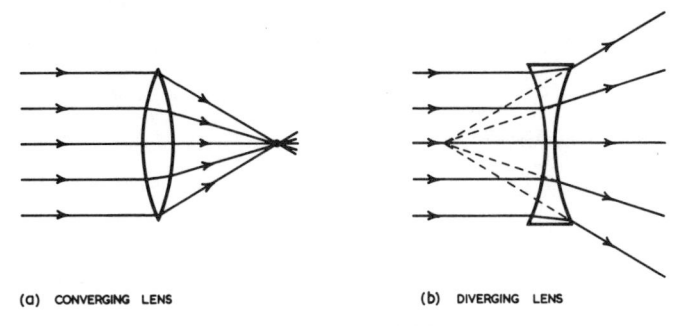

FIG. 2.1   MAIN LENSES

The action of the two main lens forms may be considered in the following way. If two glass prisms are placed base-to-base as in figure 2.2 (a), the parallel rays A and B would be deviated through the prisms to converge at a point O provided the refractive

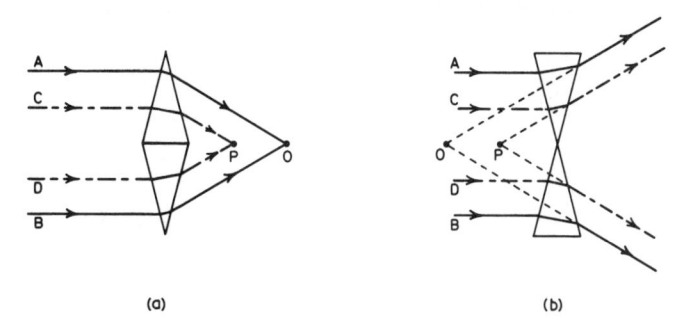

(a)   (b)

FIG. 2.2   ATTEMPTING TO FORM A LENS WITH TWO PRISMS

index was great enough. On entering the prisms the rays would be refracted towards the normal of the incident face but on leaving the prisms refraction would be away from the normal of the exit face. Similarly for a diverging lens, if the two prisms were placed apex-to-apex as in figure 2.2 (b) the rays A and B after refraction at the incident and exit faces would diverge such that they would appear to have originated from a point O.

Considering any other pair of parallel rays such as C and D incident on the prism faces of diagram (a), the point of convergence would be at P away from O. Similarly for the diverging arrangement, the emerging rays would appear to originate from a point P. The reason for the non-alignment of the two points O and P is that the rays make the same angle with the prism faces on entering and leaving its surfaces. To cause all parallel rays to converge or appear to diverge from the same point it is necessary to progressively alter the slope of the incident and exit faces of the glass medium. This may be achieved using a large number of prism faces as illustrated in figure 2.3 for the converging lens. As the number of prism faces increase, the curvature of the lens faces

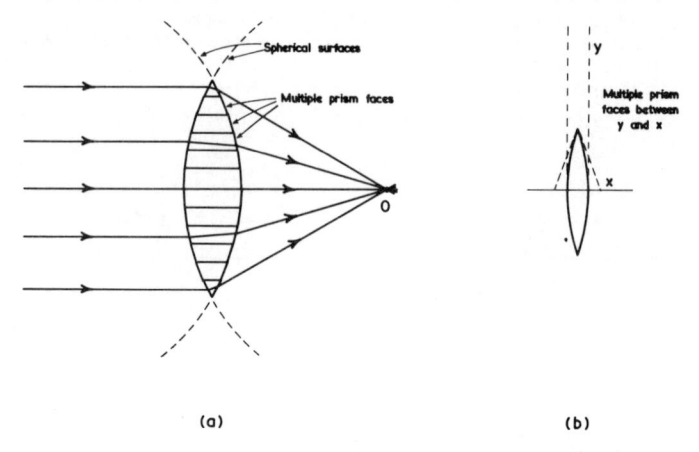

(a)   (b)

FIG. 2.3 USING MULTIPLE PRISM FACES TO PRODUCE A LENS ACTION

takes on the surfaces of **spheres.** So a lens may be considered as a very large number of pieces of principal sections of thin prisms, through whichever way the cross-section of the lens is looked at.

Each of the two main types of lenses exist in three forms as shown in figure 2.4. The upper row are all converging lenses and are capable of forming real images on their

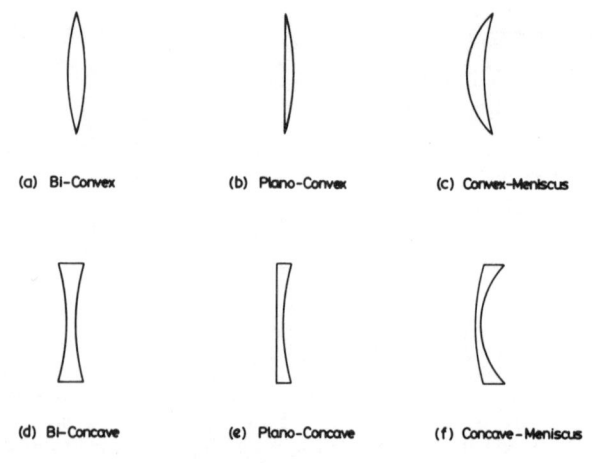

(a) Bi-Convex   (b) Plano-Convex   (c) Convex-Meniscus

(d) Bi-Concave   (e) Plano-Concave   (f) Concave-Meniscus

FIG. 2.4 SIX SHAPES OF A SIMPLE LENS

own. The lower three are diverging lenses and may be found along with converging lenses in camera lenses made up of several elements. A converging lens is always thicker at its centre than at its edges; the reverse is true of a diverging lens.

LENSES

Lens terms—The following terms are in common use and refer to figures 2.5. and 2.6.

Principal Axis—This coincides with the axis of revolution of the lens and with spherical surfaces will pass through the centres of curvature, figure 2.5.

Secondary Axis—This is any line other than the principal axis passing through the optical centre, figure 2.5.

Optical Centre—That point on the principal axis through which a ray can pass and emerge with the same or a parallel path, figure 2.5.

Aperture—The diameter AB of the lens, figure 2.5.

Principal Plane—The plane through O perpendicular to the principal axis, figure 2.5.

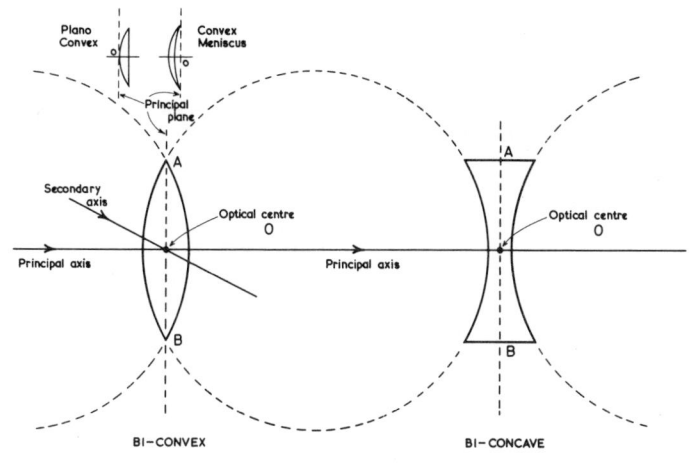

FIG. 2.5 LENS TERMS

Principal Focus—For a converging lens [figure 2.6(a)] all rays arriving parallel to the principal axis are refracted so that they pass through a point F_1, called the principal focus. Since light may be incident on either side of the lens there is a second point F_2. Thus a lens has two principal foci F_1 and F_2, and the distance $OF_1 = OF_2$.

For a diverging lens [figure 2.6(b)], parallel rays arriving on the principal axis are refracted such that they appear to originate from a point F_1 which is the principal focus. F_2 is the other principal focus and distance $OF_1 = OF_2$.

Focal Length—The distance OF_1 is the focal length, figure 2.6 (a) and (b).

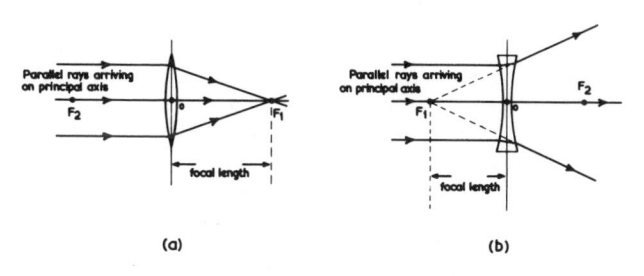

FIG. 2.6 (a) & (b) LENS TERMS

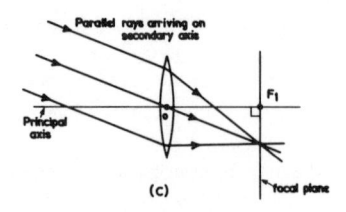

FIG. 2.6 (c) LENS TERMS

Focal Plane—A parallel beam of light arriving on a secondary axis, figure 2.6 (c), is refracted through a point which lies on a plane through F_1 perpendicular to the principal axis and is called the focal plane.

IMAGE FORMED BY A LENS

Only parallel rays of light have been considered in the diagrams used so far. When light falls on a lens from a point on a distant object the rays are substantially parallel. Such is the case for light rays coming from the sun. Thus, if we reconsider figure 2.6 (a) and (c), light from the sun would be refracted by the lens to form a small sharp image in the focal plane at F_1. It is this effect which is used in a burning glass and the schoolboy quickly learns how to burn holes in a piece of paper by positioning the lens so that it is at a distance OF_1 from the paper. The more strongly the curvature of the lens the shorter is the distance OF_1, *i.e.* the shorter the focal length, figure 2.7.

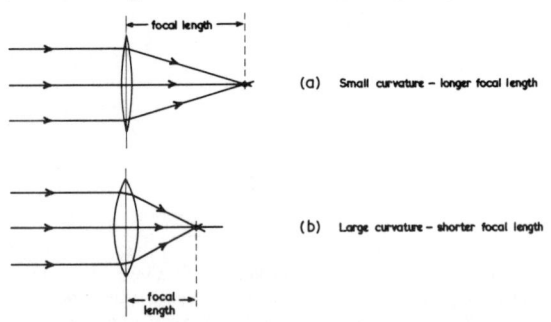

FIG. 2.7 EFFECT OF LENS CURVATURE ON FOCAL LENGTH

It is now necessary to consider the position of a sharp image of an object when the object-to-lens distance is not large in which case the arriving rays are no longer parallel, *i.e.* the lens is collecting diverging rays of light like those reflected from a point on an object having an irregular surface (see page 3). Consider an object AB, figure 2.8 (a). An image of the object may be focused on a screen placed on the opposite side of the lens to which the object lies. It is possible to predict the size and position of the image if the focal length (f) and the object-to-lens distance is known. The position of the image may be determined by graphical means in the following way:

Two rays are drawn from the top of the object.

(a) A ray parallel to the principal axis which is refracted through the principal focus F_1 of the lens.

(b) A ray through the optical centre, which passes undeviated through a thin lens.

A third ray is sometimes convenient (shown dotted) passing through the other principal focus F_2 which is refracted to emerge parallel to the principal axis.

These rays all intersect at A'. It will be seen that if lines are drawn for all rays from point A striking the lens, they would intersect at A'. Similarly, lines could be drawn

LENSES

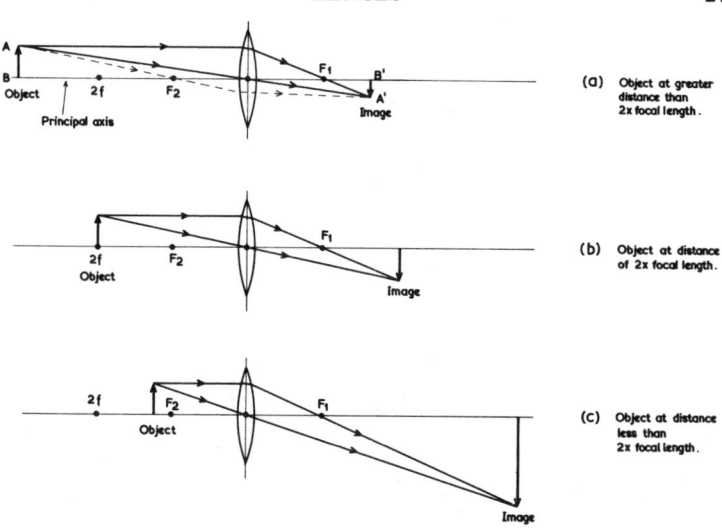

FIG. 2.8 IMAGES FORMED FOR VARYING OBJECT-LENS DISTANCES

(a) Object at greater distance than 2x focal length.

(b) Object at distance of 2x focal length.

(c) Object at distance less than 2x focal length.

from any point on the object between A and B and they would form image points in the plane $A'B'$.

The image formed in the manner described for a convex lens is a *real* image; it is always inverted (top-to-bottom and side-to-side) and may be diminished or magnified. If the object-to-lens distance is greater than 2 × focal length, $(2f)$, the image is diminished, figure 2.8 (a). When the distance is exactly $2f$, the image is the same size as the object, figure 2.8 (b). Placing the object between $2f$ and f results in an image that is a magnified version of the object, figure 2.8 (c).

When the object-to-lens distance is less than one focal length as in figure 2.9 a virtual image is formed on the same side of the lens as the object. The image is

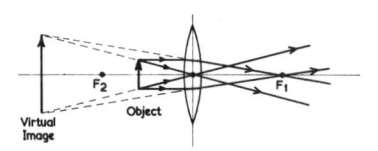

FIG. 2.9 OBJECT-LENS DISTANCE LESS THAN ONE FOCAL LENGTH

magnified but not inverted and can be seen by the eye when viewing from the opposite side of the lens to which the object is positioned. A sharp image cannot, therefore, be formed on a screen when *the object is less than one focal length away from the lens.*

Summarising, when an object is at a great distance from a convex lens an inverted, real image is formed at the focus; this is the case of the burning glass producing an image of the sun. As the object approaches the lens, the image recedes from it and grows until the object is at $2f$ the image is the same size as the object. As the object comes closer to the lens, the image recedes farther and becomes larger; this is the case of the cine projector or slide lantern. Finally, when the object moves within the focal length, the image becomes virtual, magnified and on the same side of the lens as the object; this is the case of the magnifying glass.

A diverging lens, figure 2.10, always forms a virtual image that is a diminished version of the object. The image is always nearer the lens than the object and closer to the lens than the principal focus (F_2 or F_1); also, it is non-inverted.

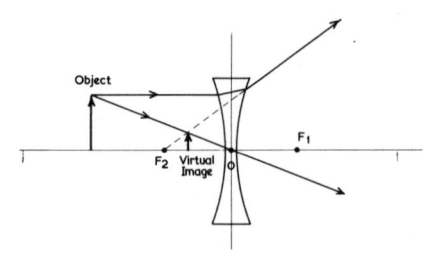

FIG. 2.10 IMAGE FORMED BY CONCAVE LENS

CALCULATING THE IMAGE POSITION

By convention the distance from the object to the lens is given as u and the distance from the image to the lens is given as v. Distances actually travelled by the light are taken as positive; distances said to have been travelled by a virtual ray are taken as negative. The focal length (f) of a converging lens is given as positive whereas for a diverging lens f is given as negative. The distances u, v and f are all measured along the principal axis.

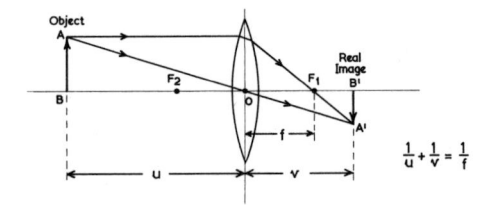

(a) CONVEX LENS

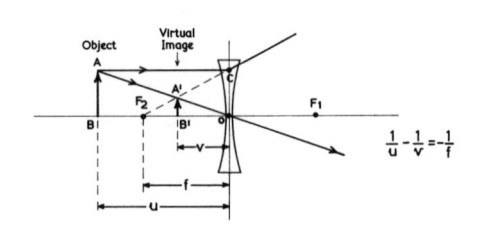

$$\frac{1}{u} + \frac{1}{v} = \frac{1}{f}$$

(b) CONCAVE LENS

$$\frac{1}{u} - \frac{1}{v} = -\frac{1}{f}$$

FIG. 2.11 CALCULATING THE IMAGE POSITION

Since in figure 2.11(a) the triangles ABO and $A'B'O$ are similar, it may be proved that:

$$\frac{1}{u} + \frac{1}{v} = \frac{1}{f}*$$

*See Appendix D, page 243.

LENSES

This is the general formula which is true for the convex lens. The formula is only true for the concave lens when the conventions outlined above are employed, *i.e.* f and v are taken as negative in which case we have:

$$\frac{1}{u} - \frac{1}{v} = -\frac{1}{f}.$$

MAGNIFICATION

A lens which forms an image three times as high as the object is said to produce a magnification of 3.

$$\text{Magnification} = \frac{\text{Height of image}}{\text{Height of object}} \quad \text{(same units)}$$

Width, of course, is magnified proportionally to height.
Also, since the triangle ABO and $A'B'O$ in figure 2.11(a) are similar.

$$\text{Magnification} = \frac{v}{u}$$

Consider now the following examples:
Example (1)
A golfer 2 metres high is 50 metres away from a convex lens having a focal length of 100 mm. Find the focused image height.

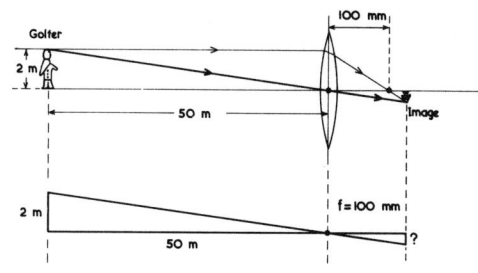

FIG. 2.12 FINDING THE IMAGE HEIGHT

$$\frac{1}{f} = \frac{1}{u} + \frac{1}{v} \text{ or } \frac{1}{v} = \frac{1}{f} - \frac{1}{u}$$

$$\therefore \frac{1}{v} = \frac{1}{100} - \frac{1}{50 \times 10^3} \text{ mm.}$$

$$\frac{1}{v} = 0.01 - 0.00002$$

$$\frac{1}{v} = 0.00998$$

$$v = 100.2 \text{ mm.}$$

Since magnification $= \dfrac{\text{height of image}}{\text{height of object}} = \dfrac{v}{u}$

$$\text{height of image} = \frac{100.2}{50 \times 10^3} \times 2 \times 10^3$$

$$\simeq 4 \text{ mm}$$

Example (2)

The scanned area of a 1″ vidicon tube is 12·7 mm wide by 9·25 mm high. An area of a control panel which is 11·35 cm high is required to be monitored by a CCTV system. The panel is 1·791 metres away from the camera lens. Find the focal length of the lens required so that the panel height dimension just fills the scanned area of the camera tube.

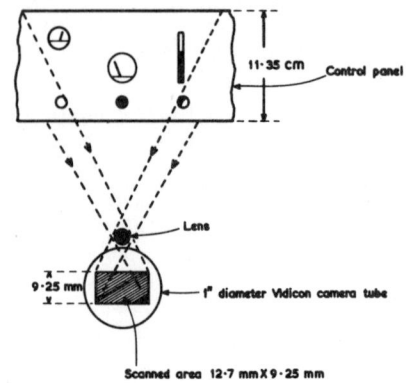

FIG. 2.13 FINDING THE FOCAL LENGTH OF THE LENS REQUIRED

$$\text{Magnification} = \frac{\text{height of image}}{\text{height of object}} = \frac{v}{u}$$

$$\therefore v = \frac{u \times \text{height of image}}{\text{height of object}}$$

$$= \frac{1 \cdot 791 \times 10^3 \times 9 \cdot 25}{11 \cdot 35 \times 10}$$

$$= 146 \text{ mm.}$$

Now $\frac{1}{f} = \frac{1}{u} + \frac{1}{v}$

$$\frac{1}{f} = \frac{1}{1791} + \frac{1}{146}$$

$$\frac{1}{f} = 0 \cdot 000558 + 0 \cdot 006849$$

$$\frac{1}{f} = 0 \cdot 007407$$

$$\therefore f = \underline{135 \text{ mm.}}$$

Formulae more practical than

$$\frac{1}{f} = \frac{1}{u} + \frac{1}{v}$$

are available for working out lens requirements. These can be found from figure 2.14

LENSES

which represents a simplified plan view for CCTV lens requirements. A t.v. camera only produces an output signal from that part of the scene image that falls on the scanned area of the light-sensitive surface of the camera tube face. The diagram shows the geometrical projection from the field of view width (W) to the scanned area width (w).

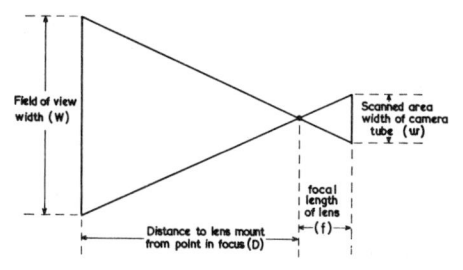

FIG. 2.14 SIMPLIFIED PLAN VIEW FOR CCTV LENS REQUIREMENTS

From similar triangles it can be said that:

$$\frac{f}{w} = \frac{D}{W} \qquad f = \frac{Dw}{W} \quad \text{...............................(i)}$$

where f = focal length of lens
w = width of scanned area on camera face
D = distance from lens mount to subject
W = width of field of view at subject (at distance D).

The approximation of equation (i) assumes that D is large compared with f which is usually the case. The measurements should all be in the same units. However, if D and W are in the same units, e.g. feet or metres, and f and w are in the same units, inches or millimetres, then the result will be the same. As the scanned area of a 1″ vidicon is 12·7 × 9·25 mm or 0·5 × 0·375 in., these widths can be instantly included in the formula instead of w.

$$\text{So } f = \frac{D}{W} \times 0.5 \quad \text{or} \quad \frac{D}{2W} = f \text{ in inches} \quad \text{...................(ii)}$$

$$\text{or } f = \frac{D}{W} \times 12.7 \text{ in mm} \text{..(iii)}$$

Example (1)

A test card measuring 12″ × 9″ is to be used for setting up the camera geometry controls. Find the distance that the card must be placed from the lens mount to fully utilise the scanned area of a 1″ vidicon tube when a lens having a focal length of 20 mm is used.

Using equation (i)

f = 20 mm
w = 12·7 mm
W = 12″
D = ? (in.)

$$D = \frac{fW}{w}$$
$$= \frac{12 \times 20}{12.7}$$
$$\simeq \underline{18.9 \text{ in}}$$

or using (iii)

$f = 20$ mm.
$w = 12.7$ mm
$W = 303.6$ mm
$D = ?$ (mm)

$$D = \frac{Wf}{12.7}$$

$$= \frac{303.6 \times 20}{12.7}$$

$$= 478 \text{ mm}$$

$$\simeq 18.8 \text{ in.}$$

Example (2)
A CCTV camera is to be used for surveillance in a factory site. If a 17 mm lens is used what will be the width of the viewing area at a distance of 60 metres from the lens mount?

Using equation (iii)

$$f = \frac{D}{W} \times 12.7$$

$$\therefore W = \frac{D}{f} \times 12.7$$

$$= \frac{60 \times 10^3 \times 12.7}{17} \text{ mm}$$

$$= 44.82 \text{ metres}$$

A table giving the width of the field of view for standard lens focal lengths is shown in figure 2.15. To find the height of the field of view multiply

$$W \text{ by } 0.73 \left(\frac{9.25}{12.7}\right)$$

Distance (D) →	2'	4'	5'	10'	20'	50'	100'	150'	200'
Focal length (f) ↓	W	W	W	W	W	W	W	W	W
12.5 mm	2.03'	4.06'	5.08'	10.16'	20.32'	50.8'	101.6'	152.4'	203.2'
25 mm	1.02'	2.03'	2.54'	5.08'	10.16'	25.4'	50.8'	76.2'	101.6'
50 mm	0.51'	1.02'	1.27'	2.54'	5.08'	12.7'	25.4'	38.1'	50.8'
75 mm	0.34'	0.68'	0.85'	1.69'	3.39'	8.47'	16.93'	25.4'	33.87'
100 mm	0.25'	0.51'	0.64'	1.27'	2.54'	6.35'	12.7'	19.05'	25.4'
135 mm	0.19'	0.38'	0.47'	0.94'	1.88'	4.7'	9.4'	14.1'	18.8'

FIG. 2.15 TABLE SHOWING FIELD OF VIEW WIDTH (W) FOR VARYING DISTANCES (D) FOR 1" VIDICON TUBE

LENSES

LENS COMBINATIONS

Lenses are often used in pairs, in contact with one another or separated by air. Each lens may have a different focal length so it is necessary to know what the effective focal length of the combination will be.

With two convex lenses *in contact*, the effective focal length (f_e) may be found from:

$$\frac{1}{f_e} = \frac{1}{f_1} + \frac{1}{f_2}$$

where f_1 and f_2 are the focal lengths of the individual lenses.

Example

Find the effective focal length of two convex lenses which are in contact if the focal lengths are 28 mm and 35 mm respectively.

$$\frac{1}{f_e} = \frac{1}{f_1} + \frac{1}{f_2}$$

$$= \frac{1}{28} + \frac{1}{35}$$

$$= 0.0357 + 0.0286$$

$$= 0.0643$$

$$f_e = \underline{15.55 \text{ mm.}}$$

When the lenses are not in contact, finding the effective focal length becomes more involved and the details need not concern us here. As far as the technician is concerned, the effective focal length will be found engraved on the side of the lens and it is this value which is used in calculations concerning field of view widths, etc. A simple t.v. camera lens may contain only a few glass sections but in a zoom lens there may be twenty sections (or elements), some of which are in contact and others not.

POWER OF A LENS (DIOPTRE)

The shorter the focal length of a lens the more it causes the rays to bend. This is referred to as the **power of the lens** and is measured in dioptres.

A lens having a focal length of 1 metre is said to have a power of 1 dioptre. Thus the power of a lens in dioptres is given by:

$$\frac{1}{\text{focal length in metres}} = F \text{ dioptres}$$

e.g. the power of a 50 mm lens is

$$\frac{1}{50 \times 10^{-3}} = \frac{1000}{50}$$

$$= \underline{20 \text{ dioptres.}}$$

The shorter the focal length the greater the power. The power of a converging lens is positive and that of a diverging lens is negative. Thus, equal power lenses in combination have no effect. A converging lens combined with a diverging lens of less power results in a converging (positive) unit.

Opticians use the dioptre instead of focal length as time is saved in working out reciprocals.

DEPTH OF FIELD

The light coming from a point on an object is not imaged by a lens as a spot of infinitely small size but rather as a disc of light of finite size. The larger the disc, the less sharply is it defined. These discs of light are called 'circles of confusion' and for good definition in a television or film camera they must be kept below a certain size.

Figure 2.16 shows a point object O_1 focused by the lens to produce a small disc I_1 in the plane AB. In a t.v. camera the tube face may be arranged to lie in this plane but in a

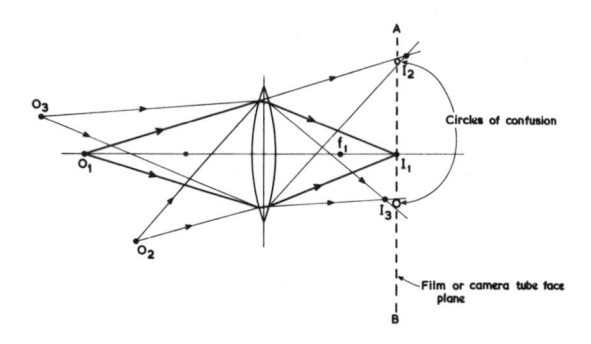

FIG. 2.16 SHARP FOCUS OCCURRING IN DIFFERENT PLANES FOR VARIOUS SUBJECT-LENS DISTANCES

film camera the film will be located here. A point object O_2 closer to the lens will be sharply focused at I_2 behind the plane AB, whereas a point object O_3 farther away from the lens will form a sharp focus at I_3 just in front of the plane AB. Objects O_2 and O_3 are not in sharp focus, but the focusing may be acceptable if the circles of confusion are not too large.

Photographers often take 0·0025 mm as the maximum value for the circle of confusion. Smaller circles of confusion are required in a negative that will be printed by enlargement as they are enlarged together with the image.

In a t.v. system, the smallest detail that can be resolved is the size of one element and the greatest circle of confusion that can be tolerated may be taken to be the size of an element on the light sensitive face of the camera tube. For example, with a 1" vidicon having a scanned area height of 9·25 mm, 585 lines are actively employed in scanning the image in a 625-line system.

Therefore the height of each element is:

$$\frac{9 \cdot 25}{585} = 0 \cdot 0158 \text{ mm}.$$

This assumes that there is no gap between successive scanning lines. In practice this is not so, thus smaller circles of confusion are indicated.

The total distance through which an object may be moved along the lens axis before the maximum circle of confusion is reached is known as the **depth of field**. This is illustrated in figure 2.17. If the point-object shown is moved along the line AB, the object will remain in acceptable focus over the distance CD which is the depth of field. The depth of field depends upon the largest circle of confusion that can be tolerated, increasing with an increase in circle diameter. Depth of field is also affected by the

LENSES

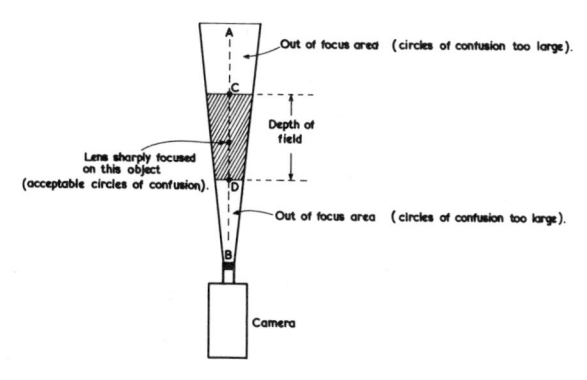

FIG. 2.17 DEPTH OF FIELD

properties of the lens system and this will be explained later.

To achieve maximum depth of field a lens must be focused on a certain object plane. The distance of this plane from the lens is known as the **hyperfocal distance,** figure 2.18. It is the object distance for which the diameter of the circles of confusion in the image plane just equal the maximum permissible value when the lens system is focused on infinity.

In figure 2.18, an object at O_1 is brought to a sharp focus in the image plane, *i.e.* where the camera tube face is located. If the object is moved towards the lens, the circle

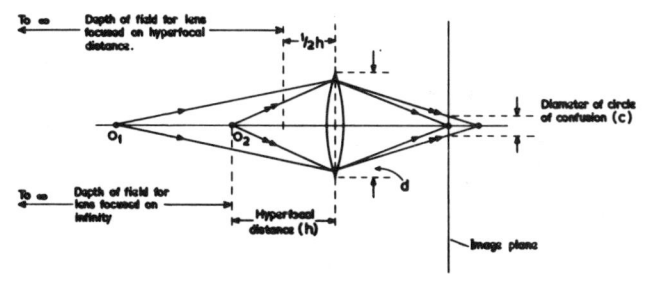

FIG. 2.18 HYPERFOCAL DISTANCE

of confusion grows until, when the object is in position O_2, the maximum circle of confusion is produced in the image plane. The distance O_2 to the lens is the hyperfocal distance. Thus, when the lens is focused to infinity, objects lying between O_2 and infinity are in acceptable focus. If the object is placed at the hyperfocal distance and brought to focus, images are in focus from half the hyperfocal distance to infinity.

This may be extended further, for if the lens is focused on an object at half the hyperfocal distance, images are in focus for all objects lying between the hyperfocal distance and one third of it.

It may be shown that the hyperfocal distance $(h) = \dfrac{f.d}{c}$ (same units)

where f = focal length of lens
d = diameter of lens
c = diameter of circle of confusion

An alternative expression which is more practical is:

$$h = \frac{(\text{focal length})^2}{\text{lens stop number} \times c}$$

Before giving examples to show the application of this expression, an explanation of 'lens stop number' will be given.

LENS STOP NUMBER (f-number)

Inside a t.v. camera lens there is a diaphragm for controlling the amount of light that passes through the lens system. It consists of a series of flat metal blades interleaved to form a central hole, the size of which may be adjusted by turning a ring on the lens barrel.

The operation of the diaphragm (iris or stop) is illustrated in figure 2.19. The object AB is placed nearer the lens than the position required to form a sharp image on the

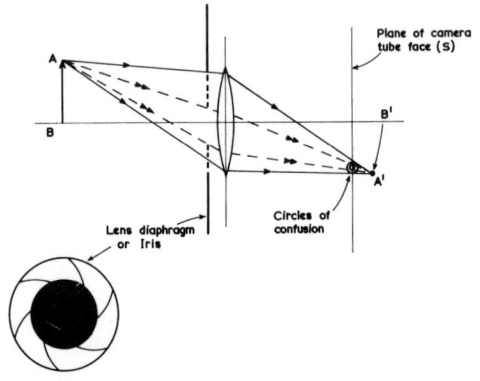

FIG. 2.19 USE OF THE LENS IRIS STOP

camera tube face. With the iris fully opened, light from the object fills the lens with light producing a circle of confusion which is unacceptable. If the iris is not fully opened (shown dotted) only light striking the centre of the lens succeeds in reaching the image space, but note that the circle of confusion is greatly reduced. Note also that the image will be slightly curved. A is slightly farther from the lens than B resulting in an image A' nearer the lens than that formed at B' by rays from B.

The amount of light passing into a lens system is referred to as its aperture, stop or f-number which is defined as:

$$\frac{\text{Aperture ratio}}{\text{(f-number)}} = \frac{\text{focal length of lens}}{\text{diameter of aperture}}$$

The aperture sizes are stated in fractions of the lens focal length having an aperture equal to the focal length, e.g.

$$\frac{f}{1}, \frac{f}{1\cdot 4}, \frac{f}{2}, \frac{f}{2\cdot 8}, \frac{f}{4}, \frac{f}{5\cdot 6} \text{ etc.,}$$

which are commonly stated as $f1$, $f1\cdot4$, $f2$, $f2\cdot8$, $f4$, $f5\cdot6$ etc.

LENSES

The reason for choosing these particular fractions may be seen from the following: The brightness of the image is proportional to

$$\frac{(\text{diameter of aperture})^2}{(\text{focal length})^2}$$

thus the brightness ratios for the f-numbers given would be

$$\frac{1}{(1)^2} : \frac{1}{(1.4)^2} : \frac{1}{(2)^2} : \frac{1}{(2.8)^2} : \frac{1}{(4)^2}$$
$$\text{or } 1 : \tfrac{1}{2} : \tfrac{1}{4} : \tfrac{1}{8} : \tfrac{1}{16}, \quad \text{etc.}$$

Thus the scale chosen for marking variable apertures *consecutively halves the brightness of the image*. Therefore, as the f-number increases, the illumination decreases but the circle of confusion is reduced.

Lenses are usually specified by stating their focal length and the maximum f-number, *e.g.* 10 mm f/1·4. In this case with maximum aperture, the lens diameter

$$= \frac{10}{1.4} \text{ mm} \simeq \underline{7.14 \text{ mm.}}$$

Example

Determine the hyperfocal distance of a 10 mm t.v. camera lens when using aperture settings of (a) f/1·4; and (b) f/5·6.

Using $h = \dfrac{(\text{focal length of lens})^2}{\text{lens stop no.} \times c}$ and assuming that $c = 0.016$ mm.

(a) $h = \dfrac{100}{1.4 \times 0.016}$ mm

$= \underline{4.464 \text{ m.}}$

(b) $h = \dfrac{100}{5.6 \times 0.016}$ mm

$= \underline{1.12 \text{ m.}}$

These results are illustrated in figure 2.20 (a) where the camera is focused at infinity. Clearly the depth of field is greatest when using f/5·6. If the focus is set at the hyperfocal distance then the depth of field extends from half the hyperfocal distance to infinity, figure 2.20 (b). In general, if a lens system is focused on a distance

$$\frac{h}{k}$$

(where h is the hyperfocal distance and k is any number), images are in focus for all objects lying between

$$\frac{h}{(k-1)} \quad \text{and} \quad \frac{h}{(k+1)}.$$

(a) FOCUS AT INFINITY

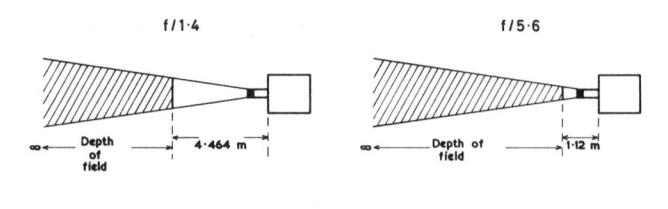

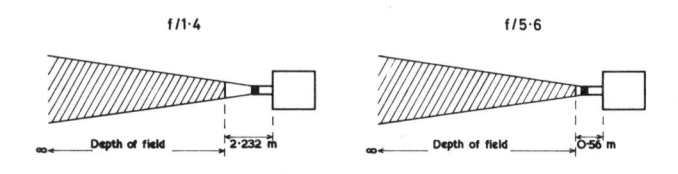

(b) FOCUS SET AT THE HYPERFOCAL DISTANCE

FIG. 2.20 EFFECT OF APERTURE SETTING ON DEPTH OF FIELD (10 mm LENS)

Thus if the focus is set to 0·56 metre, then with f/5·6 objects are in focus from

$$\frac{h}{(2-1)} \text{ and } \frac{h}{(2+1)}$$

i.e. from

$\frac{1·12}{1}$ m and $\frac{1·12}{3}$ m, whereas for f/1·4 objects are in focus from between

$$\frac{h}{(8-1)} \text{ m and } \frac{h}{(8+1)} \text{ m } i.e. \text{ from } \frac{4·464}{7} \text{ m and } \frac{4·464}{9} \text{ m.}$$

This is shown in figure 2.21.

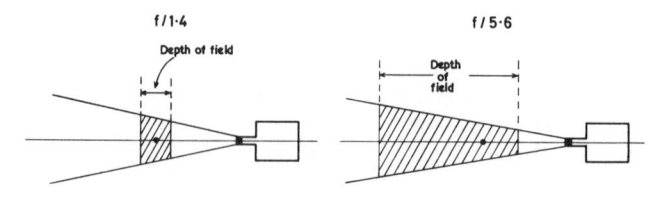

FIG. 2.21 FOCUS SET TO 0·56 METRE

FOCUSING

In the previous example we have discussed focusing the camera lens at infinity. the hyperfocal distance and a specific distance and we have seen that the depth of field is different. How is it then that in some photographic cameras there is no focus adjustment yet pictures remain acceptably sharp? Fixed-focus photographic cameras represent a compromise in design for by setting the focus at the hyperfocal distance, images are quite sharp between half the hyperfocal distance and infinity. Although a fixed-focus lens may be suitable for some CCTV applications, focusing is not crisp enough for close range studio work. Also, since a small aperture is needed to obtain depth of field, studio lighting intensities need to be high.

An adjustable focus lens has greater flexibility than a fixed-focus lens, allowing the camera operator to focus sharply on various object distances without the need for reducing the lens aperture (which may be troublesome with the varying lighting conditions met with in industry). Also, for studio work, the operator can focus on to a face and cause distractions in the distance to appear as a background blur.

The focus may be adjusted by a focus ring on the lens barrel which moves the lens in and out from the camera tube faceplate. The distance for sharp focus is usually indicated by an engraved marker which identifies the distance in feet or metres. In more refined (and expensive) cameras, focusing may be effected by moving the camera tube and its yoke assembly along the lens axis whilst maintaining a fixed focus for the lens (if a focusing ring is fitted to the lens it may be set to infinity). This method is adopted when a lens turret containing several lenses of different focal lengths is used. It has the advantage of greater range of focus and allows the focusing control to be fitted at the rear of the camera which is more convenient for the operator.

For close-up work, *i.e.* 1 metre or less from the camera, objects may not produce a sufficiently sharp image for a particular lens assembly and its range of focus adjustment. In these circumstances other techniques may be used. One method is to fit a supplementary lens by screwing it on to the front of the main lens. A positive supplementary lens shortens the focal length of the lens to which it is fitted and allows the camera to be taken nearer to the subject whilst maintaining a sharper image than is normally possible. Positive supplementary lenses are classified in various 'powers', the power (dioptres) of a lens being the reciprocal of the focal length in metres (see page 33). Powers of $\frac{1}{2}$, 1, 2 and 3 dioptres are available which have focal lengths of 2, 1, 0·5 and 0·33 metres respectively. For example, if the main lens focus is set to infinity, a + 2 supplementary attachment will produce a sharp image of an object 0·5 metre away. Tables are usually supplied with the supplementary lens indicating the object distances for various focusing distances on the main lens scale. Another method is to use an extension tube which is a hollow metal cylinder, available in various lengths. One end of the tube screws into the camera faceplate and the other end is threaded to receive the camera lens. The extension tube physically moves the lens nearer to the subject causing the plane of sharp focus to move forward so that it can be made to coincide with the face of the camera tube by the normal focus adjustment.

LENS ANGLE

For studio work and some industrial applications it is useful to be able to alter the angular view of the lens system. For example, when a camera is being used for factory surveillance, the security operator may want a general or wide view of the factory site for most of the time but when there is some suspicious activity occurring a restricted or narrow view is needed.

A television camera 'sees' a segment of the scene in front of it having width-to-height proportions of 4:3 (the aspect ratio), figure 2.22. Thus, a scene must have this aspect ratio if it is to exactly fill the active area of the camera tube face. Objects which do not bear these proportions must either be shot to lie within these viewing requirements or part will be lost beyond the active area of the camera tube face.

Figure 2.23 (a) shows a test picture placed at a suitable distance from a camera so that its image dimensions just fill the active area of the camera tube. As indicated in the diagram, the monitor displays exactly what the lens 'sees' in this viewing situation. If the angle of view of the lens is altered as in figure 2.23 (b) to give a narrower view only a small portion of the test picture (but still of aspect ratio 4:3) is seen by the lens. This smaller portion of the scene, however, now fills the monitor screen. Thus, altering the lens angle produces an apparent change in the camera-to-subject distance and the size of the subject. How is the lens angle changed? We saw on page 32 that by altering the focal length of a lens, the field of view width (and height) was changed. This is a direct result of a change in lens angle. Altering the focal length changes the lens angle. Figure

40   INDUSTRIAL AND COMMERCIAL CCTV

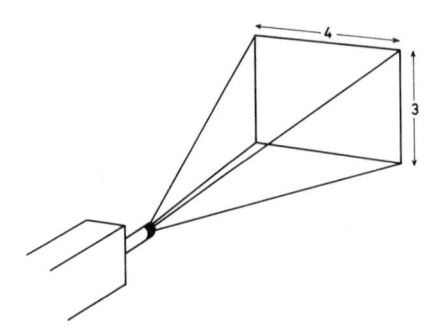

FIG. 2.22   SCENE PROPORTIONS 'SEEN' BY LENS

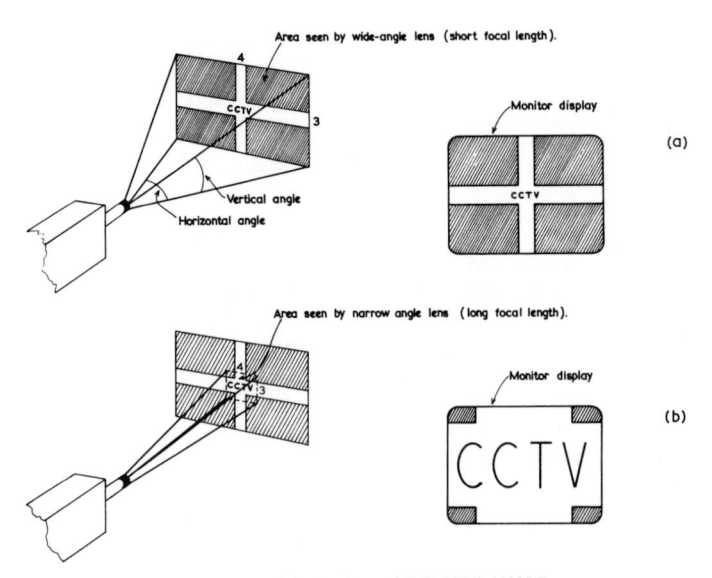

FIG. 2.23   EFFECT OF CHANGE IN LENS ANGLE

2.24 illustrates how the lens angle is derived. In this diagram only light rays which originate from the extremities of the scene are shown and these are assumed to pass through the centre of the lens system. The rays produce an image on the camera tube of width (w).

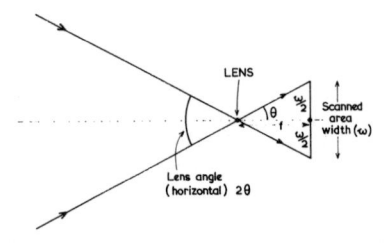

FIG. 2.24   DERIVING THE LENS ANGLE

LENSES

If f is the focal length

$$\text{Tan } \theta = \frac{w}{2f}$$

$$\text{or } \theta = \text{Arc Tan } \frac{w}{2f}$$

$$\therefore \text{Lens angle } (2\theta) = 2 \text{ Arc Tan } \frac{w}{2f}$$

This gives the horizontal angle of the lens, but if the scanned area height is substituted for w the vertical angle may be obtained.

EXAMPLE

Find the horizontal and vertical lens angles for a 10 mm lens when used with a vidicon tube having an active area of 12·7 mm by 9·25 mm.

$$\text{Horizontal angle} = 2 \text{ Arc Tan } \frac{12\cdot7}{2 \times 10}$$

$$= 2 \text{ Arc Tan } 0\cdot635$$

$$= \underline{64\cdot83°}$$

$$\text{Vertical angle} = 2 \text{ Arc Tan } \frac{9\cdot25}{2 \times 10}$$

$$= 2 \text{ Arc Tan } 0\cdot4625$$

$$= \underline{49\cdot64°}$$

The graph in figure 2.25 shows how the lens angle varies with focal length when working into a 1″ vidicon tube. A lens of long focal length, say 200 mm, has an angle of view of about 4°, whereas a short focal length of 125 mm provides a lens angle of approximately 59°. It should be appreciated that if the dimensions of the scanned area are altered (when working with camera tubes of other sizes) the lens angle changes for a lens of given focal length.

In studio work, lens angles from about 50° for wide shots to 5° for narrow shots are typical extremes. CCTV cameras are often supplied with a 'normal' lens ($f = 25$ mm)

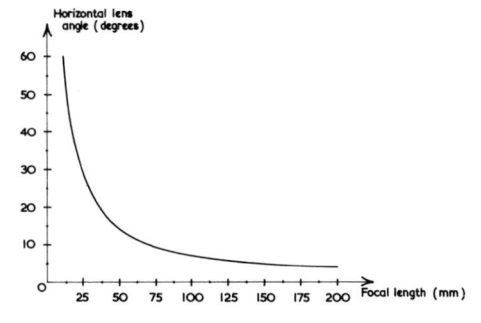

FIG. 2.25 GRAPH SHOWING HOW LENS ANGLE VARIES WITH FOCAL LENGTH OF LENS (1″ VIDICON TUBE)

which with a 1" vidicon tube has a lens angle of about 28°. A 'normal' lens is one with an angle which approximates to the angle subtended from the edges of a t.v. screen to the viewer's eyes. For the purpose it is assumed that the viewer's distance from the screen is such that he can just discern the finest detail of the television system. When using a 'normal' lens, the viewer's impression of relative size and distance will more closely correspond with those presented to the camera.

LENS DEFECTS

Lenses suffer from a number of defects which cause distortion of the image or unwanted colour in the image. Only the principal aberrations and the measures taken to reduce them will be described.

(a) Spherical Aberration

If the surface of a convex lens is truly spherical, rays from a point on an object will not converge at a single point in the image space. This is illustrated first in figure 2.26(a) for a parallel beam falling on a convex lens. The point of focus is seen to vary with the

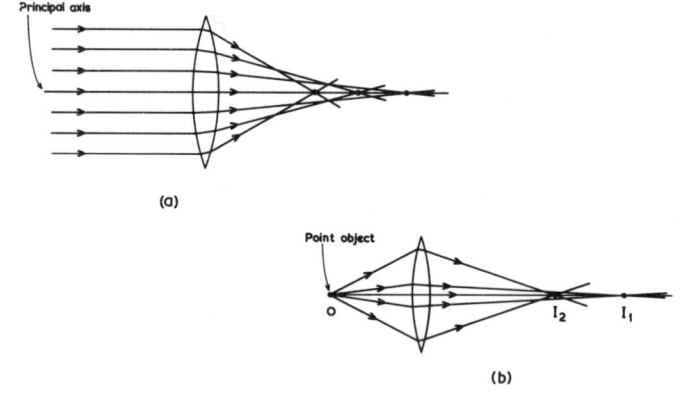

FIG. 2.26 SPHERICAL ABERRATION

height of the incident rays from the principal axis of the lens, moving closer to the optical centre with increasing height. Consider now light coming from a point object O as in figure 2.26(b). If spherical aberration is present, those rays close to the axis are brought to focus at I_1 whereas those rays refracted at the edges of the lens focus at I_2. The sharpest image is somewhere between I_1 and I_2 and is not a point but a patch of light (the circle of least confusion).

Spherical aberration can be eliminated by giving the refracting surfaces parabolic curvature (this also applies to spherical mirrors). However, parabolic surfaces are difficult to form for small lenses since they must be very accurately ground. Spherical aberration is at its worst for high angles of incidence, so for a single lens the deviation should be shared fairly equally between its surfaces. A plano-convex lens produces less aberration if the object is at infinity and light strikes the curved surface first.

Two converging lenses can be combined to reduce spherical aberration as the deviation is shared between four surfaces instead of two.

Another method is to use a combination of convex and concave elements because the aberration of a concave lens is of opposite sign to a convex lens. Such a combination can be given the required focal length and arranged for the aberrations to cancel out.

It is clear from figure 2.26 (b) that if a lens iris were used to block the rays entering the edges of the lens a more definite image would be obtained. Thus, stopping-down a lens reduces spherical aberration but will reduce the image brightness.

(b) Astigmatism

Rays of light from a point object which strike the lens obliquely are not brought to a single point but to two focal lines in the image space as illustrated in figure 2.27. This is because the curvature of the lens is different in the vertical and horizontal planes from

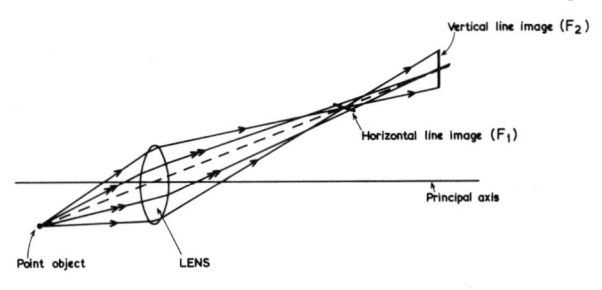

FIG. 2.27 ASTIGMATISM

the point of view of the incident light. At intermediate points between F_1 and F_2 the image is elliptical and at one point circular (the optimum focus point).

Astigmatism is not present for light arriving on the principal axis, but increases rapidly with the obliqueness of the incident light. When forming an image of a real object spherical aberration must be accompanied by astigmatism since some of the rays must strike parts of the lens obliquely. Astigmatism may be reduced by stopping-down the lens to block the oblique rays.

(c) Chromatic Aberration

When a parallel beam of white light which is a mixture of coloured lights is passed through a lens as in figure 2.28 (a), the red rays are brought to a focus F_r, the yellow rays at F_y and the violet rays at F_v. This occurs because the refractive index for the glass varies with the colour of the incident light being greater for violet than for red wavelengths. A higher refractive index will cause a greater amount of bending for the violet rays; these will be brought to focus closer to the lens than the yellow or red rays.

Diagram (b) shows the effect (greatly exaggerated) on imaging an object which is reflecting white light. I_v, I_y and I_r are the respective images of the object formed by light of these colours. This effect is called **chromatic aberration.** With a concave lens the same

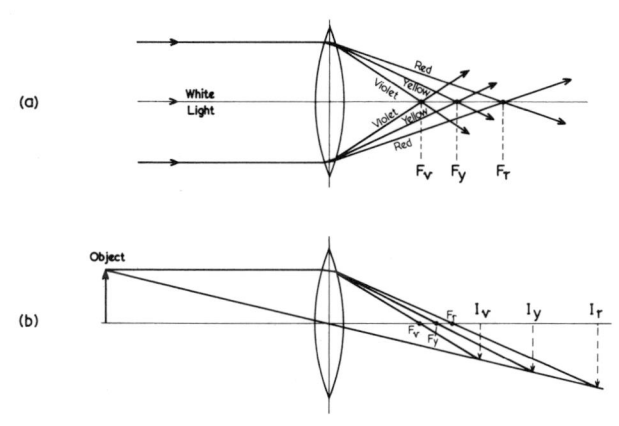

FIG. 2.28 (a) & (b) CHROMATIC ABERRATION

(c)

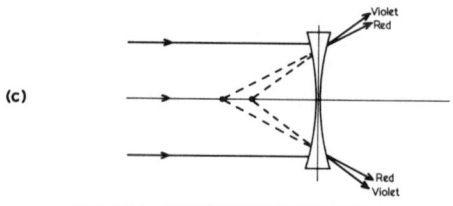

FIG. 2.28 (c) CHROMATIC ABERRATION

reasoning applies and the violet light is refracted more than the red, figure 2.28 (c).

For a simple lens, chromatic aberration is far greater than spherical aberration and it is essential to reduce the defect. To produce an **achromatic lens,** *i.e.* one which does not produce separate colour images, a combination may be used consisting of a powerful convex lens made of crown glass (low dispersive power) and a weak concave lens made from flint glass (high dispersive power). This is shown in figure 2.29 where the two faces in contact have the same curvature and may be cemented together with

FIG. 2.29 ACHROMATIC DOUBLET

Canada balsam to reduce reflections losses. The combination is known as an **achromatic doublet,** where the increasing convergence of the violet rays in the convex lens is cancelled by the increasing divergence of the violet rays in the concave lens.

(d) Flare Spots

The amount of light that passes through a lens to form an image varies from about 0·5 to 0·95 of the light which is incident on the lens. Some of the light energy is lost by absorption in the lens material. The largest proportion is lost as a result of reflection at the lens surfaces. In a complex lens unit where there are many reflecting surfaces the overall loss may be serious. Apart from loss of light energy, the reflected energy may find its way into the image space. The idea is illustrated in figure 2.30 where the incident light ray is reflected (shown dotted) at points X, Y and Z and finally emerges from face

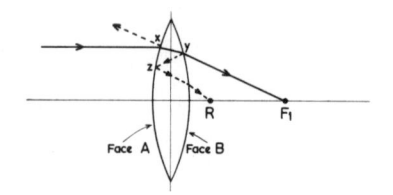

FIG. 2.30 REFLECTION WITHIN LENS CAUSING FLARE SPOTS

LENSES

B of the lens. With other rays arriving at X from different directions, a bright spot would be produced around R on the principal axis and a diffused spot in the image plane. It follows that for other incident points on face A of the lens, similar diffused spots may appear in the image plane and are known as **flare spots.** These diffused spots are usually scattered over the image space and have the effect of degrading the shadow areas of the picture.

Reflections at the lens surfaces may be reduced by coating them with a very thin film of magnesium fluoride. Lens treated in this way are termed 'coated' or 'bloomed' and can be recognised by their bluish appearance when held up against the light. Care must be exercised when cleaning a bloomed lens (see page 52) as the coating is extremely thin (a quarter of a wavelength of particular light radiations).

LENSES FOR CCTV

Lenses used to produce images of high quality for CCTV are corrected as far as possible for the various aberrations and defects affecting them.

SINGLE FOCAL LENGTH LENSES
(a) Long focal length (narrow angle)

A narrow angle lens enables the camera operator to select small detail from a distant scene that would otherwise be quite impossible to reach. The small detail fills the entire screen of the picture monitor. Because of the magnification of distant detail, cameras using narrow-angle lenses are more sensitive to camera movement; following distant action requires considerable skill on the part of the operator. Depth of field is reduced as the lens angle is made smaller, making focusing more difficult.

Lenses with a long focal length are inconvenient for use with t.v. or photographic cameras, *e.g.* a 500 mm lens requires a lens about 20 inches from the face of the camera tube, resulting in a clumsy arrangement. A **telephoto lens** provides a solution because it has a long focal length, but the distance between the lens and the camera tube face (the back focus) is shorter than the focal length.

A telephoto lens consists of a front convex lens unit and a back concave lens unit with a separation (d) which is less than the focal length of the convex unit, figure 2.31. If

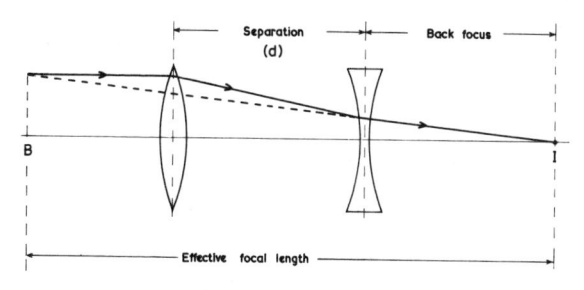

FIG. 2.31 PRINCIPLE OF TELEPHOTO LENS

f_1 is the focal length of the convex lens and f_2 is the focal length of the concave lens, the effective focal length may be found from

$$\frac{1}{f} = \frac{1}{f_1} + \frac{1}{f_2} - \frac{d}{f_1 f_2}$$

With $f_1 = +100$ mm, $f_2 = -100$ mm and $d = 50$ mm

$$f = \underline{200 \text{ mm.}}$$

Thus the combination of figure 2.31 is effectively the same as a convex lens of 200 mm focal length in the plane at B. Now, the back focus may be found from

$$\text{back focus} = \frac{f(f_1 - d)}{f_1}$$

$$= \frac{200(100 - 50)}{100} \text{ mm}$$

$$= 100 \text{ mm}.$$

The length of the lens mounting is the back focus $+ d$, i.e. 150 mm. Thus, in this case the telephoto principle results in an effective focal length of 200 mm but a lens mounting length of 150 mm. The ratio of effective focal length to the back focus is called the **telephoto magnification** which in this case is 2.

Figure 2.32 shows the format for lenses of long focal length adopting the telephoto principle. These practical lens arrangements are corrected for the lens aberrations previously described.

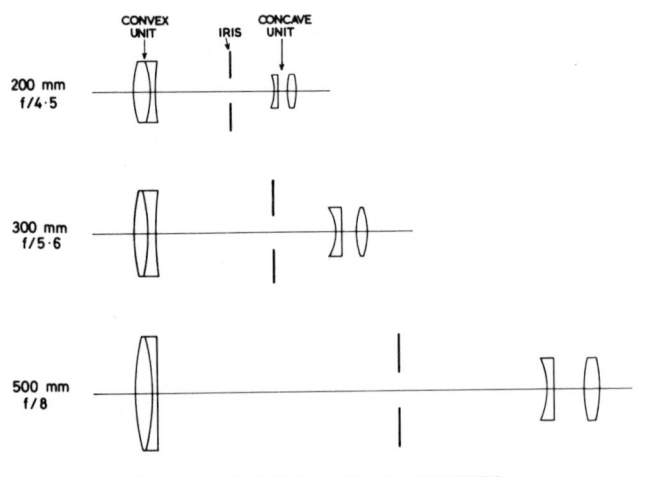

FIG. 2.32 TELEPHOTO LENS ARRANGEMENTS

(b) Short focal length (wide-angle)

A wide-angle lens gives the impression of a more distant view and is particularly useful in small studios where it gives the illusion of a spacious setting. More of the scene is shown with a wide-angle lens compared with a 'normal' lens, but the size of the detail is reduced in proportion. Thus, detail is more difficult to discern and the effect is one of a general viewpoint.

Depth of field is greater than with a lens of long focal length and movement of the subject is easier to follow. With widening lens angle, camera operation becomes easier being less susceptible to spurious movement of the camera. Lens flares are more of a problem with wide-angle lenses.

A problem with wide-angle (short focal length) lenses is that sometimes the lens cannot be brought close enough to the face of the camera tube to give a sharp image. To deal with this, an 'inverted telephoto' construction may be employed to produce a 'retrofocus lens' which has a long back focus compared with its focal length.

(c) Zoom Lens

A zoom lens is a complex optical arrangement providing a variable focal length which can be continuously altered. It has all of the advantages of a set of turret mounted, fixed focal lenses and at the movement of one control can provide any focal length within its extremities, *e.g.* 20–100 mm. Zoom lenses are designed to provide various ratios of maximum focal length; typical ratios are 3:1, 5:1, 8:1 and 10:1.

With a zoom lens the operator is able to change the field of view from a wide angle to a narrow angle ('zooming in'), smoothly and at a speed which suits the situation. On the other hand the view may be changed from a narrow angle to a wide angle ('zooming out') equally as smooth.

To change the focal length of an optical system, the position of a lens element or a group of elements has to be altered which causes a movement of the focal plane position. Since the focal plane is where the camera tube face is located, the movement would cause the image to go out of focus. This effect is reduced by what is called 'compensation'.

OPTICAL COMPENSATION

Here a stationary element or group of elements is sandwiched between two elements or groups of elements which are coupled so that they move together, as in figure 2.33. This diagram shows the movement of the image plane (greatly magnified) as the focal length is altered from one end of the zoom range to the other. The designer

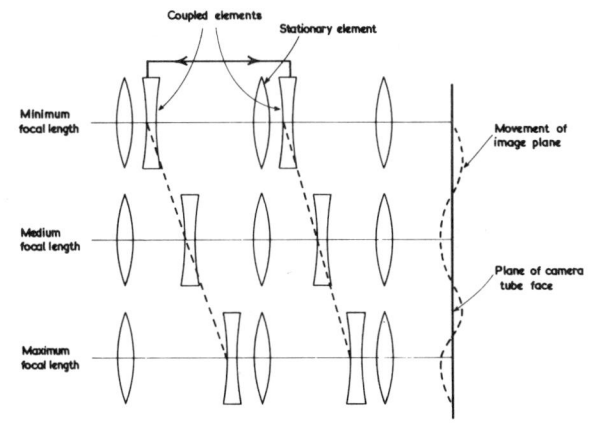

FIG. 2.33 OPTICAL COMPENSATION IN ZOOM LENS

limits this movement to within the depth of focus so that there is no noticeable deterioration of focus during zooming operations. Good results are obtained with this method provided the lens specification range is not too extreme.

MECHANICAL COMPENSATION

With mechanical compensation a second group of elements moves in the opposite direction on a non-linear cam to achieve focus in a fixed position. The idea is shown in figure 2.34. In practice, the positions of the glass components have to be controlled to within one thousandth of an inch and some of the components, particularly the cam, are machined to an accuracy of a few ten thousandths of an inch. Such precision is to be found in zoom lenses designed for the high quality demanded by the broadcast networks.

Figure 2.35 shows the complex optical arrangement for a zoom lens utilising 18 glass elements, five of which move as one unit to vary the focal length of the system. A

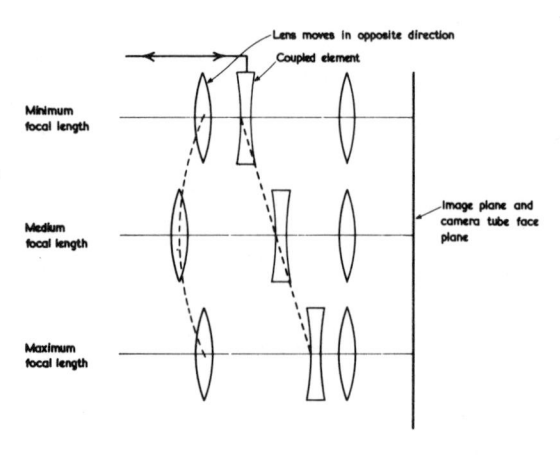

FIG. 2.34 MECHANICAL COMPENSATION IN ZOOM LENS

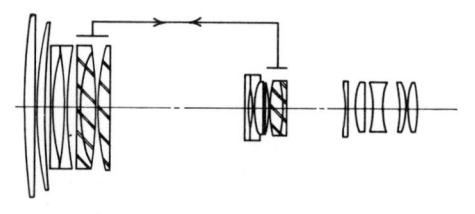

(The two moving lens units move as one in a piston which slides in the barrel on low-friction Teflon slipper pads).

FIG. 2.35 ONE FORM OF ZOOM LENS USING OPTICAL COMPENSATION (TAYLOR HOBSON OPTICS)

zoom lens requires three controls: aperture, zoom and focus. Adjustment of focus is something of a problem as the lens angle is varied. With a wide-angle setting, depth of field is good and it is difficult to judge whether or not a small detail in the wide angle view is accurately in focus. On 'zooming in' to give a narrow angle of view, depth of field is reduced. Thus, the small detail of the wide-angle view may be out of focus when featured in the narrow-angle shot. It is therefore best to focus on detail using a narrow-angle setting, then to re-check the focus on the same detail in the wide-angle view repeating as often as necessary until there is no need to re-focus at either end of the zoom range.

Figure 2.36 shows the external control rings of a zoom lens with a typical arrangement for mounting the lens. The mounting consists of a lens ring which screws into the camera faceplate and is threaded to receive the chosen lens. By screwing the ring in or out, the position of the focal plane of any lens used with the camera may be accurately set so that it coincides with the face of the camera tube. It is best to follow the lens manufacturer's instructions (usually supplied with the lens) so that the lens focuses over its intended range. Locking screws are provided which prevent the lens mounting ring being moved once it has been set. For CCTV cameras the thread commonly used is known as C-mount. If a lens with a different thread us to be used with a C-mount camera a thread adapter will be required. See photograph on page 50.

REMOTE CONTROL OF LENS ADJUSTMENTS

Quite often in CCTV systems used in industry, the camera(s) will be placed at some considerable distance from the operator and the viewing monitor; or the camera may

LENSES

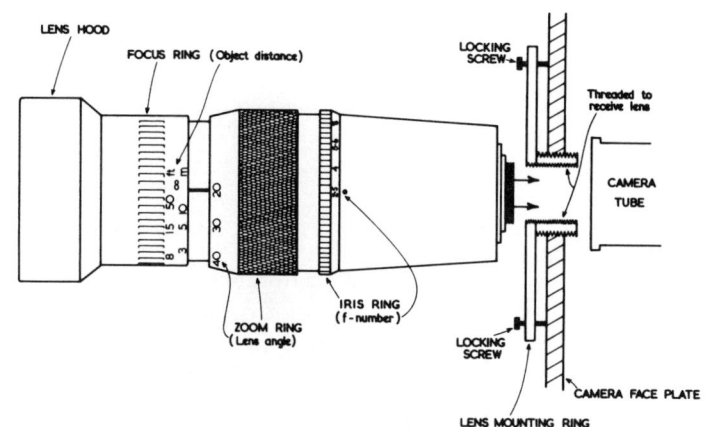

FIG. 2.36 EXTERNAL FEATURES OF ZOOM LENS WITH USUAL CCTV CAMERA LENS MOUNTING ARRANGEMENT

be mounted in a difficult place to reach. Clearly, it would be inconvenient for an operator to travel several hundred yards or to scale a high gantry in order to adjust the camera focus. In such situations motorised remote control provides the solution.

The photograph on page 51 shows a zoom lens which uses motorised control of zoom, focus and iris. The lens has three geared control rings; one each for zoom, focus and iris. Pinions engaging the control rings are driven by three d.c. motors through speed-reduction gears and limited-torque clutches. The clutches slip at the limit of travel of the lens rings in either direction. Current-reversing switches on the control panel determine the direction of motor rotation.

A basic circuit for motorised remote control of a zoom lens is shown in figure 2.37 Four diodes (D1—D4) form two separate full-wave rectifiers to provide low voltage d.c. supplies of opposing polarities. Voltages of either polarity are fed to the control motors *via* spring-loaded push-buttons. For example, when F_1 is operated, A is made positive with respect to B, but when F_2 is pushed, A becomes negative with respect to B.

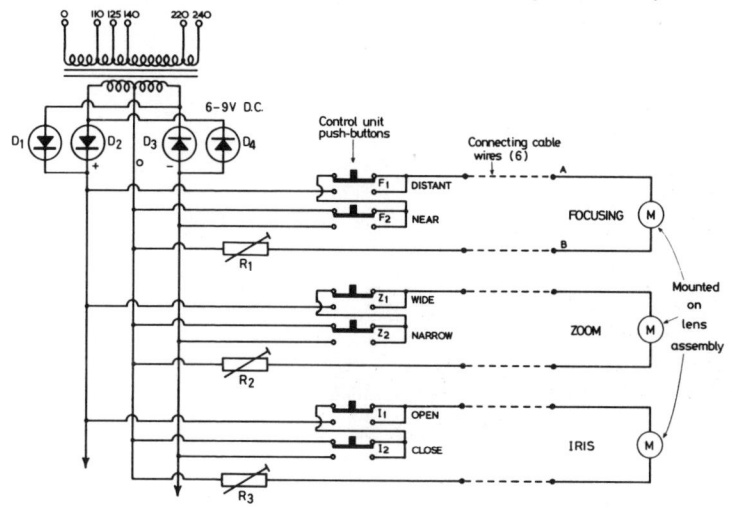

FIG. 2.37 BASIC CIRCUIT OF CONTROL UNIT FOR MOTORISED REMOTE OPERATION OF ZOOM LENS

(Photograph courtesy of Rank Taylor Hobson Optics, Leicester)

5:1 ZOOM LENS WITH FOCAL LENGTH RANGE OF 17—85 mm AT f/3·8

5:1 ZOOM LENS WITH MOTORISED ZOOM, FOCUS AND IRIS FOR REMOTE CONTROL

(Photograph courtesy of Rank Taylor Hobson Optics, Leicester)

Thus, the direction of current through the motor is altered which reverses the direction of rotation. Each motor is fed with the d.c. *via* a suppression circuit to prevent interference due to brush noise. The preset resistors ($R_1 - R_3$) ensure correct adjustment of the speed of each motor and/or compensate for the length of the connecting cables from the control unit to the camera.

VIGNETTING

Optically, the image formed by a lens system must have an edge. The idea is shown in figure 2.38. It will be seen that as the angle of the incident rays increase, eventually an angle is reached where all of the light is reflected (ray 4): So an image of these extreme

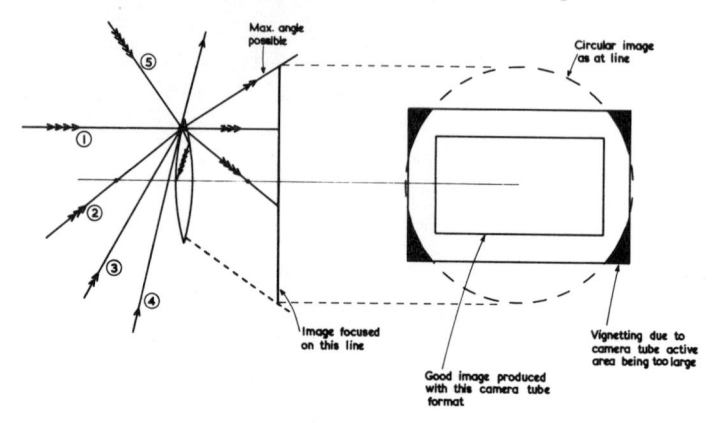

FIG. 2.38 DIAGRAM ILLUSTRATING VIGNETTING

rays cannot be reproduced and an image with a circular cut-off is formed. If it is attempted to obtain a larger 4:3 rectangular image that will fit in the circle, the corners will be cut off and the image completely black in the corners. Thus, a lens is designed to work into a certain sized format of camera tube.

Apart from vignetting due to optical cut-off with a single lens, there are various factors in a lens unit that may cause darkening of the image in the corners. 16 mm cine lenses which are converted for vidicon formats usually vignette, *i.e.* the picture darkens in the corners. This is because the converted lens is being asked to cover a larger format than the one it was designed for. Stopping-down the lens will reduce darkening in the corners but will have no effect when optical cut-off has occurred.

LENS CARE

The essential requirement is to keep the lens surfaces free from dust, fingerprints, grease and extreme changes of temperature. The external surfaces of the lens may be cleaned with a soft camel hair brush, or a clean soft cloth dipped in a small quantity of ethyl alcohol.

A lens should always be handled with extreme care to avoid dropping it and to prevent the surfaces from becoming scratched or damaged. It is unwise to attempt to dismantle a lens—an operation best carried out by the lens manufacturer. The moving parts of a lens should not be oiled as the oil will find a way inside and may cause clouding of the glass elements with a resulting degradation of picture quality. A lens which is used in a permanent camera installation out of doors must be fitted in a weather-proof housing to protect the lens (and camera) from the elements. The 'window' of the housing is kept clean by a windscreen wiper and washers which may be remotely controlled from the operator's position.

With reasonable care, a lens will have almost an indefinite working life.

CHAPTER 3

CCTV SIGNALS AND PRINCIPLES

PICTURE ELEMENTS

A TELEVISION camera lens focuses an image of the scene to be televised on to the light-sensitive face of the camera tube. The camera produces an electrical voltage at its output with an amplitude proportional to the amount of light falling on the light-sensitive area of the camera tube. At the monitor the camera voltage, after suitable amplification, is used to intensity modulate an electron beam which causes the monitor c.r.t. to fluoresce recreating the original camera image.

In order to convey the scene to the monitor by means of an electrical voltage it is necessary to break the scene down into picture elements. The idea is shown in figure 3.1

(a) SCENE TO BE CONVEYED (black cross on white background)

(b) SCENE DIVIDED INTO 48 ELEMENTS

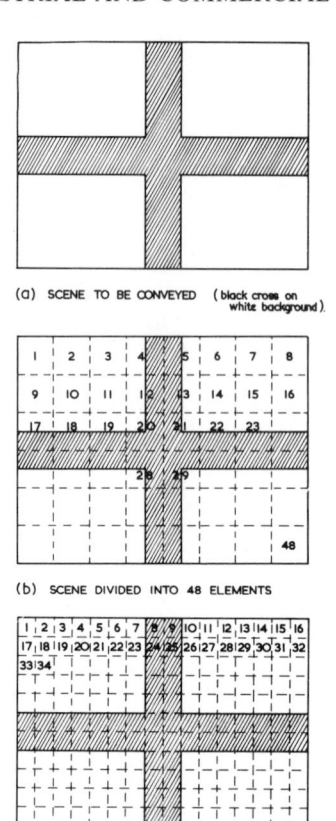

(c) SCENE DIVIDED INTO 192 ELEMENTS

FIG. 3.1 BREAKING THE PICTURE INTO ELEMENTS

where diagram (a) is the scene imaged on the face of the camera tube. Imagine that the scene image area is divided into 48 squares of equal size as in diagram (b). In a monochrome CCTV system it is the luminance information of the scene image that must be sent to the monitor. Thus, each of the 48 squares contains a certain amount of luminance information of the complete scene. We will assume that the camera produces an output voltage of 1 V for the white parts of the scene and 0 V for the black parts. Thus, squares 1, 2, 3, 14, 15 and 16, etc., which are completely white will each give rise to 1 V of output. Squares which are half white and half black will each cause 0·5 V of output, *i.e.* squares 4, 5, 12 and 13, etc., whilst squares 20, 21, 28 and 29 which are only one quarter white will each cause 0·25 V at the camera output. Now, if the camera output voltage for each square or element is responsible for causing the brightening or darkening of a corresponding square on the picture monitor, the image of the black cross will not be clearly defined. Squares 4, 5, 12, 13, 17 and 18, etc. will be reproduced as mid-greys whilst squares 20, 21, 28 and 29 as deeper greys, there being no sharp transition from white to black as intended. This is because there are insufficient squares or elements to deal with the scene detail.

If each square is subdivided into four smaller squares or elements there will be a total of 192 elements in the picture as illustrated in figure 3.1(c). As before, assume that each element gives rise to a camera output voltage which is used to brighten or darken corresponding elements on the monitor. There will now be sufficient elements to deal with the scene detail, *i.e.* the picture on the monitor will show a sharp transition from white to black on the vertical and horizontal parts of the black cross. The scene used in this example contains relatively large areas of constant light information, thus only 192 elements are required in our example to reproduce the scene on the monitor. If, however, there was information in the scene smaller than one of the squares of diagram (c), it could not be reproduced unless a greater number of elements were used. Therefore, to convey very fine picture detail a large number of elements are needed and the larger the number the finer the picture detail that can be dealt with.

SEQUENTIAL TRANSMISSION OF PICTURE INFORMATION

It would be impossible without the use of a highly sophisticated (and expensive) television system to send a signal from the camera to the monitor which *simultaneously* conveyed the information of each element constituting the scene. Thus for technical and economic reasons a *sequential* method of relaying picture information from the camera to the display device is used. In a sequential system, the luminance information contained in the area of each small picture element of the camera tube image is 'read-off' in a logical sequence. At each 'reading' the camera tube produces an electrical voltage output proportional to the luminance of each element. These sequential voltages are sent from the camera to the monitor where the image is reconstituted element by element on the c.r.t. Clearly, the elements must be read over and over again in rapid sequence in order to create a continuous picture on the c.r.t. and to capture movement within the scene as then the light falling on each camera tube element will be varying all the time.

Building up a picture on the monitor screen element-by-element in time sequence and giving the viewer the impression of 'picture completeness' works because the eye has a time lag (persistence of vision). As long as the information forming a complete picture is repeated often enough, say, 50 times a second, the eye perceives a continuous picture.

To 'read' each element of the camera tube image an electron beam (generated within the camera tube) scans the elements in a predetermined order. The idea of a simple scanning system is shown in figure 3.2(a). Here the internal electron beam traces out a series of sloping lines from left to right across the face of the camera tube

(a) Scanning the camera tube image   (b) Scanning the monitor screen

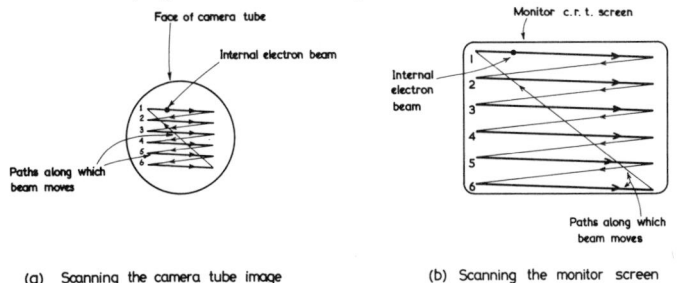

FIG. 3.2 DIAGRAM SHOWING IDEA OF SIMPLE SCANNING

commencing at line 1. At the end of each line the beam quickly returns to the left-hand side to commence the next *line scan*. As the beam is moving across the tube a force is also exerted on the beam causing a downward movement (this is responsible for the downward slope of the lines). Thus the beam traverses the full image formed on the face of the camera tube by the lens system. In doing so, each element is 'read' (explained in greater detail in Chapter 4) and an output voltage corresponding to each element is produced at the camera output. When the electron beam reaches the bottom of the tube (at the end of line 6) the beam rapidly returns to the top and repeats the scanning process. In practice the beam traces out a greater number of horizontal lines than are shown in the diagram which illustrates the basic idea. The movement of the beam from the top to bottom constitutes what is known as the 'field scan'.

At the monitor the same scanning process is adopted, figure 3.2(b), using an electron beam which scans the screen of the monitor c.r.t. As the beam travels across and down the screen, the beam is made more or less intense (by the sequential voltage output of the camera) causing a greater or smaller light output from the screen.

In both camera and monitor the beam is deflected in two directions at right angles to each other. With modern equipment the beam is deflected by magnetic fields set up at right angles to one another using two sets of deflection coils (line and field scan coils). In either the horizontal or vertical directions the movement of the beam must be at a linear rate during the *scan* and rapid later on during the return or *flyback*. This means that the magnetic field must be increased gradually in a linear manner during the scan and then rapidly reduced to its 'start of scan' value during the flyback. To achieve this, currents having a sawtooth waveshape have to be fed into the deflector coils.

Figure 3.3 shows the effect on the beam position when the deflector coils are fed with the required sawtooth currents. During the interval *a–b* the beam is moved

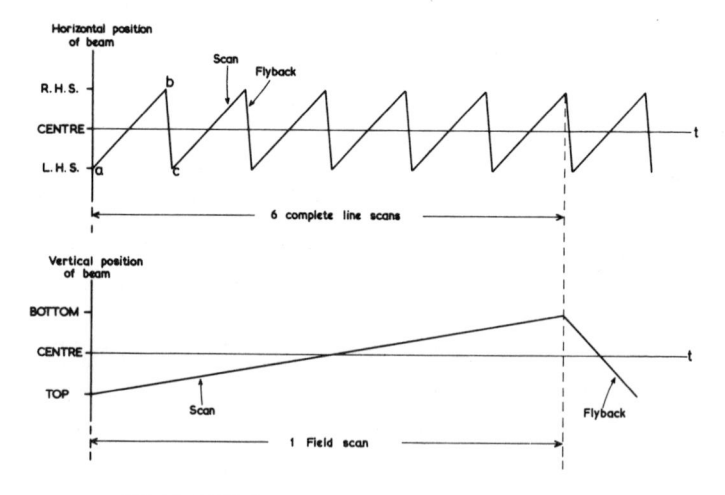

FIG. 3.3 SCANNING WAVEFORMS FOR SIMPLE SCANNING

horizontally across the screen or camera tube face with linear motion due to the effect of the *line scan current*. At the same time the beam is also under the influence of the magnetic field set up by the *field scan current* causing the beam to slowly move downwards as it traces out its path across the screen. Between instants *b* and *c* the beam quickly returns from the right-hand side to the left-hand side of the camera tube or monitor screen but is still being urged downwards by the field scan. At the end of six line scans (using the same earlier example) the field waveform commences its flyback causing the beam to return to the top of the camera tube or monitor screen so that it is ready to start the next field scan.

To generate the necessary scanning currents, scanning generators or timebases are employed. There are two timebases associated with both the camera and the monitor called the 'line timebase' and the 'field timebase'. Quite clearly there must be some co-ordination in the movement of the camera and monitor electron beams so that the image reproduced on the monitor at any particular instant is in the same relative position as the image formed on the camera tube. To co-ordinate the two beams, synchronizing pulses must be sent from the camera to the monitor. Two sets of pulses are needed, one set (line sync. pulses) to synchronize the line timebase in the monitor and the other set (field sync. pulses) to synchronize the field timebase. The synchronizing pulses may be generated within the camera or external to it, but in either case pulses must also be fed to the camera timebases so that they operate exactly in step with those in the monitor.

continued on page 59

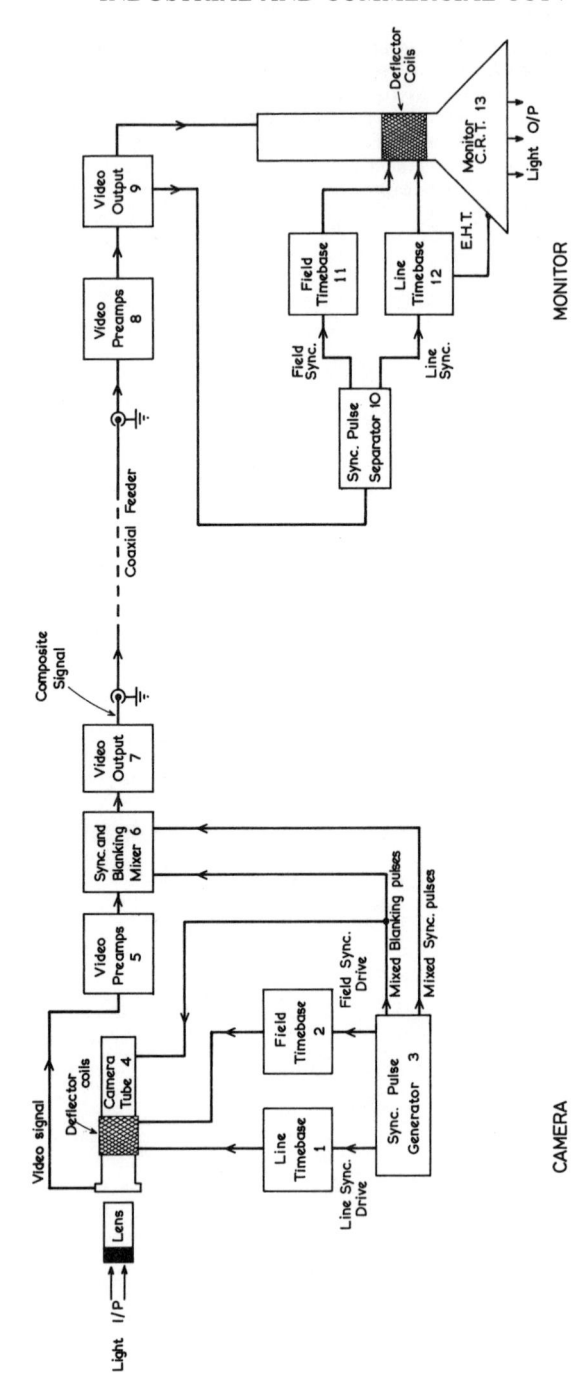

FIG. 3.4 SIMPLIFIED BLOCK SCHEMATICS OF CAMERA AND MONITOR

CCTV SIGNALS AND PRINCIPLES

Basic block diagrams of camera and monitor are shown in figure 3.4 which will now be described. In the camera, light from the scene is focused by the lens on to the face of the camera tube (4) where an electrical charge image of the scene is formed. The camera tube electron beam is deflected over the photo-sensitive face of the tube by the sawtooth currents supplied to the deflector coils from the line timebase (1) and the field timebase (2). The two timebases are fed with sync. drive pulses which are produced in the sync. pulse generator (3). These pulses ensure that the camera timebases operate at the correct frequency and in step with the monitor timebases. As the electron beam scans the charge image, it 'reads' each element in turn producing an electrical (video) voltage output which is fed to block (5). Here the video signal is amplified since the video voltage from the camera tube is quite small. In block (6) the line and field sync pulses (from block 3) are added to the video signal and will eventually be used to synchronize the timebases in the monitor. Also fed to block (6) are line and field blanking pulses generated in block (3). These pulses remove spurious signals generated by the camera tube during the line and field flyback periods. The blanking pulses are also fed to the camera tube to cut off the beam during the flyback periods. In block (7) the camera video signal and sync. pulses are amplified to provide a *composite* signal output at a standard level which is now fed to the monitor *via* the coaxial feeder.

At the monitor, the composite signal is fed to block (8) where it is amplified. The amplifiers in this block usually give frequency compensation and control of contrast. Further amplification is provided by block (9) which raises the video signal to a sufficiently large value to fully drive the monitor c.r.t. (13). Block (9) also feeds the composite signal to the sync. pulse separator (10). Here, the video information is removed from the composite signal allowing just the sync. pulses to be passed on to their respective timebases (11) and (12) to keep them synchronized. Sawtooth currents from the timebases are fed to the deflector coils which deflect the electron beam of the c.r.t. over the fluorescent coating of the screen. As the beam scans the screen, the beam is intensity modulated by the video signal applied to the c.r.t. from block (9). Thus an image of the original scene is produced on the screen of the c.r.t.

INTERLACED SCANNING

It has been explained that the scanning beams in camera and monitor must repeat the scanning process many times per second to give the impression of a continuous picture to the viewer. Persistence of vision lasts for approximately 1/25th second, thus if a picture repetition rate of 25 pictures per second were used the eye would perceive an uninterrupted picture on the monitor screen. There would, however, still be some flicker present, particularly in the bright areas of the picture. To overcome flicker it is necessary to increase the picture rate to 50 pictures per second. This speeds up the whole scanning process, *i.e.* more information has to be packed into a shorter time interval resulting in a doubling of the video bandwidth required. The transmission bandwidth available to the Television Broadcasting Authorities has always been at a premium. In the early days of monochrome transmissions alternative ways were investigated to get over the flicker problem without resorting to 50 pictures per second. This led to the introduction of *interlaced scanning,* a system which is used internationally for broadcast television and for CCTV.

In an interlaced scanning system an odd number of scanning lines must be used. An example of a 19-line system is shown in figure 3.5. The complete picture which is repeated 25 times per second is divided into two distinct fields with each field containing 9½ lines. Scanning commences at *A* with the beam tracing out lines 1, 2, 3, 4 and 5, etc. At each end of each line scan, the line timebase commences its flyback stroke causing the beam to return to the left-hand side of the monitor screen and camera tube. Due to the effects of the field scan, the beam is also being deflected downwards, hence the reason for the sloping lines as previously mentioned. Now half-way through line 10 the first (odd) field scan is completed and field flyback commences at point *B*. The beam now returns to the top of the screen and camera tube. Assuming that the field flyback is instantaneous, the other half of line 10 will now be completed as shown when the

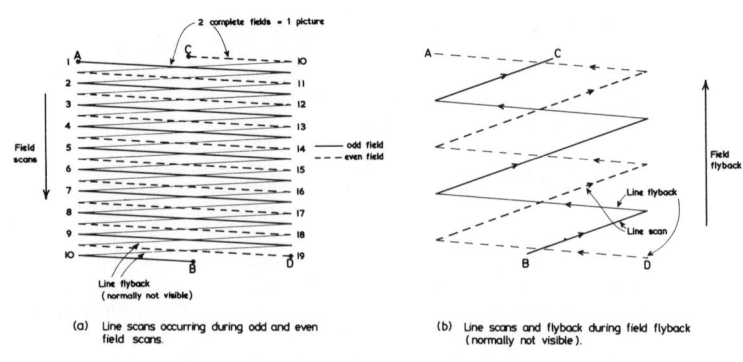

FIG. 3.5 INTERLACED SCANNING (19-LINE SYSTEM)

(a) Line scans occurring during odd and even field scans.

(b) Line scans and flyback during field flyback (normally not visible).

second (even) field scan commences at point C. Because of the half-line, scanning lines 11, 12 and 13, etc. now fit in between lines 1, 2, 3 and 4 etc. of the first field. This action continues until the end of line 19 at which point the second field scan is completed. At D field flyback occurs once more and the beam is returned to the starting point A. This action is repeated over and over again, 25 times per second. Each picture is thus composed of two interlaced fields and therefore the field timebase must be producing 50 scans per second.

In practice, of course, the field flyback is not instantaneous as was assumed in diagram (a). A finite time is required for the field flyback which is considerably longer than the line flyback period. Whilst the field timebase is in its flyback stroke, the line timebase (being continuous in operation) is still deflecting the beam in a zig-zag path across the monitor screen and camera tube. Diagram (b) shows the path taken by the beam during the flyback period of each field, but the beam is not normally visible during these periods.

If the brightness control in the monitor is turned up a little when there is no picture modulation, *i.e.* if the lens cap is fitted the scanning lines produced by the c.r.t. electron beam can be seen. The pattern of lines visible on the screen is called the 'raster'.

Interlaced scanning 'tricks' the eye into believing that every part of the scene is repeated 50 times per second when in fact each element of the scene is only being flashed up on the monitor screen 25 times per second. The difference between interlaced and non-interlaced scanning and its effect on human vision may be appreciated by considering figure 3.6 which shows a small area of picture detail (about two beam widths from top to bottom). With interlaced scanning, line x on, say, the odd

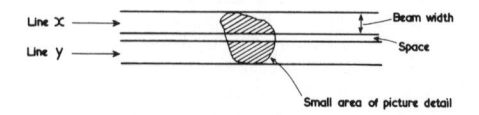

FIG. 3.6 OPTICAL 'TRICK' OF INTERLACED SCANNING

field causes the top half of the detail to brighten-up and 1/50th second later on the even field, line y causes the lower half of the detail to brighten up. Now, although it takes 1/25th second for the whole area of the detail to be shown, a part of the detail is being flashed up on the monitor screen every 1/50th second. In a non-interlaced, 25 pictures per second system, lines x and y will occur in the same field. Thus the upper and lower portions of the picture detail will brighten up in the time interval of one line but will not brighten up again until 1/25th of a second later on the following field scan.

Scanning current waveforms for interlaced scanning using the 19-line system as an example are illustrated in figure 3.7. Here it is assumed that the field flyback is

CCTV SIGNALS AND PRINCIPLES

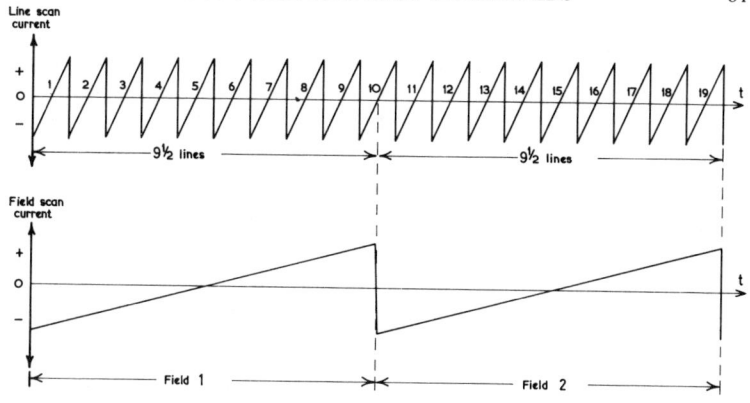

FIG. 3.7 SCANNING WAVEFORMS FOR INTERLACED SCANNING

instantaneous. In practice, the duration of the field flyback stroke is quite arbitrary (within limits) and does not have to be an exact number of line durations as at first may be thought. Provided the field flyback times is the same on both odd and even fields, satisfactory interlacing of the scanning lines will result.

FIELD AND LINE FREQUENCIES

In Great Britain and other European countries a picture repetition rate of 25 per second is used. Thus, with 2 fields to each picture the field timebase will be operating at 50 Hz. The number of pictures per second is chosen to give the field frequency synchronization with the 50 Hz mains supply. While this is not essential, it has certain advantages. In monitors and cameras that are powered from the mains, stray mains pick-up may find its way into the signal circuits causing hum bars on the monitor screen. When the field timebase frequency is not 'locked' to the mains supply any hum bars present will drift up and down the monitor screen as the phase between the timebase and the mains varies. If the field timebase is locked to the mains frequency there will be a constant phase relationship and any hum bars on the screen will be stationary, resulting in a reduction of their annoyance value. In mobile equipment where mains lock is not possible, a free-running 50 Hz field oscillator is used which may be locked by external sync. signals from a stable sync. generator. In America where the mains supply is at 60 Hz a picture repetition rate of 30 per second is adopted. It should be mentioned that for colour television it is a disadvantage to lock to the mains (synchronous operation) since the mains frequency is liable to vary. Field and line frequencies are derived from very stable generators which are independent of the mains (asynchronous operation).

Figure 3.8 shows in block form the basic idea of mains lock. Block (1) contains a free-running oscillator which is synchronized by a sample of the mains voltage after it has been suitably shaped. This positively locks the field oscillations to the mains

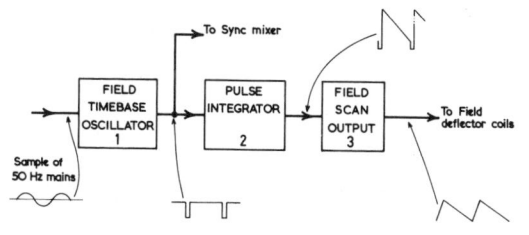

FIG. 3.8 LOCKING THE CAMERA FIELD TIMEBASE TO THE MAINS SUPPLY

supply. The output of the oscillator which in this case consists of pulses, is fed to block (2) where the waveform is integrated to produce a sawtooth drive to the field output stage (3). The final block generates the required amplitude scanning current for supplying the field deflector coils. This method of mains lock is usually adopted in cameras operating the **random interlace** mode which will be discussed later.

In an interlaced scanning system there is an odd half-line in each field, so there must be an odd number of lines in each complete picture. The actual number of lines used in a practical t.v. system determines the vertical resolution of the picture, improving the resolution with an increase in the number of lines adopted. Table 3.1 gives the essential details of a few different systems. Modern CCTV cameras operate on 625 lines but the

Line system	Picture frequency	Field frequency	Line frequency = Picture frequency × No of lines	2× Line frequency
British 405	25 Hz	50 Hz	10,125 Hz	20,250 Hz
British 625	25 Hz	50 Hz	15,625 Hz	31,250 Hz
American 525	30 Hz	60 Hz	15,750 Hz	31,500 Hz
French 819	25 Hz	50 Hz	20,475 Hz	40,950 Hz

TABLE 3.1   STANDARD FREQUENCIES

technician may occasionally meet up with 405-line equipment. Taking as an example the 625-line system, there are 625 lines (not all carry picture information) in each complete picture, *i.e.* $312\frac{1}{2}$ lines in each field. Now, since the picture is repeated 25 times per second, the line frequency

$$= \text{Picture Frequency} \times \text{Number of Lines}$$
$$= 25 \times 625$$
$$= 15,625 \text{ Hz}$$

The period of each line scan $= \dfrac{1}{\text{line frequency}}$

$$= \frac{1}{15625} \text{ sec.}$$
$$= 64 \ \mu s$$

2:1 INTERLACE RATIO

All broadcast television systems and some CCTV camera equipment provide an interlaced picture, comprising 2 fields for each complete picture or 2:1 interlace. It is possible to provide 3:1, 4:1 and even higher interlace ratios to make flicker even less noticeable but this would add to the complication of the system and is therefore not adopted.

To achieve a correctly 2:1 interlaced picture there must be an exact timing relationship between the line and field timebase oscillations in the camera, *i.e.* they must be mutually synchronized. If this is so, the monitor timebase will likewise have the correct relationship because of the synchronism that exists between camera and monitor. For good phase stability it is normal practice to derive the line and field timing pulses for the camera and monitor by frequency division from a very stable oscillator working at *twice the line frequency*. A typical arrangement is shown in Fig. 3.9 for CCTV equipment.

At the centre of the sync. pulse generating circuits is the **master oscillator.** This is a stable oscillator working at twice the line frequency. Either a crystal, L/C or blocking oscillator may be used. The output of block 1 consists of pulses at 31·25 kHz and these are fed to a ÷2 stage which produces a pulse output at 15·625 kHz. These pulses

CCTV SIGNALS AND PRINCIPLES

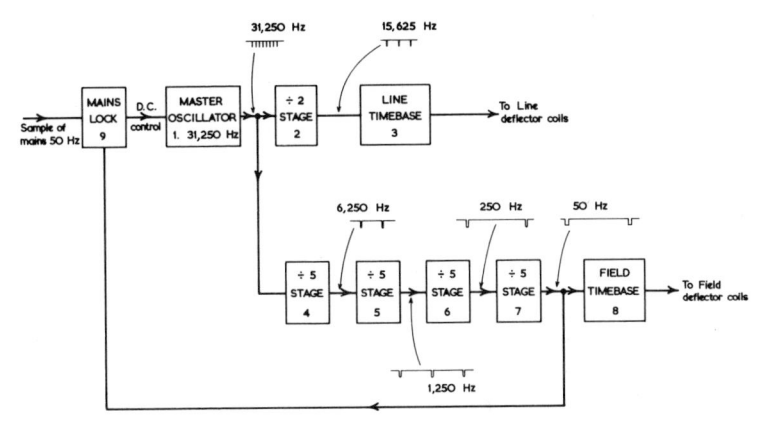

FIG. 3.9 DERIVING THE LINE AND FIELD FREQUENCIES BY FREQUENCY DIVISION FOR 2:1 INTERLACING (625-LINE CAMERA BLOCK DIAGRAM)

provide the line frequency timing and are fed to block (3) to drive the camera line timebase.

Pulses from the master oscillator are also fed to a 625-divider chain consisting of four $\div 5$ stages [(blocks (4)–(7)] employing bistable multivibrators or blocking oscillators. By dividing the output of the master oscillator by a factor of 5^4, pulses at 50 Hz may be obtained. These pulses are used for the field frequency timing and are fed to block (8) to drive the camera field timebase.

In addition to driving the camera line and field timebases, the derived 50 Hz and 15·625 kHz pulses will also be used for providing the line and field sync. pulses which are 'mixed' with the camera video output and sent to the monitor.

Since the 50 Hz and 15·625 kHz pulses are derived from a common source, there will be a constant timing relationship between the line and field oscillations. Hence, a properly interlaced picture is possible. The reason for using a master oscillator operating at twice the line frequency and not directly at line frequency is that in order to obtain 50 Hz from 15·625 kHz by frequency division implies dividing by $312\frac{1}{2}$. Unfortunately, practical dividers can only divide by whole numbers. Now 31,250 divided by 50 equals 625—a whole number as is required. In broadcast television, pulses at twice line frequency are specifically needed (for insertion into the synchronizing waveform at the beginning of each field period). Thus, pulses at twice

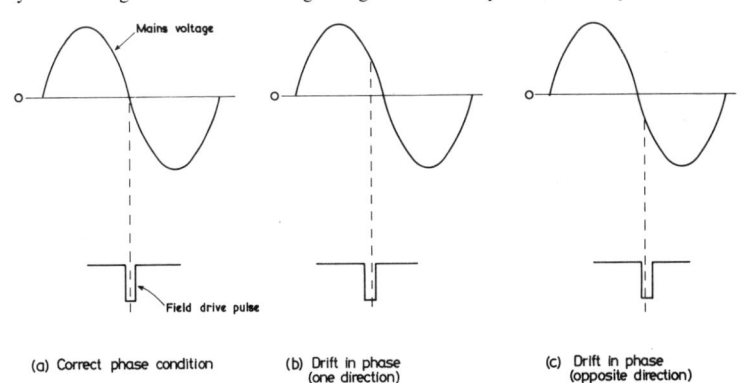

FIG. 3.10 PRINCIPLE OF PHASE LOCK

line frequency, line frequency and field frequency must be made available for high-class studio operation.

Block (9) is concerned with the locking of the field drive pulses to the 50 Hz mains. This block is actually a phase comparator and the basic principle of one arrangement is shown in figure 3.10. One input to the comparator is a sample of the 50 Hz mains supply voltage. The other input is the field drive pulses taken from the output of block (7). The circuit of the phase comparator is designed so that when the field pulse timing coincides with the zero crossing points of the mains input no 'error' voltage is fed to the master oscillator whose frequency is undisturbed. If there is a drift in phase of either the mains or the master oscillator, the field pulse timing will no longer coincide with the zero crossings of the mains input. Under this condition the phase comparator gives out an 'error' voltage which corrects the frequency of the master oscillator and HENCE the field drive pulses to restore the correct phasing condition. The 'error' signal is in the form of a d.c. voltage which can be used to vary the 'aiming potential' if the master oscillator uses a blocking oscillator. When a crystal or L/C oscillator is employed the d.c. voltage may be used to vary the capacitance of a varactor diode connected across the frequency determining circuit.

RANDOM INTERLACE

A simpler method of driving the camera timebases is to use the **random interlace** mode and is adopted by CCTV cameras used for non-critical applications in industry and commerce. In this method the line and field drive pulses are generated within the camera, unlike 2:1 interlace systems where the drives may be supplied from an external s.p.g. (sync. pulse generator).

With random interlace there is no constant timing relationship between the line and field oscillations. Thus, precise triggering of the field timebase does not occur which is essential to maintain the half-line difference between the start of successive field scans. Figure 3.11 shows one arrangement for obtaining the line and field pulses in a random interlace camera. Here separate free-running oscillators are used for line and field.

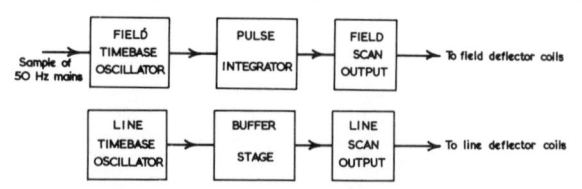

FIG. 3.11 TIMEBASE ARRANGEMENTS FOR A RANDOM INTERLACE CAMERA
(Mains lock used on field, but no mutual synchronization between line and field)

Being independent, the oscillators are liable to drift in phase or frequency relative to one another. If mains lock of the field is required it may be incorporated as previously described. The line timebase oscillator may operate directly at 15,625 Hz but sometimes a master oscillator working at 31,250 Hz is employed followed by a divide-by-two stage. The latter arrangement is adopted when the camera is to be operated in other modes, *i.e.* receiving external line and field drives.

As there is no guarantee of a 'half-line difference' in a random interlace camera there are approximately 312 lines in each field (625-line operation). There is a random chance of small vertical displacements of successive fields so a degree of interlace is possible, but it is purely a transistory condition. It should be realized that with random interlace, *identical* and not *additional* information may appear on the randomly placed adjacent pairs of scanning lines. The diagrams of figure 3.12 illustrate various degrees of interlace. Diagram (a) shows correct 2:1 interlace where there is even spacing between the lines of adjacent field scans. Partial loss of interlace is shown in (b) where the lines of successive fields tend to 'pair'. Complete loss of interlace is represented by (c) where the lines of successive field scans are superimposed. The conditions shown in

CCTV SIGNALS AND PRINCIPLES

(a) Correct 2:1 interlace   (b) Line pairing   (c) Complete loss of interlace

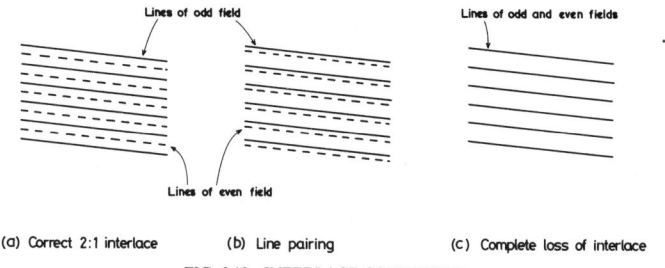

FIG. 3.12 INTERLACE CONDITIONS

diagrams (b) and (c) will result in a loss of vertical definition, but this is a defect which is normally only recognized by the 'trained eye'. A characteristic feature of random interlace is the presence of a flickering bar at the bottom of the picture on the monitor. This can be hidden from view by a slight increase in the picture height of the monitor.

COMPOSITE SIGNALS

Before looking at the composite signal, consider figure 3.13 to reinforce the ideas of basic camera operation. Diagram (a) shows a test picture to be relayed to the monitor.

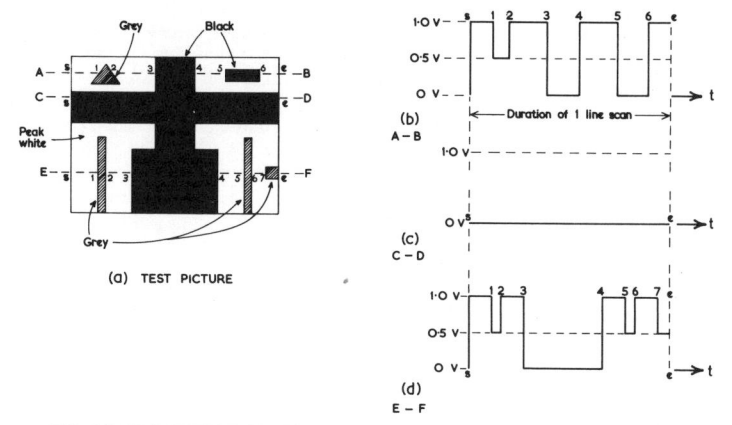

FIG. 3.13 EXAMPLES OF CAMERA VIDEO OUTPUT DURING SINGLE LINE SCANS

Again we will assume that the camera gives 1 V of output on peak white and 0 V for black. All luminance values lying between these two limits must produce an output signal between zero and 1 V. In general, the luminance values will be quite arbitrary when the camera is looking at a real coloured scene. But for this test picture only three levels of luminance are considered. It will be assumed that the grey (which is white at reduced luminance level) produces an output of 0·5 V. Diagrams (b), (c) and (d) show the video output signal during one line scan for lines corresponding to positions A–B, C–D and E–F of the test picture. Note that at the end of the line scans the video signal may assume any arbitrary value between 0 V and 1 V.

A composite signal consisting of video information during one line scan and line sync. pulses is shown in figure 3.14. This will be considered in more detail.

(a) Video Signal

The composite signal is divided on an amplitude basis into two voltage ranges. The range between black level and peak white (P) is reserved for the video or picture signal,

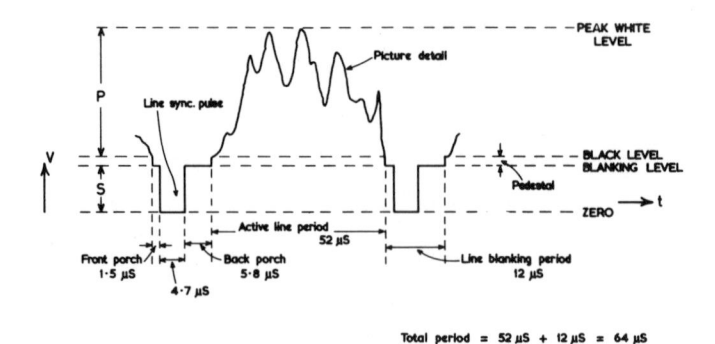

FIG. 3.14 VIDEO INFORMATION AND LINE SYNC. DURING ONE LINE SCAN (625-LINE OPERATION)

and the range between zero and blanking level (S) is taken up by the sync. pulses. For broadcast and CCTV, the ratio of P/S is approximately 2:1. This ratio (or thereabouts) is chosen so that when at some remote receiver or monitor where the signal level is weak, the timebases will fail to synchronize at the same time as the picture signal-to-noise ratio falls to such a level that the picture ceases to be of viewable quality.

The video signal can have any arbitrary value between black and peak white levels and on a normal scene each line of video information will be different from the proceeding ones. When 'scoping' the line waveform of such a scene using an ordinary c.r.o., the display will show *superimposed* lines of information. As the video content is different line-to-line, the displayed video will appear very fuzzy. The line sync. pulses, being of repetitive shape, will be quite distinct. When the scene information is repetitive line-by-line as with some test pictures, *e.g.* a 'grey-scale', the video information of the displayed waveform will also be distinct.

(b) Line Blanking Period

This is the period lasting from a short time just before the line sync. pulse until an interval after it. During the line blanking period the composite waveform is devoid of picture information. Any spurious signals generated by the camera between the end of one line of picture information and the commencement of the next line are suppressed by means of a blanking pulse (of suitable amplitude and duration) which is added to the video signal. By using a limiter to limit the added pulse at blanking level, the spurious signals are lost in the limiter. The blanking pulse does not appear in the composite waveform, but this non-picture information period is given the name 'blanking period' because of the 'blanking' pulse used to create it. The line blanking period can be divided into three sections:

(i) Front Porch

This is the brief period (1.5μs) just before the commencement of each line sync. pulse. It serves as a 'cushioning' period for the video circuits in the camera and monitor, allowing them to settle down before the commencement of the regularly occurring line sync. pulses. The level of the video signal voltage at the end of a line picture information is quite arbitrary. It may be high on peak white or low on black (see figure 3.13). Now, electronic circuits cannot change their voltage state instantaneously and the front porch allows sufficient time for the voltage level to fall from peak white to blanking level before the line sync. pulse commences. Without this interval, the sync. pulses may be late in starting following lines ending in peak white compared with the

CCTV SIGNALS AND PRINCIPLES

pulses following lines ending on black, resulting in faulty line timebase synchronization.

(ii) Line Sync. Pulse

This pulse provides the timing of the line timebase oscillations in the monitor thus synchronizing them with similar oscillations in the camera. The *leading* and downward going edge of the pulse initiates the commencement of line timebase *flyback*. A duration of 4·7 μs is allowed for the pulse and during this interval the electron beams in the camera and monitor tubes will be in a re-trace stroke.

(iii) Back Porch

The back porch provides a further 'grace' period for the line timebase in the monitor to complete its flyback before picture information commences again on the next line. This is necessary due to variations in design of the line output stage used in the different makes of monitor equipment. In addition, the back porch serves as a reference level in connection with the maintenance of the d.c. component of the video signal during its passage through the camera and monitor circuits.

(c) Pedestal or 'set-up'

The voltage range between black level and blanking level is known as the **pedestal** or **set-up**. In adjusting the camera circuit controls, the technician has to 'set-up' the level of the signal corresponding to black to give the required pedestal. Now, in the monitor the composite signal is fed to the modulating electrode of the c.r.t. and the brightness control is set so that the screen is blank on black level. The pedestal will thus ensure that the voltage levels corresponding to blanking and the line sync. pulse are not visible on the monitor screen. In some equipment, the black level is set coincident with blanking level, *i.e.* there is no pedestal. A pedestal is not normally used in studio equipment.

STANDARD OUTPUT

Cameras are designed to give a standard voltage output for the composite signal. Two standard voltage outputs are in existence in the CCTV industry and these are given in figure 3.15. The overall voltage outputs of 1·0 V or 1·5 V refer to the *camera video ouput* when the camera is *correctly terminated*. Standard voltage levels allow

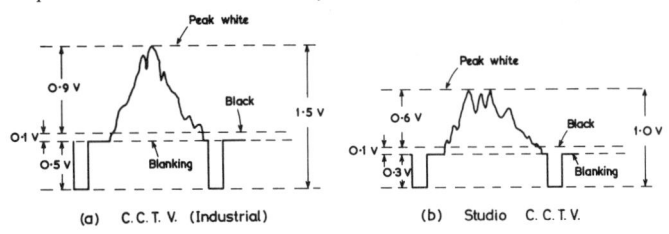

FIG. 3.15 STANDARD VOLTAGE DIMENSIONS FOR COMPOSITE SIGNAL

equipment of different manufacture to be used together. Preset controls within the camera are provided for the setting of the various levels of the composite signal and these levels must be accurately maintained for stable operation of ancillary equipment (monitors, video tape recorders and special effects generators, etc.).

FIELD SYNC. PULSES

At the end of each field scan a field blanking waveform is inserted into the composite signal waveform for suppressing any spurious signals generated by the camera. The blanking pulse has a duration equal to 20-line periods, figure 3.16. Thus in a complete picture there are 40 lines which do not bear picture information, *i.e.* there are only 625 − 40 = 585 active picture lines. As with the line blanking pulses the field blanking waveform does not show up in the composite waveform.

To initiate the flyback of the field timebase at the end of each field scan, a field synchronizing pulse is added to the composite waveform for use in the monitor. This

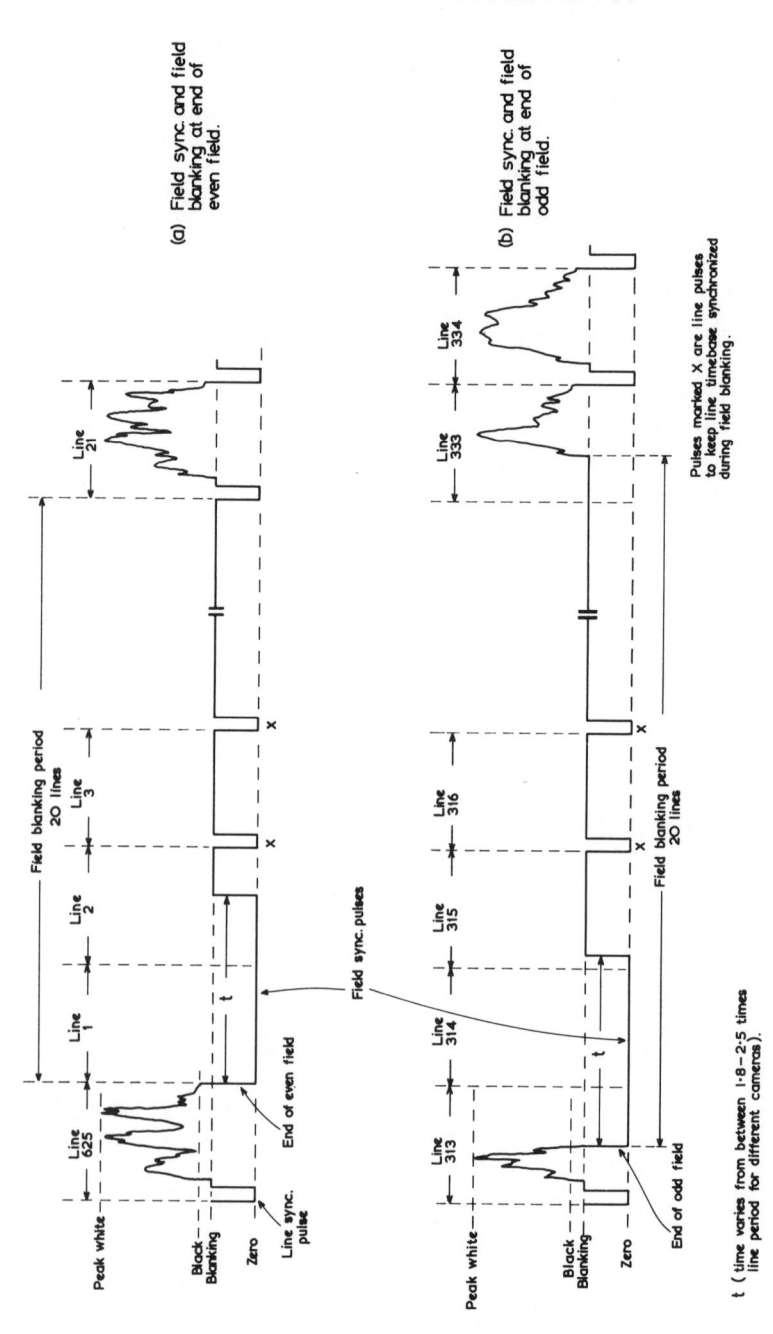

FIG. 3.16 FIELD SYNCHRONIZING WAVEFORM (CCTV CAMERAS)

pulse is of the same amplitude as a line sync. pulse but is of longer duration. For CCTV cameras a single pulse is commonly used having a duration of about 1 ms to 1·6 ms. Field timebase flyback commences sometime after the leading edge of each field sync. pulse (but before the trailing edge) and the suppression period of 20 lines gives ample time for the flyback to be completed before picture information occurs once more.

During field blanking it is advantageous to keep the line timebase fully synchronized, so that at the commencement of picture information at the start of the next field scan, horizontal displacement of the picture scanning lines does not occur. This is the purpose of the line pulses marked x which are repeated regularly at the end of each line period during field blanking. A front porch is not required for the field sync. pulse as field flyback does not commence until a time interval after the start of the field sync. pulse. This time interval is longer than the time taken for the picture signal to fall from peak white to blanking level.

It will be noted in figure 3.16 that there is a whole line period between the last line sync. pulse and the field sync. pulse at the end of an even field, but only a half-line period at the end of an odd field. This can lead to difficulties in the monitor, the circuits of which are required to produce from the pulse train field synchronizing pulses having precisely the same shape on odd and even fields. If this is not so, the interlace will be impaired to some extent depending upon the monitor circuitry. In the composite signal used for broadcast television, equalizing pulses are inserted into the wave train before and after the field synchronizing period, figure 3.17, Here, instead of using a single field pulse, the sync. period is broken up into five field sync. pulses each having a duration of $27 \mu s$. The equalizing pulses reduce disimilarities in the shape of the field sync. pulses produced in the monitor, thereby ensuring good interlace.

It will be observed that the equalizing pulses and field sync. pulses are repeated at *twice line rate* during their appearance in the composite waveform every 1/50th sec. This timing ensures that the line timebase is synchronized *every line* of the 625 lines in the system. More complex pulse generating circuits are required to produce this type of field sync. arrangement and is therefore usually found only in professional or semi-professional studio equipment.

In order to observe the field sync. period pulse arrangement on an ordinary c.r.o. it should be synchronized at picture frequency (25 Hz) and should be capable of large X expansion. Detailed examination of the field pulses is best carried out using a c.r.o. with a strobe timebase facility.

The diagram of figure 3.18 shows a form of pre-field sync. pulse equalizing used in some CCTV cameras. Here a front porch period is provided for the field sync. pulse and into this period is inserted the pulses (marked x) which are normal line sync. pulses. These pulses assist in the forming of identically shaped field sync. pulses in the monitor.

PICTURE RESOLUTION

The *resolution* or *definition* of a television system is a measure of its ability to deal with fine picture detail, *i.e.* to reproduce a sharply defined image of small detail in the scene. Normally, we are only interested in the maximum resolution of a system which gives an indication of the finest detail that the system can handle.

The definition of any reproduced picture depends upon both the *horizontal* and *vertical* resolution. Horizontal resolution is a measure of the ability to reproduce luminance changes along a horizontal line such as $x-y$ in diagram (b) of figure 3.19. In this diagram, vertical black and white stripes are shown imaged on the face of a camera tube. If the system has insufficient horizontal resolution, the luminance changes along any line such as $x-y$ will appear blurred, *i.e.* the sharp changes from black to white across the scene will not be reproduced. The finer the vertical bars, the faster the system will have to respond in order to reproduce them clearly. In any reproduced picture, *horizontal resolution* shows up on the *vertical edges of picture detail* and depends upon

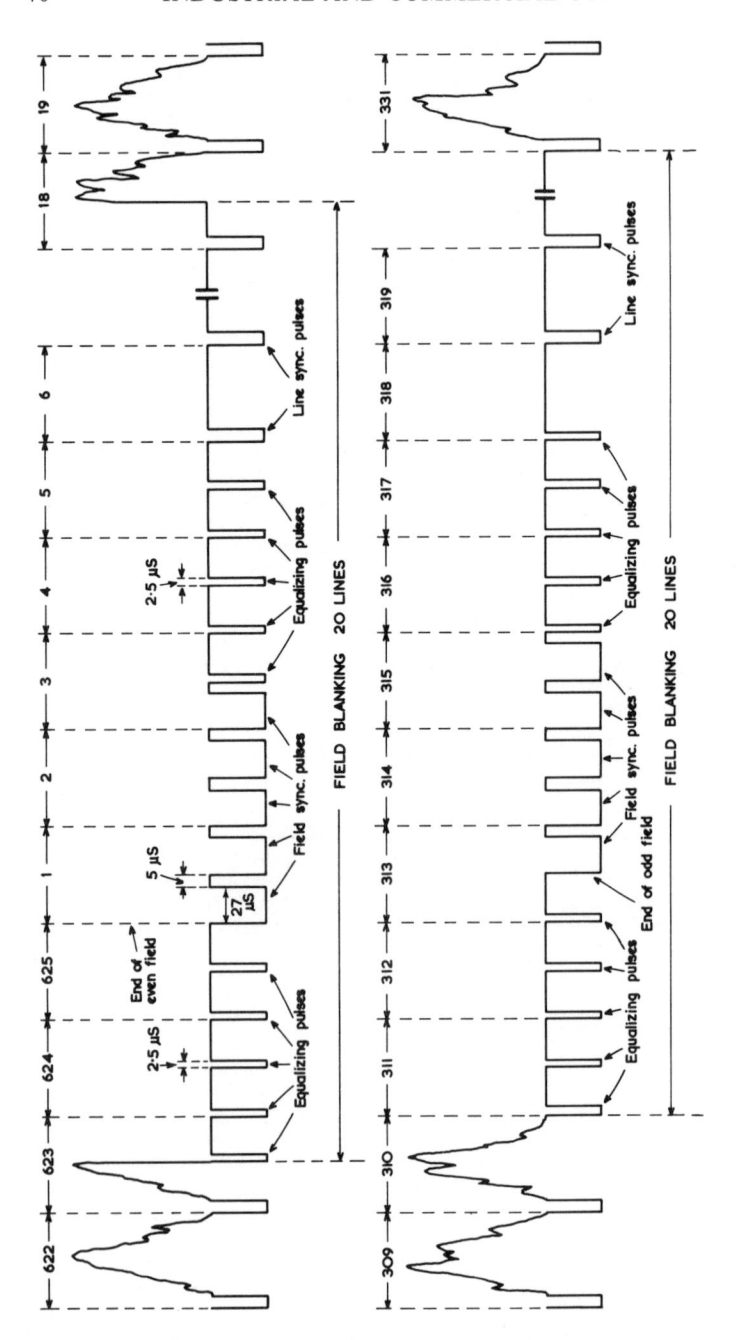

FIG. 3.17 625-LINE FIELD SYNCHRONIZING WAVEFORM (BROADCAST STANDARD)

CCTV SIGNALS AND PRINCIPLES

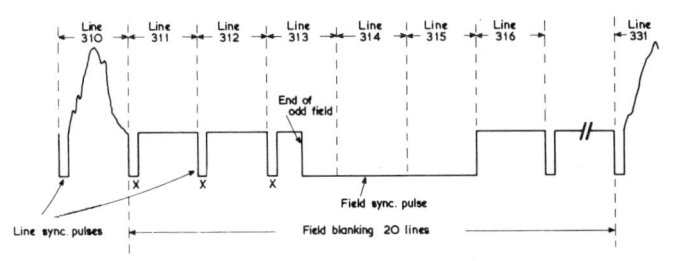

FIG. 3.18 PRE-FIELD SYNC. PULSE EQUALIZING USED IN ONE TYPE OF CCTV CAMERA

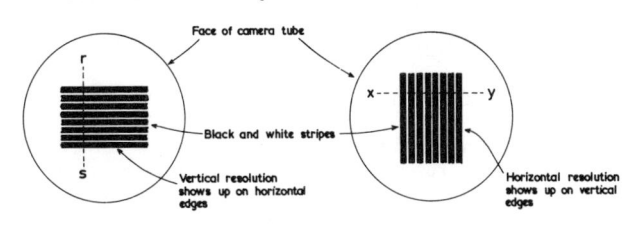

(a) Horizontal bars (b) Vertical bars

FIG. 3.19 VERTICAL AND HORIZONTAL RESOLUTION

the frequency bandwidth of the system. Vertical resolution is a measure of the ability to reproduce fine detail along a vertical line such as *r–s* in diagram (a). Here, horizontal black and white stripes are being considered. To reproduce fine detail in the vertical sense, a large number of scanning lines are required and the greater the number the better the vertical resolution. In any reproduced picture, *vertical resolution* shows up on the *horizontal edges of detail*.

MEASURING THE RESOLUTION

In the CCTV industry, resolution is measured in terms of LINES PER ACTIVE PICTURE HEIGHT and may be quoted for both the vertical and horizontal measurements in technical literature. Occasionally, definition may be quoted in terms of the highest video frequency and whilst this is useful in quantitative assessment of horizontal resolution it has no real significance as regards vertical resolution. We will consider resolution in terms of lines by looking at definition in the vertical sense which forms the basis of this measuring system.

VERTICAL RESOLUTION

It is important to realize that the size of the electron beam imposes a limitation on the smallest detail that can be resolved (other factors being maximised) in either the vertical or horizontal senses. Consider an electron beam scanning a horizontal stripe pattern imaged on the face of a camera tube as in figure 3.20(a). Here it is assumed that the diameter of the scanning beam is much smaller than the width of each stripe. Thus the beam makes several scans such as *a, b* and *c* etc. across each stripe before the luminance level changes on the following stripe. In this case the pattern can be clearly resolved as in nearly every part of the scene the beam is only 'reading' one particular imaged level of luminance. If now the size of the beam is increased so that its diameter embraces three stripes of the pattern as in diagram (b), the detail cannot be resolved. This is because in any position of the beam it will 'read' the average luminance value from parts of three stripes. Thus, detail which is smaller than the electron beam cannot

72   INDUSTRIAL AND COMMERCIAL CCTV

(a) Scanning beam diameter smaller than picture detail dimensions (detail may be resolved).

(b) Scanning beam diameter larger than picture detail dimensions (detail cannot be resolved).

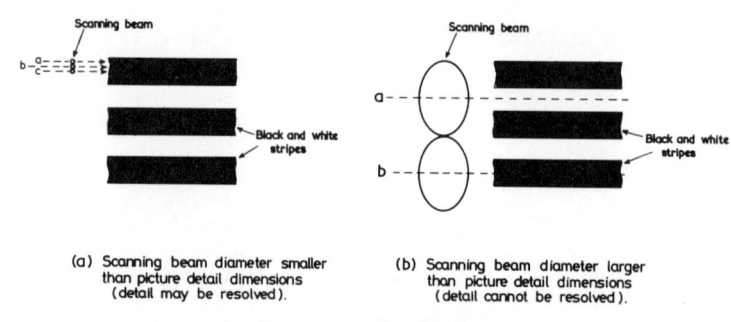

FIG. 3.20 DETAIL SMALLER THAN BEAM DIMENSIONS CANNOT BE RESOLVED

be resolved. The ultimate size of detail that can be resolved is that equal to the diameter of the electron beam.

If the scanned height of a small camera tube is, say, 1 cm, it is evident that the diameter of the scanning beam must be very small in order to accommodate 625 separate scanning lines as in a 625-line system. Now, if the ultimate size of picture detail that can be resolved is equal to the diameter of the beam, the maximum number of resolvable elements disposed vertically down the scene image is 625, *i.e.* equal to the number of scanning lines. This assumes that every lines carries picture information which is not so as we have seen. In a 625-line system there are only 585 active lines thus the maximum vertical resolution would appear to be 585 LINES PER ACTIVE PICTURE HEIGHT. In technical literature the words 'per active picture height' are dropped. In practice the maximum vertical resolution is less than 585 lines and the reason for this will now be considered.

The vertical resolution obtainable from a television system is affected by the relative positioning of the scanning beam in relation to the horizontally disposed picture detail, figure 3.21. In these diagrams we are considering the scanning of an ultimate size stripe pattern. If the centre of the beam corresponds to the centre of each

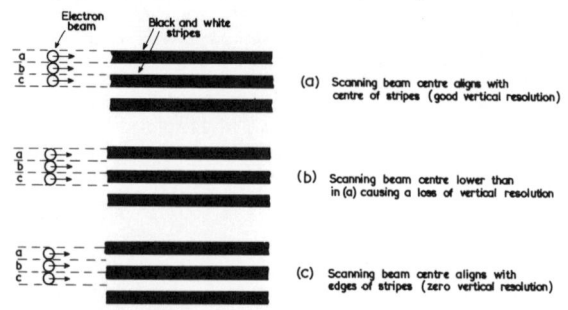

(a) Scanning beam centre aligns with centre of stripes (good vertical resolution)

(b) Scanning beam centre lower than in (a) causing a loss of vertical resolution

(c) Scanning beam centre aligns with edges of stripes (zero vertical resolution)

FIG. 3.21 DIAGRAMS SHOWING HOW POSITION OF SCANNING BEAM IN RELATION TO ULTIMATE STRIPE PATTERN AFFECTS VERTICAL RESOLUTION

stripe as in diagram (a), the pattern can be clearly resolved. However, if the beam commences its first scan *a* slightly lower down as in diagram (b), it will be 'reading' mainly all of one stripe and a little of the following stripe in any position. Thus there will be some loss of definition. When the beam is disposed as in dagram (c) there will be complete loss of vertical definition as in every position the beam will 'read' the average luminance level of black and white which is grey.

Now, the position of the scanning beam in relation to such a test pattern (or similar picture detail) will depend upon how accurately optical positioning and scan geometry

CCTV SIGNALS AND PRINCIPLES

is set and maintained. Statistical and subjective testing suggest that the effective number of lines is about 0·7 of the total number of active lines. This factor is known as the 'Kell' factor (named after an early television worker). The Kell factor cannot be precisely determined therefore values other than 0·7 may be ascribed to it.

Assuming a Kell factor of 0·7, the maximum vertical resolution obtainable in a 625-line system is:

$$0.7 \times 585 \text{ lines}$$
$$= \underline{409.5 \text{ lines}}$$

The corresponding figure for a 405-line system is approximately 264 lines. These figures assume a perfect 2:1 interlaced picture, thus random interlace cameras will have a poorer vertical resolution.

HORIZONTAL RESOLUTION

It would seem an obvious aim in designing a television system to make the horizontal resolution the same as the vertical resolution. Thus with a maximum vertical resolution of 409·5 lines per active picture height, the *horizontal resolution required to match this figure is also 409·5 lines per active picture height*. A horizontal resolution of 409·5 lines or approximately 410 lines means that the object size resolved is 1/410 of the effective active picture height.

Now the horizontal resolution is determined by the video frequency bandwidth. To find out what bandwidth is needed to equate the horizontal resolution to the vertical resolution consider figure 3.22. This diagram shows a checker-board pattern consisting of alternate black and white squares which is required to be reproduced on the screen of

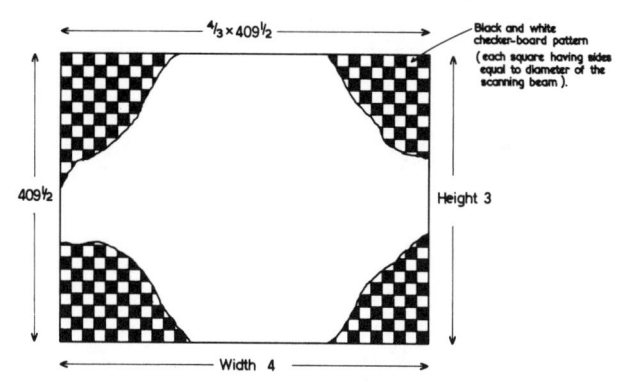

FIG. 3.22   ULTIMATE CHECKER-BOARD PATTERN TO DETERMINE HORIZONTAL RESOLUTION

a monitor. There are 409·5 squares along the vertical sides of the board, this number being chosen as it represents the maximum practical vertical resolution. Each square has sides corresponding to the diameter of the scanning beam and thus represents the ultimate pattern for testing the resolution of a television system.

In television the ratio of Picture Width:Picture Height (called the *aspect ratio*) is 4:3. Therefore the number of squares disposed across the board in a single horizontal line is

$$\frac{4}{3} \times 409.5 = 546.$$

Each line scan this number of squares has to be scanned by the beam in a time period of 52μs (the active line period). To reproduce such a pattern, the video waveshape

(a) Checker pattern

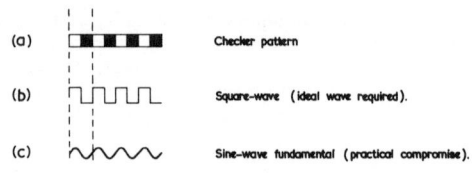

(b) Square-wave (ideal wave required).

(c) Sine-wave fundamental (practical compromise).

FIG. 3.23 WAVEFORM REQUIRED TO PRODUCE PATTERN

required is ideally a square wave, see figure 3.23(b). To handle this waveshape too great a demand would be placed on the system in terms of bandwidth. A reasonable technical compromise is to accept a sine wave response as in diagram (c). Note, however, that each cycle corresponds to *two* squares. Thus the number of cyclic changes across each picture line

$$= \frac{546}{2} = 273.$$

Therefore the periodic time of one cyclic change $= \frac{52}{273}$ μs.

$$= 0{\cdot}19 \ \mu s.$$

Thus the frequency of the cyclic changes $= \dfrac{1}{0{\cdot}19 \times 10^{-6}}$ Hz.

$$= 5{\cdot}26 \ \text{MHz}$$

We can state the horizontal resolution either in terms of the highest video frequency (5·26 MHz) or in lines (409·5). In a 625-line system there are thus

$$\frac{409{\cdot}5}{5{\cdot}26}$$

lines per MHz of bandwidth = 78 lines per MHz of bandwidth (160 lines/MHz for 405-line).

In technical literature figures for horizontal resolution may be quoted in excess of 409·5 lines. Why is this so when 409·5 lines would appear to be the maximum horizontal resolution? The reason is that we have assumed there is no space between the scanning lines. In practice there is a gap between adjacent line scans, thus the beam diameter is smaller than we have assumed. Because of this many more smaller elements can be accommodated across the width of the picture. Provided, therefore, the video circuits can work fast enough, finer detail can be resolved. Thus a camera with a video bandwidth of 8 MHz is theoretically capable of a horizontal resolution of $8 \times 78 = 624$ lines (a typical figure for a CCTV camera of good quality). Such a camera is capable of providing better resolution horizontally than vertically, of course.

Strictly speaking the resolution of a monitor picture is subject to limitations imposed by the acuity of the eye. Under the very best conditions, the eye can see an object with a viewing angle of one minute of an arc. It may be shown that at a viewing distance of four times the picture size (measured diagonally) the useful maximum resolution is about 610 lines under the best conditions. At distances farther away from the monitor screen, say, 10 times the picture size the maximum useful resolution is reduced to 245 lines. Thus, the benefits to be gained by using resolutions in excess of about 600 lines can only be realized at close viewing distances. For CCTV in industry a

CCTV SIGNALS AND PRINCIPLES

viewing distance of about 2–3 times the screen size would appear to be typical, but occasionally it may be much greater.

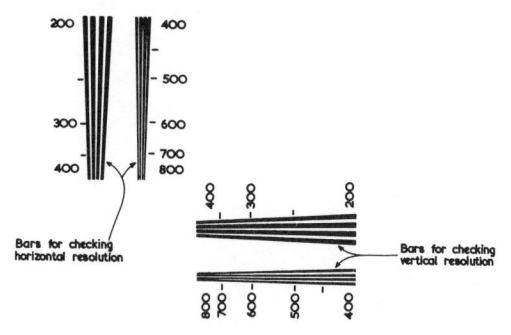

FIG. 3.24 BARS WITH FIGURES (CORRESPONDING TO LINES PER ACTIVE PICTURE HEIGHT) FOUND ON CCTV TEST CARDS

MODULATION

Some cameras provide, in addition to the composite video output, a modulated r.f. output. The idea of the generating circuits is shown in block schematic form in figure 3.25. An LC oscillator generates a carrier wave having a frequency lying in Band 1. Usually the carrier frequency may be adjusted to one of three Band 1 channels by

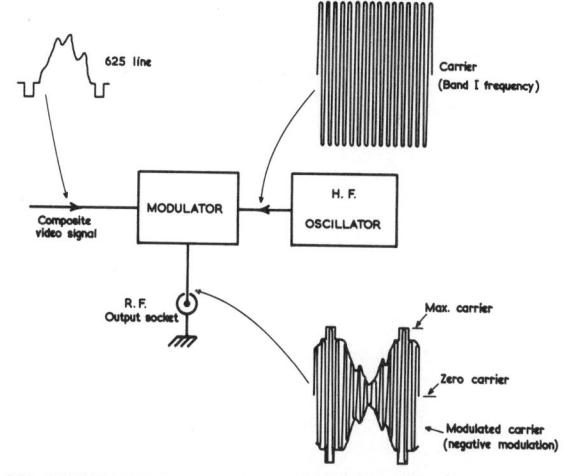

FIG. 3.25 MODULATED R.F. CARRIER FACILITY PROVIDED BY SOME CAMERAS

preset tuning. In the modulator stage the composite video signal amplitude modulates the carrier to provide a modulated r.f. output of about 10 mV r.m.s. across a 75 Ω termination. To conform with 625-line vision transmissions, negative modulation of the carrier is employed, *i.e.* peak white causes the carrier to assume its minimum amplitude. This facility allows a camera to be used with an ordinary domestic receiver. If the camera operates on 625 lines, the t.v. receiver must select the v.h.f. tuner in the 625-line mode to utilize the modulated r.f. facility.

Details of the modulation levels for the vision carrier transmitted by the televison broadcast authorities in this country are shown in figure 3.26. There is no pedestal, thus black and blanking levels are the same and these produce about 77% modulation. The

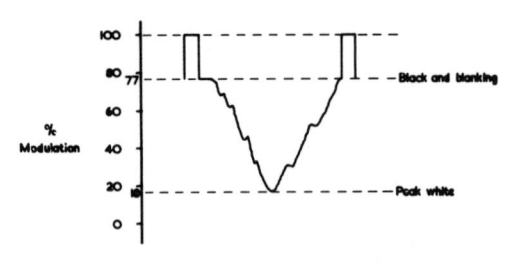

FIG. 3.26 MODULATION LEVELS FOR BROADCAST TELEVISION VISION CARRIER (625-LINE)

sync. pulse tips cause the carrier to assume its maximum value, whilst peak white corresponds to about 18–20% modulation. The reason why the carrier is not reduced to zero is that some residual carrier must remain in order to create the 'intercarrier sound' used in domestic receivers.

CHAPTER 4

TELEVISION CAMERA TUBES

CAMERA tubes are available using two different operating principles. In one type, the electrode on which the optical image is formed is PHOTOEMISSIVE, *i.e.* it is coated with a material which liberates electrons on exposure to light. Camera tubes adopting this principle have been extensively used in high-definition studio work. In other camera tubes instead of using a photoemissive layer, a PHOTOCONDUCTIVE material is employed, the electrical resistance of which is lowered when light falls on the material. The change of resistance is normally slow, resulting in some smearing in the reproduced images of moving objects. Research into photoconductive materials has now considerably reduced the 'lag'. The photoconductive principle is used in the VIDICON and PLUMBICON camera tubes. Because of its low cost, simplicity of operation and its small size, the vidicon camera tube is usually chosen for monochrome CCTV systems.

THE VIDICON CAMERA TUBE
(a) Construction
The general layout of a vidicon camera tube is shown in figure 4.1. The tube is about 6 inches long and has a diameter of about 1 inch. At one end of the tube is the

(a) Construction

(b) Details of target

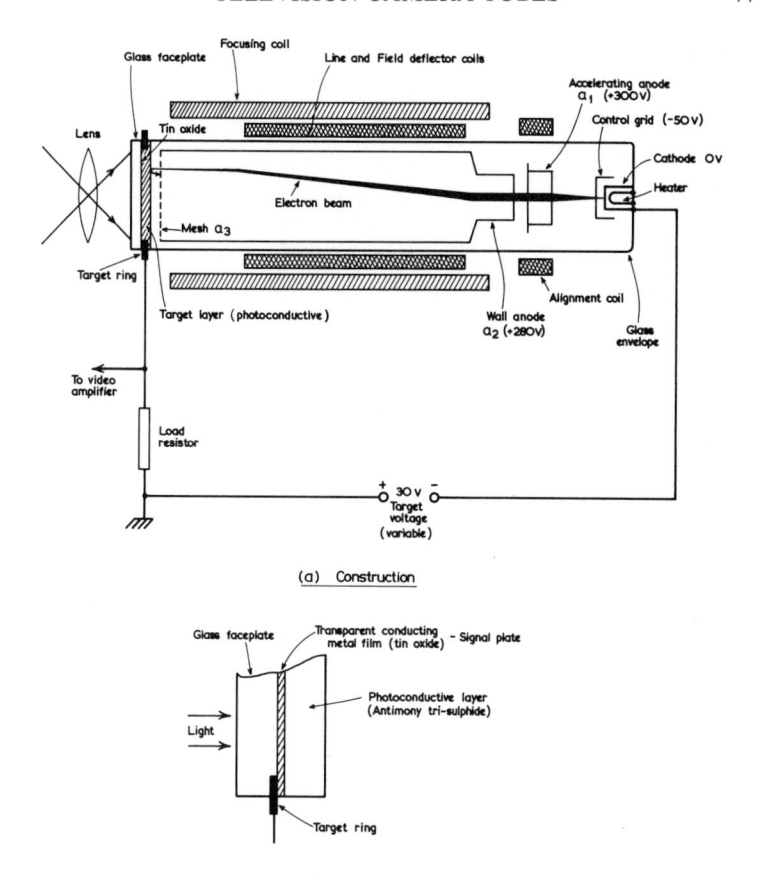

FIG. 4.1 THE VIDICON CAMERA TUBE

photosensitive target layer and at the other end an electron gun assembly. The electron gun fires a narrow beam of electrons along the highly evacuated glass envelope. Electrons are emitted from the cathode which is raised to the required emitting temperature by the heater which may be fed with either a d.c. or an a.c. voltage. Electrons are accelerated towards the target by the accelerating anode which is held at a constant voltage, some 300 V positive with respect to the cathode. Before reaching the accelerating anode the electrons must first pass through the control grid, a disc with a small aperture in its centre. The control grid is fed with a variable voltage which is negative with respect to the cathode by about 50 V. Varying the grid voltage controls the density of the electron beam passing through the grid aperture. The control which adjusts this voltage is called the 'beam control' and will be required to be set up by the technician (or operator).

On leaving the accelerating anode the beam passes through the wall anode which extends for most of the length of the tube. Here the beam is focused by the focus coil (mounted externally to the tube) which has a direct current flowing in it. The electron beam must be brought to a sharp focus on the target which is achieved either by varying the potential on the wall anode or by varying the current in the focus coil. The control which is used for this adjustment is called the 'electrical focus' (not to be confused with the 'optical focus' used to focus the scene image).

To provide horizontal and vertical deflection of the beam so that it scans the target image, suitable currents are fed into the line and field deflector coils which, like the focus coil, are mounted outside the glass envelope of the tube.

Situated at about 2–3 mm from the target is a very fine metal mesh through which most of the arriving electrons pass. The electric field set up between the mesh and target provides a uniform decelerating field for the electron beam so that the electrons strike the target with low impact. This is necessary to reduce secondary emission from the target which can lead to spurious signals being produced. In an *integral mesh* vidicon, the mesh is electrically connected to the focusing (wall) anode. Some improvement in resolution can be obtained by connecting the mesh to higher positive potential than the wall anode (about 1·5 times higher). This arrangement, which is in increasing use, is known as a 'separate mesh vidicon'. It is important to note at this stage that the electron beam *strikes the target at right angles to the surface wherever it lands*. This is known as 'orthogonal' scanning of which more will be said later.

The target which is shown in greater detail in figure 4.1(b) consists of a layer of photo-conductive material such as Antimony Trisulphide. When light falls on this material, the resistance between opposite faces decreases. On one side of the photo-conductive layer is deposited a transparent, conductive metal film of tin oxide. This is electrically connected to the 'target ring' which is sealed into the glass envelope. the target ring serves as the output electrode. To develop an output signal voltage a load resistor is required which is connected to the target ring. Light from the scene passes through the optically flat, polished faceplate and the transparent tin oxide film on to the photoconductive layer.

(b) Producing an Output Signal

For explanatory purposes we may consider that between any two points on opposite faces of the photoconductive layer a small capacitor is formed. One 'plate' of the capacitor is formed by the signal plate and the other 'plate' which is floating is produced by the electron beam when it arrives at the small area under consideration. Between the plates is the dielectric formed by the photoconductive material. The layer structure may therefore be regarded as being composed of innumerable small capacitors each isolated from the other with one plate commonly connected *via* the signal plate. In parallel with each of the capacitors so formed between points on opposite faces is the axial resistance of the photoconductive layer. The idea is shown in figure 4.2(a) where just a few capacitors and their associated 'leakage resistances' are shown.

It will be assumed that no light is allowed to fall on the target layer, *i.e.* the lens cap is fitted. When the scanning beam falls on any point such as B, D, F and H, etc., that point will assume cathode potential. This will cause each of the capacitors to charge up to the target voltage, *i.e.* to 30 V, after the first scan. If all the capacitors were perfect they would hold their charge. However, due to the axial resistance of the target layer each capacitor will discharge a little between scans. This will cause a minute current to flow in the load resistor when the electron beam is on the next and subsequent scans and the charge lost is replaced. This very small current constitutes the **dark current** of the tube (typically 0·01 μA). For successful operation of the vidicon, the lateral resistance of the target must be large compared with axial resistance so that current flow is parallel to the tube axis. This means that the photoconductive layer must be very thin.

Let us now allow light to fall on the target as shown in figure 4.2(b). High intensity light falling on point A causes the resistance R_1 between A and B to fall. This action

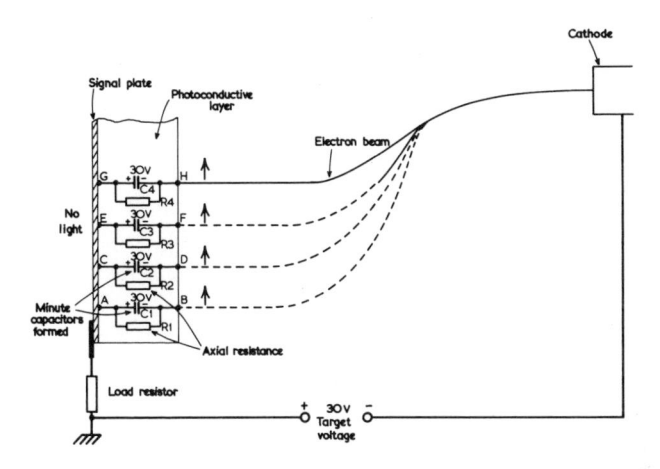

(a) State of charge of a few layer capacitances after first scan by electron beam at points B, D, F and H.

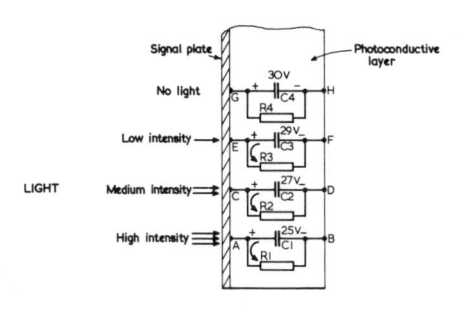

(b) When light falls on the photoconductive layer, resistances R1, R2 and R3 become lower. C1, C2 and C3 discharge, the charge lost being proportional to the intensity of the light falling on those points A, C and E.

FIG. 4.2 DIAGRAMS SHOWING INITIAL CHARGING OF LAYER CAPACITANCES AND EFFECT OF VARYING LIGHT INTENSITY

causes C_1 to discharge to, say, 25 V. Light of medium intensity falling on point C causes the resistance of R_2 to fall but the reduction in resistance is not so great as for R_1. As a result C_2 discharges to, say, 27 V. Low intensity light likewise causes C_3 to discharge but not so much as for C_2, say, to 29 V. C_4 does not discharge as no light falls on point G. Thus it can be seen that the charge lost by each capacitor is proportional to the intensity of the light falling on the points under consideration.

Now, when the electron beam subsequently scans the target, each capacitor is recharged back to 30 V. Figure 4.3(a) shows the beam striking point B causing C_1 to recharge to 30 V. As the capacitor charges current flows in the load resistor in the

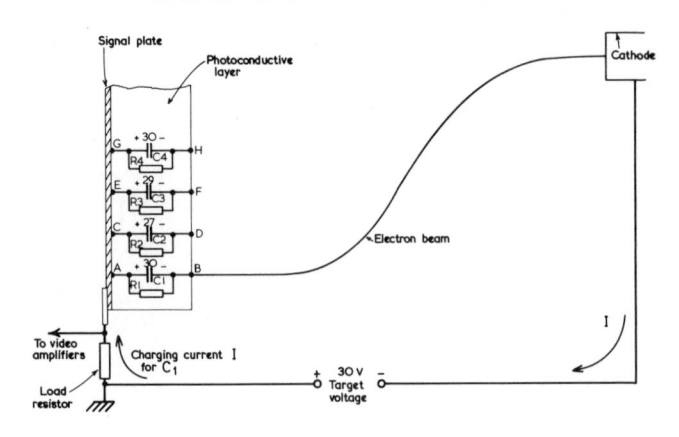

(a) Subsequent scan of electron beam causes C_1, C_2, and C_3 to charge. The charging current (proportional to the light) flows in the load resistor producing a signal output voltage.

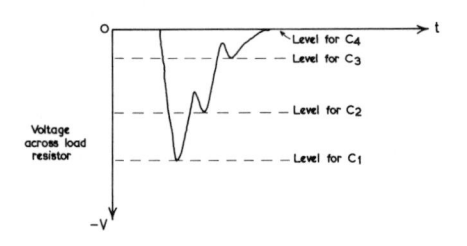

(b) Output signal voltage across load resistor.

FIG. 4.3 DIAGRAMS SHOWING HOW CURRENT IS PRODUCED IN LOAD RESISTOR TO DEVELOP THE VIDEO OUTPUT SIGNAL

direction shown. The voltage developed across this resistor constitutes the output signal. As the beam scans the target area each capacitor is recharged in sequence, the charging current of which flows in the load resistor. The charging current and hence the output voltage is proportional to the intensity of the light falling on the individual capacitors. The peak current flowing in the load resistor is typically of the order of 0·25-0·4 μA. Figure 4.3(b) shows the changing voltage across the load resistor resulting from the recharging of a few of the layer capacitors.

IMAGE BURNS

If the vidicon photoconductive layer is exposed to a scene emitting bright light, the scene may still be seen faintly in the background when the camera is pointed in another direction. This is called 'image sticking'. Sometimes the image will gradually fade away depending upon various operating conditions. Often in industrial use, a camera may be viewing a static scene over long periods with the result that objects of high luminance permanently burn the photoconductive layer. On the monitor the 'burn' appears as a bright-up. It is thought that the axial resistance is permanently lowered in those areas receiving continuous light of high luminance. Thus, each time the beam scans the 'burnt' area (even though the lens cap may be fitted) the associated layer capacitors have to be recharged producing a large video output from the camera and causing corresponding bright-ups on the monitor screen. It is important, therefore, to prevent bright lights from falling on the camera target and to avoid unnecessary contrasty viewing situations. The lens cap should always be fitted when the camera is not in use whether the camera is switched ON or OFF.

FOCUSING THE ELECTRON BEAM

In common with other camera tubes, a long focus coil is used for focusing the electron beam in a vidicon tube. A long magnetic field is produced by a d.c. flowing in the coil which extends for most of the length of the tube, figure 4.4. Electrons are brought to a focus by the axial component of the magnetic field. The mechanism of

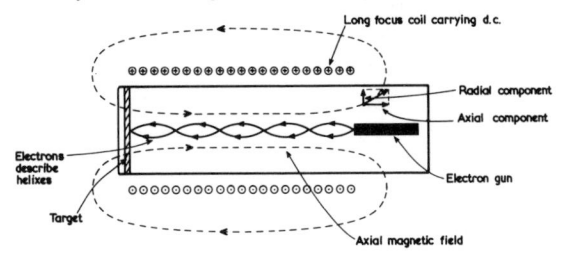

FIG. 4.4 FOCUSING WITH A LONG MAGNETIC LENS

focusing is thus quite different from a short coil or permanent magnet (which at one time was used for focusing in domestic television receivers) where focusing is mainly due to the radial component of the magnetic field.

Electrons enter the magnetic field from the gun arrangement at a small angle to the axial field or parallel to it and with varying velocities. Those electrons making a small angle with the axial field describe helixes about an axis parallel to the tube axis. All the electrons take the same time to complete one loop of the helix and they all meet at some other point after going through a complete loop. As the electrons enter the field at random velocities, helixes of different diameters are described but these have circumferences which pass through a common point on the tube axis. Thus, points of focus occur at regular intervals along the tube axis as shown. One of these points must be made to coincide with the plane of the target layer. Now the spacing between the points of focus is proportional to

$$\sqrt{\frac{V}{H}}$$

where V is the accelerating potential and H the magnetic field strength. Thus by varying the current in the coil (to alter the field strength) or by varying the potential on a_2 of the vidicon a point of focus can be made coincident with the target layer.

ORTHOGONAL SCANNING

In low-velocity camera tubes such as the vidicon, the line and field deflector coils are surrounded by a long focus coil as in figure 4.5. An electron beam travelling parallel to the tube axis is subjected to two magnetic fields when it enters the vicinity of the deflection system. There is an axial field (H_F) due to the focus coil and a tranverse field (H_D) set up by the deflector coils (just one pair are considered here). The electron beam thus pursues a helical path along an axis parallel to the resultant (H_R) while under the influence of the combined fields.

On leaving the region of the combined fields, the electron beam is subjected to the axial field of the focus coil only. The beam enters this field at an angle and is again deflected to perform helixes about an axis parallel to the tube axis. Thus the effect of the two fields is to laterally displace the beam in the direction of the transverse field without altering its initial direction. The amount of lateral displacement is

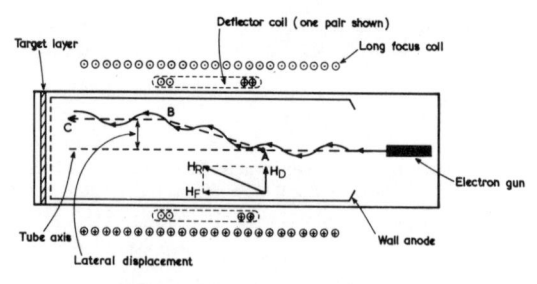

FIG. 4.5 ORTHOGONAL SCANNING

proportional to the strength of the transverse field (set up by the current in the deflector coils) and can be reversed by reversing the direction of the current in the coils. In this form of magnetic deflection where there is an axial field, the beam is deflected towards one of the deflecting coils and not at right angles to it. Thus the line and field deflector coils are disposed as in figure 4.6.

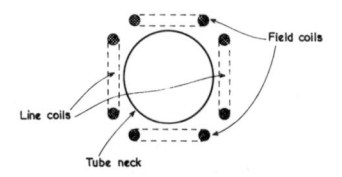

FIG 4.6 DISPOSITION OF SCANNING COILS IN CAMERA

This is termed 'orthogonal scanning', *i.e.* the beam strikes the target layer at right angles to the surface wherever it lands. This ensures that electrons which are scattered or repelled remain in the vicinity of impact. A sharper focused beam is thus possible using this method of scanning as the beam will have a circular cross-section at the point of impact, unlike ordinary scanning where the beam will have an elliptical cross-section. Figure 4.7 illustrates the difference between orthogonal scanning in a low velocity camera tube and ordinary scanning as used in the monitor c.r.t.

(a) Orthogonal scanning in low velocity camera tube.

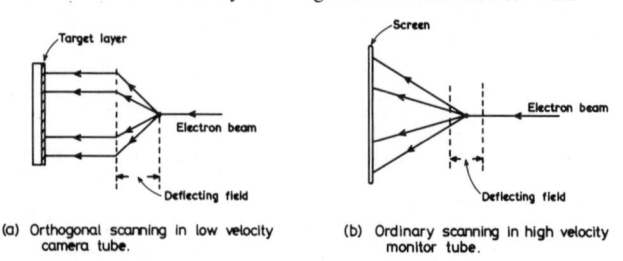

(b) Ordinary scanning in high velocity monitor tube.

FIG. 4.7 COMPARISON OF ORTHOGONAL AND ORDINARY SCANNING

VIDICON LAG

Time is taken for the target layer to change its resistance when varying intensity light falls on the target layer. This causes some smearing of reproduced moving images. The promptness in the change of the target conductivity depends upon the scene illumination, changing faster with an increase in the scene illumination. Thus to reduce smearing on moving objects a large light input to the camera is required. The slow response of the vidicon is one of its main disadvantages and is thus more suitable in situations where there is little rapid movement in the scene.

CAMERA TUBE CONTROLS

An example of the circuits associated with the main camera controls are shown in figure 4.8.

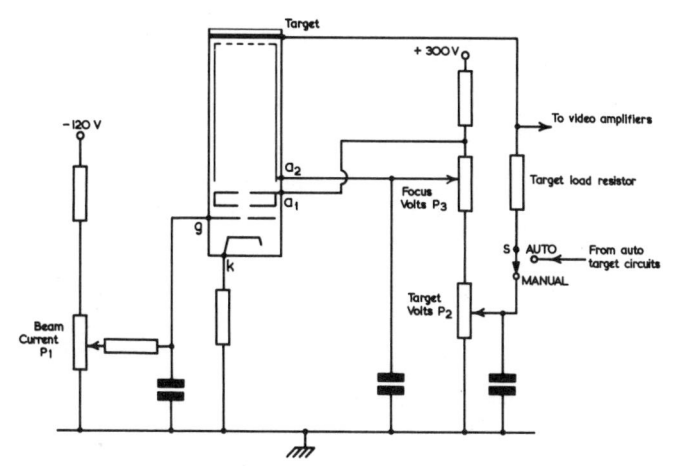

FIG. 4.8 CAMERA TUBE CONTROLS

Beam Current Control P_1

As previously explained, this control varies the grid-cathode bias of the camera tube. With small bias the beam current is high and with large bias the current is small. If the beam current is too small, there will not be sufficient electrons to recharge the 'layer capacitors' to cathode potential in the brightest areas of the scene. This will cause 'clipping on whites', *i.e.* areas of high but different brightness levels will all be interpreted as being of the same brightness level and the picture will appear 'flat'. Too much beam current will cause some loss of focus and shorten the life of the tube and for this reason manufacturers usually specify a maximum beam current (of the order of 0·3 to 0·4 μA).

If P_1 is adjusted starting at the zero beam current position, then as the current is increased a negative or inverted picture will appear on the monitor screen. As the control is advanced further the picture will change to a normal black-and-white one, but at this stage the picture may appear 'flat'. Thus the control should be set slightly higher until in the brightest areas of the picture the peak whites are not limited.

Target Volts Control P_2

This control alters the sensitivity of the tube. As the target voltage is increased, the 'layer capacitors' are able to charge to a higher voltage (assuming that there is sufficient beam current available). Thus, for a given scene illumination a larger video output signal will be obtained across the target load resistor with a higher target voltage setting. Unfortunately, as the target voltage is increased, the dark current of the tube increases. This, however, is not uniform and causes shading of the picture, *i.e.* ranging from black on one side to grey on the other side of the monitor screen. The dark current also increases with temperature, thus cameras working under high ambient temperatures may require special cooling arrangements. Manufacturers usually specify a maximum target voltage of about 80–100 volts. In some circuits 'target limit' is applied; here a suitably biased diode prevents the target voltage from exceeding a permitted maximum value.

As the target voltage alters the amplitude of the video output signal, automatic target control circuits may be employed to maintain a steady video output with varying scene illumination by automatically adjusting the target potential. The changeover from 'manual' to 'auto' may be by means of a switch (such as S in the diagram) or by a semi-permanent link. In practice the beam and target control adjustments are interdependent. Thus to obtain the optimum setting for each control they should be adjusted alternately so that after each adjustment one is getting closer to the optimum setting.

Focus Volts P_3

The a_2 electrode of the vidicon tube is held at a potential of about 30–50 volts negative of the accelerating anode a_1. The electric field set up between these two electrodes produces an electrostatic lens which, in conjunction with the long magnetic field from the focus coil, is responsible for focusing the electron beam. Adjustment of the focus is usually done by varying the a_2 potential while keeping the current in the focus coil constant. There may be a pre-set focus coil current adjustment, but this acts as a coarse adjustment keeping optimum focus at the centre of P_3 travel. Note that the electrical focus is concerned with focusing the electron beam on the target. Although it has a similar effect to the optical focus in so far as clarity of detail on the monitor is concerned, it cannot be used to correct for maladjustment of the optical focus.

As P_3 is adjusted about its correct setting the picture on the monitor will go 'in' and 'out' of focus. If the control is 'rocked' about its optimum setting, a rotation of the picture on the monitor about a point near the picture centre may also be observed. This results from a change in the pitch of the helix described by the electron beam as the a_2 potential is altered.

THE PLUMBICON CAMERA TUBE

The plumbicon is very similar in construction to the vidicon. It uses the same low-velocity electron beam and orthogonal scanning. The tube is intended for use in monochrome and colour television broadcast work, also industrial and scientific applications. The main difference between the plumbicon and vidicon is in the make-up and operation of the target layer, see figure 4.9.

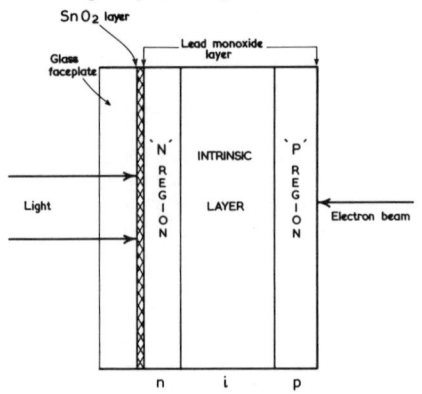

FIG. 4.9 DETAILS OF TARGET IN PLUMBICON TUBE

In the pumbicon the layer on which the scene is imaged consists of microcrystalline lead monoxide, hence the name 'plumbicon'. The lead monoxide layer is made up of a thin p-type region, a thick intrinsic lead monoxide region and a thin n-type region. The target plate consists of a thin transparent film of conductive tin oxide (SnO_2) located between the glass faceplate and the n-region of the light sensitive layer. During operation the target plate potential is held positive to the cathode as for the Vidicon.

The lead monoxide layer forms a p-i-n diode (or many p-i-n diodes in parallel) and during operation it is reversed biased. Thus, with no light falling on the camera tube, only a very small leakage current (the 'dark' current) flows in the p-i-n diode and the

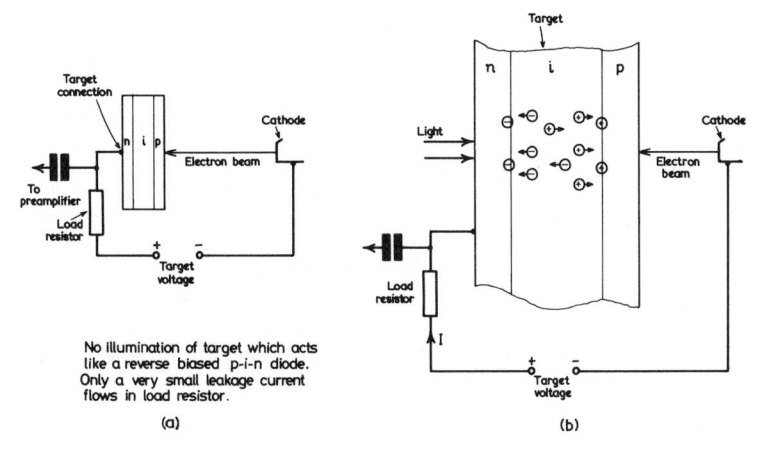

FIG. 4.10 DIAGRAMS ILLUSTRATING BASIC OPERATION

load resistor, figure 4.10(a). The dark current is of the order of 0·003 μA which is many magnitudes smaller than in the vidicon. When light falls on the lead monoxide layer, bonds are broken in the intrinsic layer releasing holes and electrons which find their way to the conductive 'n' and 'p' regions, figure 4.10(b). The number of charge carriers released is dependent on the intensity of the incident light. As a result of this action the potential of the 'p' region increases. When the electron beam scans the target layer the beam neutralizes the charge until the 'p' region drops to cathode potential. This action causes a current to flow in the load resistor which produces the output signal.

Altering the target voltage has little effect on the magnitude of the output signal voltage. If the voltage is increased it merely increases the dark current. Thus, varying the target voltage will not alter the tube's sensitivity as it does in the vidicon. Therefore if automatic sensitivity is desired, a lens with an automatic iris adjustment is required. The main advantages of the plumbicon tube are:

(a) The low level and uniformity of the dark current.
(b) Low lag characteristic.
(c) High sensitivity, *i.e.* ability to work at lowlight levels (see figure 4.11).

In the standard plumbicon tube, very high peaks of light result in the production of large charges on the 'p' region which require a number of scans to neutralize them. If

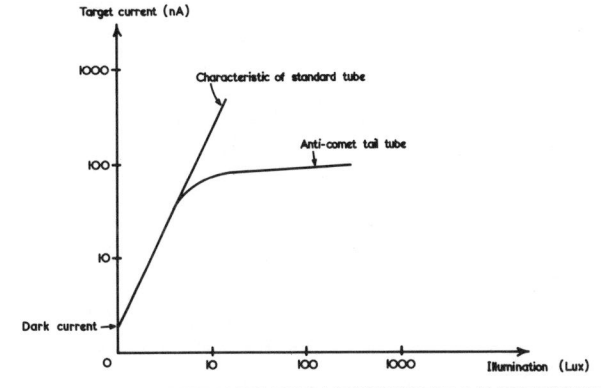

FIG. 4.11 TARGET CURRENT-LIGHT CHARACTERISTIC FOR A PLUMBICON TUBE

the source of the peak light is moving, the effect is to produce a flare (comet-tail) trailing after the object on the monitor screen. Also, stationary objects appear to 'bloom' on the screen. These effects are reduced in the 'Anti-Comet Tail' Plumbicon tube.

In this tube the electron beam is not suppressed during flyback. Instead the beam current is increased, the action being controlled by the focus lens. The increase in beam current neutralizes the high charges. To ensure that only the peak charges are neutralized and not the entire layer area, the cathode potential has to be raised simultaneously by a few volts. The effect of this action is to neutralize all charges exceeding a preset value so that the remaining charges can be dealt with by the normal beam scans. Thus light above a certain intensity will cause the target current to saturate, see figure 4.11.

Because of the velocity distribution of the electrons in the beam there is a lag in recharging the target. This lag can be reduced by increasing the dark current of the tube. The method used is to illuminate the rear of the target by means of a light pipe which receives light from a weak external light source, figure 4.12.

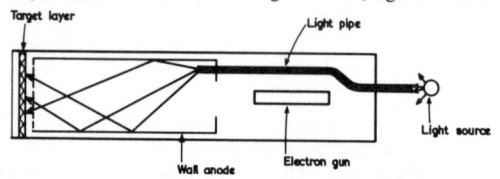

FIG. 4.12 LIGHT PIPE IN ANTI-COMET TAIL PLUMBICON TUBE TO ILLUMINATE REAR OF TARGET THEREBY INCREASING DARK CURRENT AND REDUCING LAG

GAMMA CORRECTION

The gamma (γ) of a device may be defined as the slope of its transfer characteristic when it is plotted on log-log graph paper. If a linear relationship exists between the input and the resulting output of a device the gamma is unity. Now, the relationship between the intensity of the light (L) falling on the face of a vidicon camera tube and the resulting output signal voltage (V_0) is a non-linear one, see figure 4.13(a). For a typical vidicon it obeys the law $V_0 = kL^{0.6}$ where 0.6 is the gamma. As illustrated in figure 4.13(b) equal increments in light falling on the camera tube do not produce equal increments in the video output signal. The effect is to compress variations in the

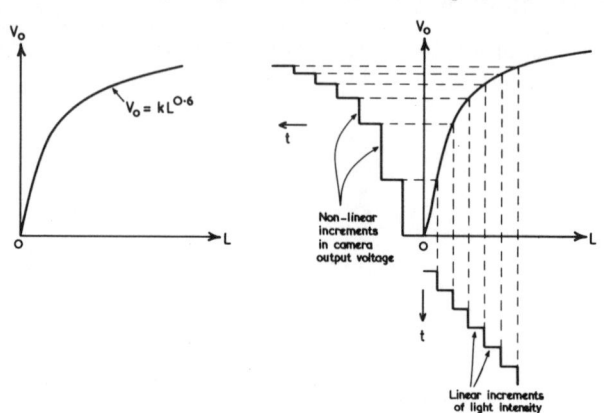

(a) Transfer characteristic of Vidicon tube.

(b) Application of linear staircase light variations.

FIG. 4.13 VIDICON TUBE GAMMA

highlights of a scene but to expand variations in the lowlights. Quite clearly if the video output signal was processed by a linear system between the camera tube output and the monitor c.r.t., the resulting picture on the monitor would not be of the correct tonal value and considerable correction for the non-linearity of the camera would be required.

Fortunately, the c.r.t. used in the monitor also exhibits non-linearity but with opposing effects to that of the camera tube. The transfer characteristic of a typical monitor or t.v. receiver c.r.t. is shown in figure 4.14. Here the light output (L) bears the relationship

$$L = kV_D^{2 \cdot 2}$$

with the signal drive voltage (V_D), where 2·2 is the gamma. For a linear staircase drive waveform this non-linear relationship results in the highlights being stretched and the lowlights compressed.

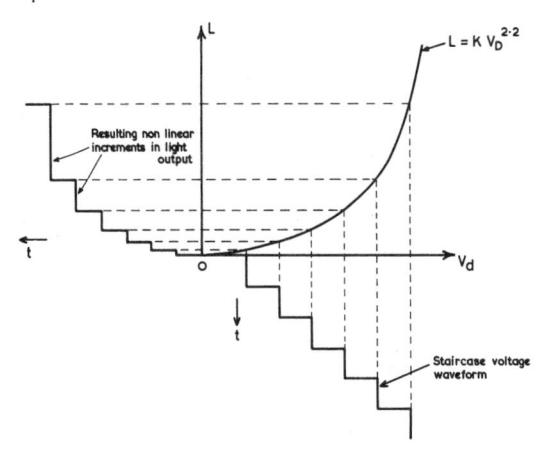

FIG. 4.14 GAMMA OF TYPICAL MONITOR C.R.T.

The overall gamma of a number of devices is the product of the individual gammas. Thus for the vidicon tube and the c.r.t. (with the values given) the overall gamma is 0·6 × 2·2 = 1·32. At first it may be thought that the ideal overall gamma would be unity, but for direct viewing a gamma of about 1·2 to 1·3 is preferred. Thus, when using a vidicon camera tube, no gamma correction is required—which is most fortunate. For a plumbicon tube the gamma lies between 0·95 and unity, *i.e.* the transfer characteristic is almost linear. When used with a monitor c.r.t. the resulting overall gamma is, say, 0·95 × 2·2 = 2·09. This gamma value indicates, for the overall system, expansion in the highlights and therefore some correction is required.

Gamma correction may be applied in one of the video stages following the camera tube output by operating the amplifier in a non-linear manner. To correct for an overall gamma of 2·09, the correcting amplifier should have the relationship

$$V_0 = kV_i^{\frac{1}{2 \cdot 09}}$$

where V_i and V_0 are the input and output voltages of the amplifier, figure 4.15. Quite commonly the correction is performed by diodes which conduct at different levels to shape the amplifier characteristic as required. By carrying out the gamma correction in the camera circuits, no correction is required in the monitor which thus can be efficiently used with cameras employing different tubes.

(a) Correcting amplifier

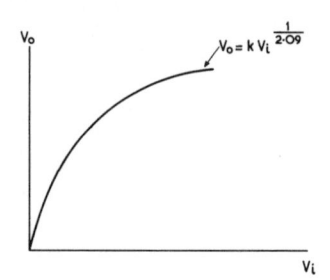

$V_o = k V_i^{\frac{1}{2 \cdot 09}}$

(b) Characteristic of amplifier for example given in text.

FIG. 4.15 GAMMA CORRECTION

APERTURE DISTORTION

The limit of resolution in any camera tube is the size of the scanning beam. Consider diagram (a) of figure 4.16 where the scene to be televised contains sharp transitions from black to white. If the scanning beam is of extremely small cross-

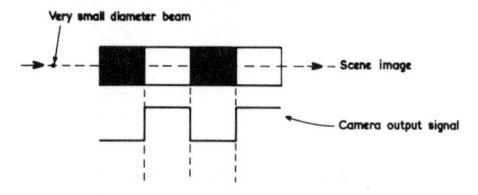

(a) Camera output for beam of very small dimensions

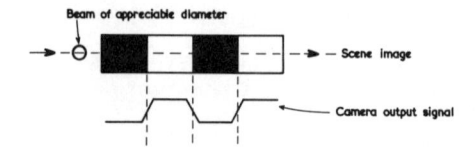

(b) Camera output for beam of appreciable dimensions

FIG. 4.16 APERTURE DISTORTION

section, the camera output signal rapidly changes level each time the beam crosses the boundary line between the black and white areas. Such an output waveform contains components up to a very high frequency and represents the ideal case.

In practice the beam has appreciable cross-section and takes time to cross the boundary line. When the beam is disposed centrally on the boundary line, the camera output is intermediate between black and white. This results in an output waveform with flanks that are less steep as shown in diagram (b). Thus the effect of a beam with finite width is to attenuate the high frequency components of the output signal. This is called 'aperture distortion' and may be compensated for in the camera circuits by special techniques which have the effect of sharpening abrupt changes in the video waveform (see Chapter 5, page 100).

SILICON DIODE TARGET VIDICON

Out of all the many photoconductive materials that exist only antimony trisulphide and lead oxide have (until more recently) given really good performance in a wide range of television camera applications. Even then, the preparation of these materials requires very closely controlled physical structures and impurity levels. When high quality, single crystals such as silicon and germanium became available (as a result of development of the transistor industry) these were naturally considered for television camera tube targets. However, in their intrinsic form these crystals have a resistivity which is far too small to satisfy the dark current and resolution requirements of a television camera. The obvious structure to adopt was an array of reverse-biased p-n diodes because both low dark current and lateral isolation of the picture elements could be achieved. Silicon was the material chosen for fabricating the diodes since it can work at high and low temperatures without loss of photo-sensitivity. Also, silicon targets cannot be damaged even by a focused image of the sun and no permanent scan burn occurs.

Most tube manufacturers produce silicon diode-array vidicons under their own trade names (Sidicon, Sivicon, Tivicon and S1 Vidicon are the more common). Except for the target area, the construction of a silicon diode vidicon is the same as the ordinary antimony trisulphate vidicon. The techniques used in the fabrication of the silicon diode target are similar to those used in the manufacture of planar integrated circuits. Figure 4.17 shows a cross-section of one form of silicon diode target which consists of a regular arrangement of minute, discrete p-type regions diffused into a thin

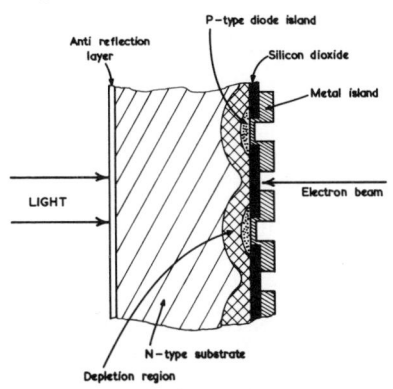

FIG. 4.17 SILICON DIODE-ARRAY TARGET

wafer of n-type silicon. A silicon dioxide masking is used to define the p-regions and also serves to insulate the n-type silicon (between the p-regions) from the scanning beam. Between the substrate and each p-type area a p-n diode is formed. Each diode is capped by a metal island, covering as much of the adjacent oxide as possible; this increases the electron beam capture area. For normal television applications, *e.g.* 600 lines, about half a million such diodes are needed in the scanned area of the tube. Since the scanned area of a 1″ vidicon is only 12·7 mm by 9·5 mm, integrated microcircuit engineering is called for.

During operation the conventional positive target potential is applied to the n-type substrate, which is common to all of the diodes. In the unilluminated state, the electron beam stabilizes the p-type contact of each diode at cathode potential. The diode-array is therefore reverse biased and the depletion capacitance of each diode is charged to the target voltage. When light falls on the target, hole/electron pairs are generated in the n-type substrate. The holes drift into the depletion region causing the diode 'capacitors' to discharge in proportion to the intensity of the incident light (the electrons are collected at the positive target contact). As each diode is subsequently scanned by the electron beam the diode 'capacitors' recharge. The recharging currents which flow in the target and load resistor constitute the video signal as in the conventional vidicon.

An anti-reflective coating on the light incident side of the target is used to increase the sensitivity at chosen wavelengths; this is necessary because the reflectivity of untreated silicon is very high.

The dark current of a silicon diode tube is of the order of 15 nA which is less than the antimony trisulphide vidicon but a little higher than the plumbicon. The most common picture blemishes produced by the silicon diode-array are small white spots caused by high leakage diodes scattered over the array.

Compared with the antimony trisulphide and lead oxide tubes, the silicon diode vidicon offers a much higher photo-sensitivity particularly in the red and infra-red sectors of the spectrum. Other points in its favour are its low lag (freedom from image sticking) and the ability to withstand light overloads.

SPECTRAL RESPONSE

The spectral response of a typical silicon diode target is shown in figure 4.18. All camera tubes have a varying sensitivity to lights of different wavelengths and the

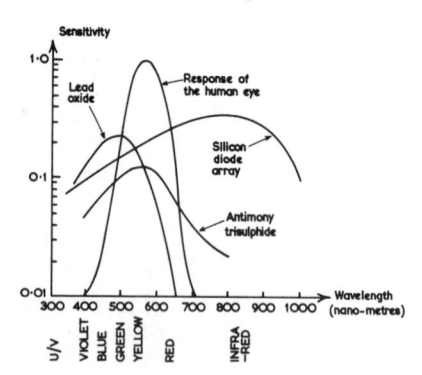

FIG. 4.18   SPECTRAL RESPONSE OF PHOTOCONDUCTIVE TARGETS

responses of the ordinary vidicon and lead oxide tubes are given for comparison. For normal television broadcast and CCTV pictures the response required approximates to that of the human eye. The actual response obtained in a particular tube design depends entirely upon the material used for the target and its processing during manufacture. Although the silicon diode vidicon is responsive over the entire visible light band, it has an enhanced response in the infra-red region which makes it suitable for use in i.r. television. Typical applications of i.r. television include observation of nocturnal animals, of hospital patients requiring constant observation and for security/surveillance applications. Infra-red vidicons have also been used to detect hot spots in furnace walls and for monitoring temperature patterns at the tips of cutting tools.

The sensitivity of a silicon diode vidicon cannot be significantly altered by varying the target potential (within the normal operating range); thus if automatic sensitivity control is required it has to be carried out in the lens system using an automatic iris. With the silicon diode tube the relationship between light input and video signal output is a linear one, thus a gamma of unity is obtained.

CHALNICON TUBES

These tubes, developed by Toshiba, first became available in 1972. The Chalnicon, pronounced with a hard 'k' (kaélnikon), is a photoconductive type camera tube. It employs a multilayer structure target consisting of Cadmium Selenide (CdSe) and other chemical compounds. The name 'Chalnicon' comes from the fact that CdSe is one of the 'Chalcogenides' which is the chemical name of sulfides, selenides or tellurides.

Except for the target structure which is shown in figure 4.19 the construction of the Chalnicon is similar to the ordinary vidicon. Magnetic focusing and deflection is

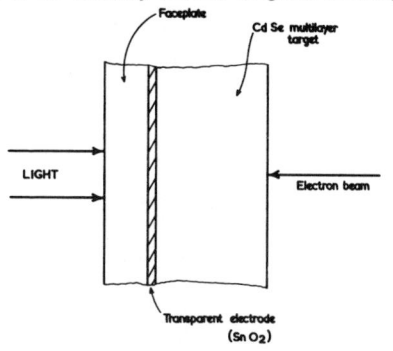

FIG. 4.19 CHALNICON TARGET STRUCTURE

employed using a separate mesh tube. The inner surface of the glass faceplate is coated with a thin, transparent, conductive film of SnO_2 to which the positive target voltage is applied. Deposited over this film is a layer of CdSe and this is followed by multilayers of Chalcogenides. The thickness of all the photoconductive layers is only a few micrometres.

As in the ordinary vidicon, the action of the electron beam is to stabilize one surface of the target layer at cathode potential. When light falls on the other surface of the target, after passing through the glass faceplate and SnO_2 layer, carriers in the N-type CdSe layer are excited in proportion to the quantity of light received. These excited carriers drift through the multilayers by the electric field applied to the target. The carriers raise the surface potential of the layer. Thus the light image is converted into an image of potential distribution on the target surface. When the surface is subsequently scanned by the electron beam, the surface potential of each picture element is reduced to cathode potential by neutralizing the positive charges with the electron beam. The discharge current due to this action flows in the load resistor located in series with the target connection. A video signal is thus developed across the load resistor which is passed on to the video amplifier in the usual way.

The principle advantages of the Chalnicon are its very high photo-sensitivity, low dark current, wide spectral response (see figure 4.20), unity gamma and virtual freedom from image sticking (provided the target voltage is set correctly). It is available in 18 mm and 25 mm diameters. Applications for this camera tube include Security Observation in low illumination, Industrial or Educational CCTV and Slow Speed Scan Pick-up and Data Processing.

HIGH-SENSITIVITY CAMERA TUBES

The quality of the picture attained in a television system gets worse as the scene illumination falls, in the same way as the eye discerns less of its surroundings in fading light. It is fortunate in broadcast and ordinary CCTV applications that the declining response of the eye and the system electronics are not too badly matched. Thus if the light is sufficient for footballers to play it will suffice for televising the match.

When a television picture is to be produced where the available light is insufficient for the unaided human eye, *e.g.* in night-time military/security surveillance or X-ray

92 INDUSTRIAL AND COMMERCIAL CCTV

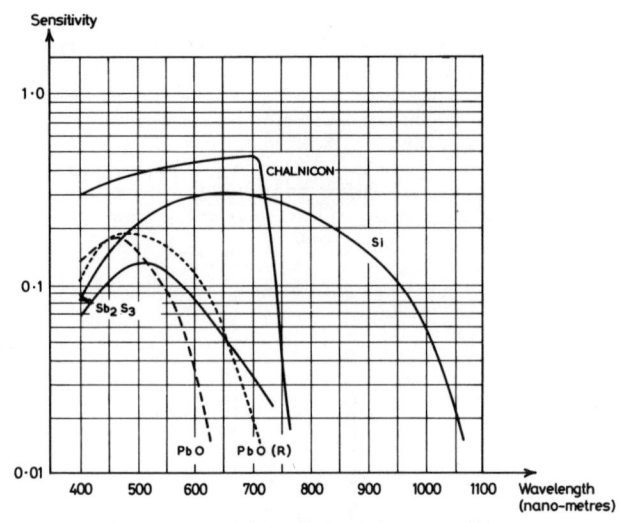

Si = Silicon diode Vidicon; Sb_2S_3 = Antimony trisulphide Vidicon; PbO = Lead oxide Vidicon and PbO R = Lead oxide Vidicon (extended red sensitivity).

FIG. 4.20 SPECTRAL RESPONSE OF CHALNICON

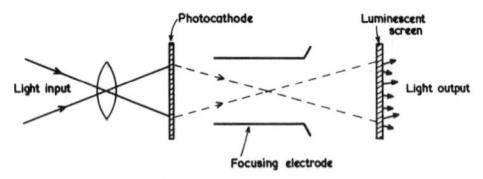

FIG. 4.21 AN IMAGE INTENSIFIER (BASIC ACTION)

fluoroscopy, an **Image Intensifier** may be used. Clearly, for night-time military work, extra light cannot be provided without disclosure to the enemy.

An image intensifier consists essentially of a photo-cathode, a focusing system and a luminescent screen as shown in figure 4.21. When light falls on to the photo-cathode from the scene, electrons are emitted from each point on its surface in proportion to the quantity of light received. Due to the liberation of electrons an electronic image of the scene is formed on the photo-cathode surface. The emitted or photo-electrons are accelerated by high voltages and focused on to a fluorescent screen deposited on the output window of the device. High voltages are required for accelerating the electrons, applied to the focusing electrode(s) and the fluorescent screen. The potential difference between the photo-cathode and screen may be of the order of 18 kV. On striking the screen the photo-electrons give up their kinetic energy to the phosphor coating which fluoresces to create an optical image once again. This image may be viewed directly or, alternatively, the light from the output window may be directed on to the photo-cathode of a second image intensifier section. The light output from the second intensifier's luminescent screen may then be directed on to the photo-cathode of a third image intensifier section, and so on. In this way, the overall gain is increased but in practice three-stage intensifiers are about the limit owing to picture deterioration as a result of noise.

Until more recent years, the only way of relaying the image from one image intensifier section on to the photo-cathode of the next was by means of lenses which are

TELEVISION CAMERA TUBES

(Reproduced by courtesy of English Electric Valve Co. Ltd, Chelmsford, Essex)

Left to right: IMAGE INTENSIFIER; 30 mm LEDDICON (PLUMBICON); ANTIMONY TRISULPHODE VIDICON; and SILCON DIODE VIDICON

bulky, heavy and expensive. The new technology of fibre optics has provided an alternative and acceptable solution to the image coupling problem. Image Intensifiers are now manufactured with fibre optic faceplates at input and output windows, as well as using this technique of coupling between sections. Essentially, a fibre optic faceplate is an array of parallel light-pipes, which convey light from one side of the plate to the other. The spread of light is confined to the diameter of the individual light-pipe.

Each individual pipe of a fibre optic faceplate is made of a core of high refractive index glass enclosed in a sheath of low refractive index glass as shown in figure 4.22.

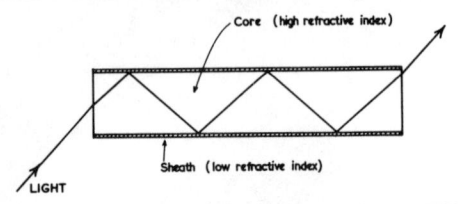

FIG. 4.22 PATH OF LIGHT IN LIGHT-PIPE (FIBRE-OPTICS)

The light is reflected many times as it travels along the fibre but suffers little attenuation due to the phenomenon of total internal reflection (see Chapter 1, page 7). During manufacture the individual fibres, which are only a few micrometres in diameter, are fused together to form a vacuum-tight bundle.

A three-stage image intensifier is shown in figure 4.23. The high voltages required for accelerating and focusing the photo-electrons may be developed by built-in e.h.t.

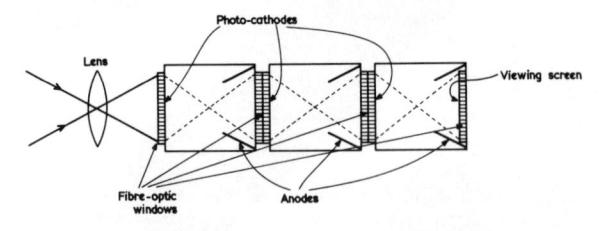

FIG. 4.23 THREE-SECTION IMAGE-INTENSIFIER

multiplier circuits. Nearly all types of camera tubes can be manufactured with fibre-optic faceplates to allow the coupling of an image intensifier; in some camera tubes the intensifier is built in. Different materials may be used for the photo-cathode to give peak sensitivity at the wavelength of the primary source of illumination. At night, the light emanating from the stars and filtering through the clouds of an overcast sky is sufficient to produce acceptable pictures when using an image-intensifier tube.

OTHER CAMERA TUBES

Only the principal photoconductive camera tubes have been described, thus the list is not an exhaustive one. Photoemissive tubes such as the Image Orthicon and Image Isocon have not been dealt with; the Isocon tube is now obsolete and the Image Orthicon is used only in small numbers. Solid-state pick-up devices are being developed by some companies, but at the time of writing the resolution of those demonstrated has been poor. The silicon-diode target array already described is a hybrid device, but represents a step on the way to a fully integrated solid-state image sensor. This would have a diode array, digital scanning circuits and video amplifier all fabricated on the same semiconductor substrate. Within a few years it is believed that solid-state pick-up devices will be able to compete with vacuum-type camera tubes.

CHAPTER 5
CAMERA CIRCUIT OPERATION

IN this chapter we shall be considering the various circuit techniques to be found in 625-line cameras of the type used in industrial and commercial CCTV applications. A block diagram of a 625-line camera is shown overleaf in figure 5.1. This diagram may be used in conjunction with the circuits that follow so that an understanding of the way in which the circuits are interconnected may be obtained.

VIDEO SIGNAL STAGES

The stages handling the video signal developed across the target load resistor should ideally have a gain-frequency characteristic that is substantially flat from d.c. up to the highest video frequency. This is shown in figure 5.2(a) where $f\mu$ (the upper video frequency) may lie, say, anywhere in the range of 4 MHz–10 MHz depending on

(a) Gain-frequency response   (b) Phase-frequency response

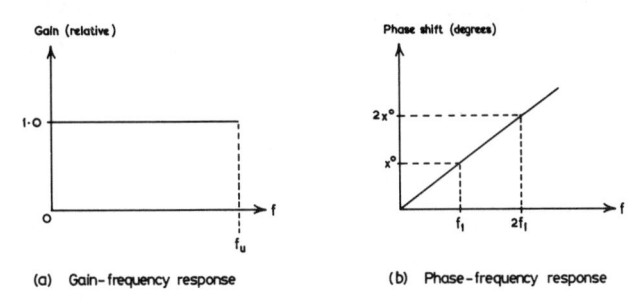

FIG. 5.2 IDEAL GAIN AND PHASE RESPONSE OF CAMERA VIDEO STAGES

the quality of the camera. Equal in importance as the frequency response is the phase response of the video stages. The video signal is a complex wave pattern and as with any complex wave it may be broken down into a fundamental and harmonics of the fundamental. In passing the video signal through an amplifier the relative phases of the various frequency components must be maintained otherwise phase distortion will result which distorts the picture image. To prevent phase distortion, any phase shift introduced by the video stages should be proportional to frequency as shown in figure 5.2(b).

The performance of a video amplifier may be assessed by testing with a square wave. Figure 5.3 shows the output waveshape (for square wave input) from a video amplifier when it is exhibiting various defects in its frequency and phase characteristics.

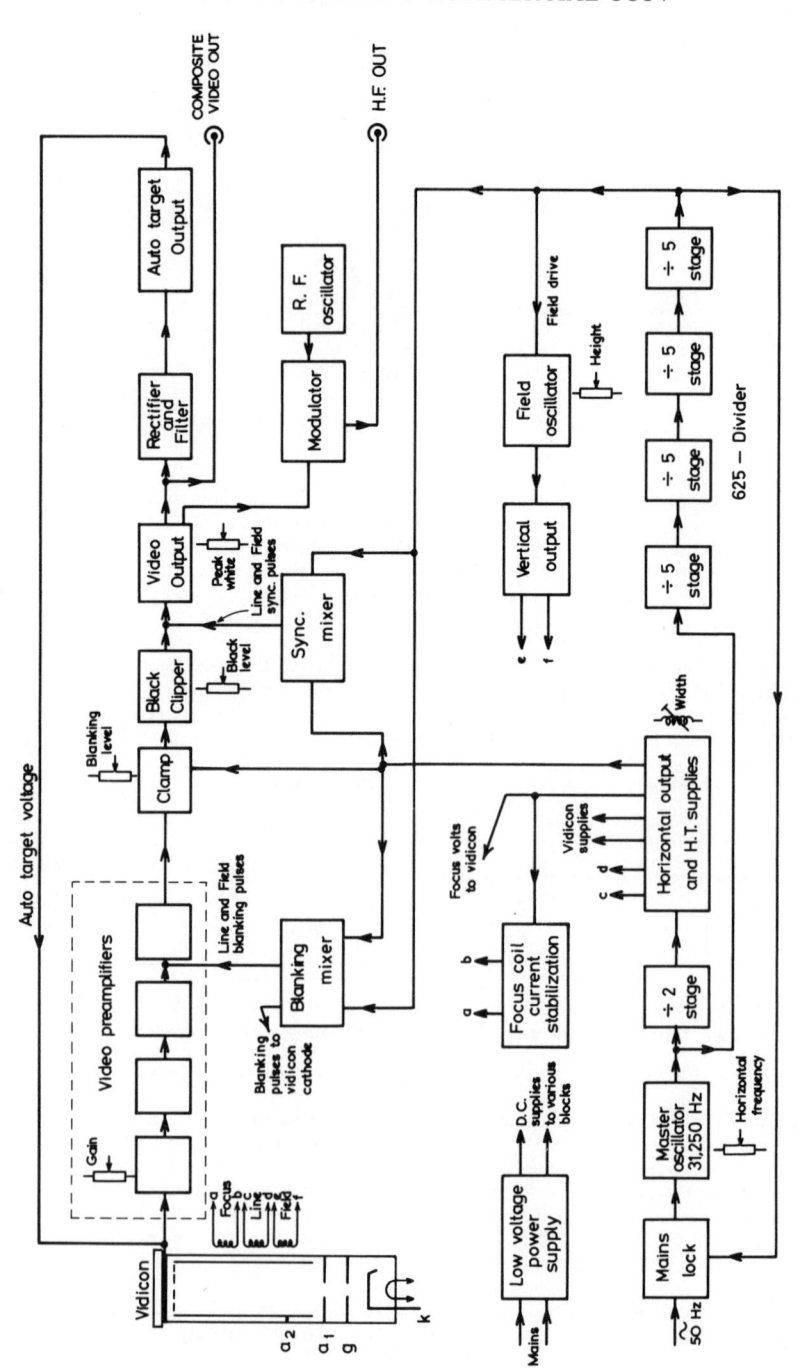

FIG. 5.1 BLOCK SCHEMATIC OF 625-LINE CAMERA (2:1 INTERLACE)

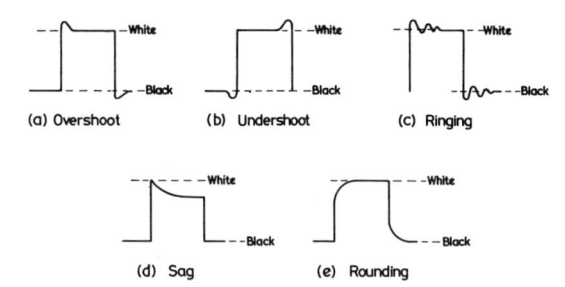

FIG. 5.3 VARIOUS GAIN-FREQUENCY AND PHASE-FREQUENCY DEFECTS WHICH CAUSE DISTORTION OF THE PICTURE

The causes and effects on the reproduced picture of these defects will now be briefly considered.

OVERSHOOT

In attempting to achieve a flat response up to high video frequencies, peaking coils are sometimes included in the video stages. Although these coils improve the picture definition by decreasing the rise time of the stage in which they are fitted, they do unfortunately disturb the phase characteristic at high frequencies causing overshoot, figure 5.3(a). Here, the signal overshoots its steady level after executing the sudden change from black level to white level or *vice versa*. The effect of this on a picture is to cause white borders following black vertical lines of detail or black borders following white vertical lines.

UNDERSHOOT

This defect may be caused by an amplifier with an erring phase characteristic resulting in some of the frequency components of the signal arriving 'early' relative to others at the amplifier output. Here, there is a momentary change in direction of the waveform just prior to the start of a step, figure 5.3(b). The effect on a picture is to cause black borders on the left-hand side of white vertical lines or white borders prior to black vertical lines.

RINGING

If the frequency-correcting circuits give a sharp cut-off, *i.e.* are of high Q, ringing may be produced as shown in figure 5.3(c). Transients in the video signal cause damped oscillations which on the picture show up as vertical black and white bars on the right-hand side of vertically disposed picture detail.

SAG

An amplifier with a frequency characteristic which falls off towards the l.f. end of the video band and is accompanied by a leading phase angle will produce the output waveform shown in figure 5.3(d). This defect usually results in a shading of the picture (going from white to black) superimposed on the scene detail.

ROUNDING

If the amplifier has a falling characteristic towards the h.f. end of the video band, rounding of the output waveform occurs as shown in figure 5.3(e). The effect of this on the picture is to cause loss of fine detail, the overall effect being that of a blurred picture.

An amplifier which is being overloaded or exhibits frequency and phase distortion in the m.f. range, say, around line frequency, may cause STREAKING of the picture. On the monitor, the important white picture detail is followed by black streaks and black detail by white streaks. The streaks may only last a fraction of the line period or extend right across the screen.

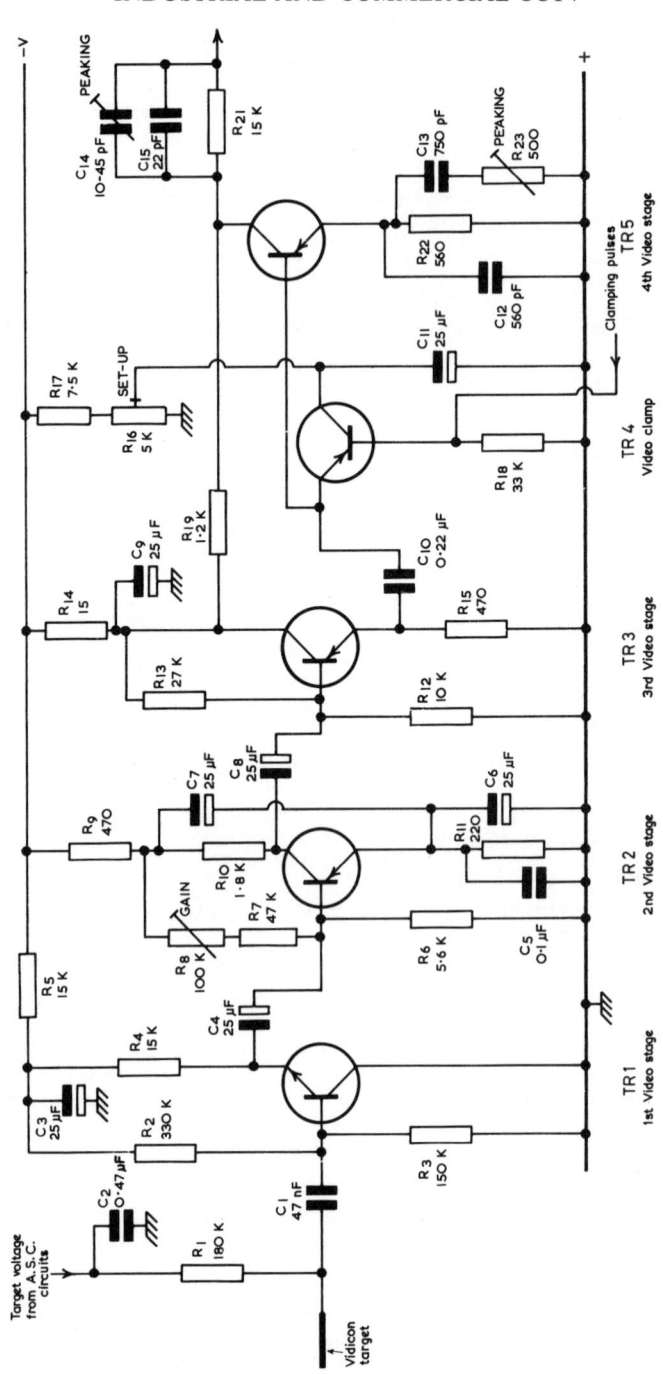

FIG. 5.4 PREAMPLIFIER STAGES PYE 'LYNX' CAMERA

CAMERA CIRCUIT OPERATION

PREAMPLIFIER STAGES

These stages are concerned with amplyifying the low level video signal from the target electrode of the camera tube to provide an output signal of the required standard level (see page 67). The amplifier must perform this function without introducing too much noise and must possess a frequency characteristic to match the camera specification. Most of the frequency-correction circuits are incorporated in the preamplifier and the design will, as far as is practically possible, ensure freedom from the defects outlined above. Four or five transistor stages are normally required in the preamplifier which will be mounted close to the vidicon target connection. Screening of the preamplifier will ensure stable operation and freedom from spurious signal pick-up.

Figure 5.4 shows the preamplifier stages of the Pye 'Lynx' Camera. R_1 is the target load resistor which is grounded at one end by C_2. If we assume a vidicon peak current of 0·2 μA, the video signal appearing across R_1 is of the order of 30—40 mV. This voltage is passed *via* C_1 to the base of TR_1 which is connected as an emitter-follower. This stage matches the high output impedance of the vidicon target to the low input impedance of TR_2. The first stage plays a vital part in obtaining a good signal-to-noise ratio. This is achieved by operating TR_1 at low currents, a factor enabling a good noise figure to be obtained with little loss of gain. The signal is a.c. coupled by C_4 to TR_2 which is connected in the common-emitter mode. This stage provides large voltage amplification, the degree of amplification being controlled by R_8 (gain control) which sets the base bias. The amplified signal from across R_{10} is coupled *via* C_8 to a second emitter-follower stage TR_3. This stage provides a low impedance drive to the clamp transistor TR_4.

The clamp circuit restores the d.c. and low frequency components lost in the a.c. couplings of the first three stages. To restore the d.c. component of a waveform either a d.c. clamp or d.c. restorer circuit may be used. A d.c. restorer would be unsatisfactory to deal with the waveform of figure 5.5 as it would be unable to distinguish between the black level (a) and the negative-going signal peaks (b). On the other hand, a d.c. clamp

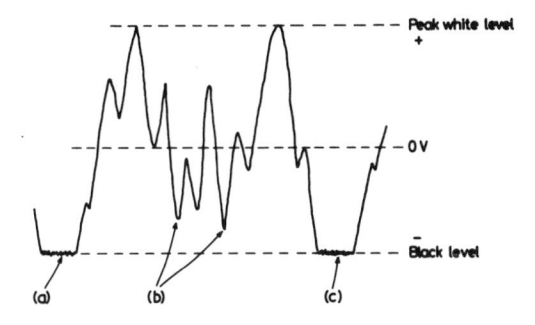

FIG. 5.5 THE VIDEO SIGNAL WITH D.C. COMPONENT REMOVED

operated by pulses occurring at regular intervals during periods (a) and (c) etc. has a fixed reference level (the black level). Thus the d.c. component introduced by the clamp will be referenced to black level and not a varying level set by the negative peaks of the video signal.

The clamp is operated by negative-going line pulses applied to the base of TR_4 which functions as a switch. Each pulse closes the switch allowing C_{10} to charge up to the potential set by R_{16} (the 'set up' control) since the collector and emitter of TR_4 are practically at the same potential when TR_4 is conducting. During the interval between clamping pulses, TR_4 is OFF and C_{10} is able to discharge. However, the potential of C_{10} remains substantially constant as the discharge path for C_{10} is *via* the base-emitter circuit of TR_5 which provides a suitably long time-constant. As a result of this action,

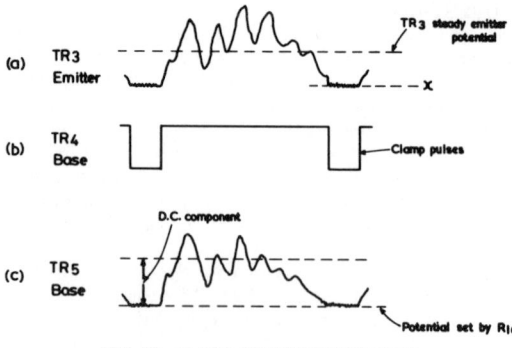

```
                    TR3 steady emitter
                         potential
(a)  TR3
     Emitter
                                          x

(b)  TR4
     Base                          Clamp pulses


         D.C. component
(c)  TR5
     Base
                              Potential set by R₁₆
```

FIG. 5.6   CLAMP CIRCUIT WAVEFORMS

the level x of the video waveform shown in diagram (a) of figure 5.6 is clamped to the potential set by R_{16} as in diagram (c).

The restored video signal is applied to TR_5 which, together with the following stage (not shown), forms a two-stage d.c. coupled amplifier. TR_5 is a 'high peaker' stage which lifts the h.f. response. In the emitter circuit R_{22} provides attenuation of the low frequencies by negative feedback, whilst C_{12} and C_{13}, R_{23} respectively raise the high and mid-band frequency response. Control over the lift is given by R_{23}. In addition, C_{14} provides control over the h.f. lift and forms part of the coupling circuit to the next stage.

The preamplifier stages of a Sony Camera are shown in figure 5.7. Output from the vidicon target is developed across R_1 and coupled through C_2 to the input of the first preamplifier stage. The output lead from the vidicon target is kept as short as possible to reduce the h.f. losses and to minimize the pick up of spurious signals. To match the high impedance of the vidicon target an f.e.t. is used in the first stage. This stage has a good noise-figure which is instrumental in achieving a good signal-to-noise ratio. The amplified video signal from across R_3 is applied to TR_2, the first of four d.c. coupled common-emitter amplifiers. These stages together with TR_1 provide most of the video gain of the camera. Mid-band and h.f. compensation is applied in the emitter circuits of TR_2, TR_3 and TR_4 by R_8/C_6 (m.f. and h.f.), C_8 (h.f.) and C_9 (m.f. and h.f.) respectively.

Overall negative feedback is used from TR_5 collector to TR_1 gate. R_{20} determines the degree of feedback and is used for setting up the preamplifier gain. C_{12} sets the frequency response. This capacitor together with R_{19}/C_{11} form a potential divider in the feedback path for the video signal resulting in more feedback at low frequencies compared with high video frequencies. The use of overall n.f.b. assists in keeping the amplifier gain stable over long periods.

Restoration of the d.c. component (lost in C_2) is carried out in a later stage in this camera.

APERTURE CORRECTION

In some cameras a correcting circuit is used to reduce the effects of aperture distortion which was discussed in Chapter 4. Blocks for these circuits have not been shown in figure 5.1 as this diagram is typical of cameras used for non-critical CCTV applications.

Aperture correction is applied to the video signal usually after preamplification but before sync. and blanking waveforms are added. Only the principle of operation will be dealt with and one example is shown in figure 5.8.

Waveform 1 is the video input signal to the correcting circuit and is taken to be rectangular. This signal is amplified in block 1 and applied to the input of a delay line in

CAMERA CIRCUIT OPERATION 101

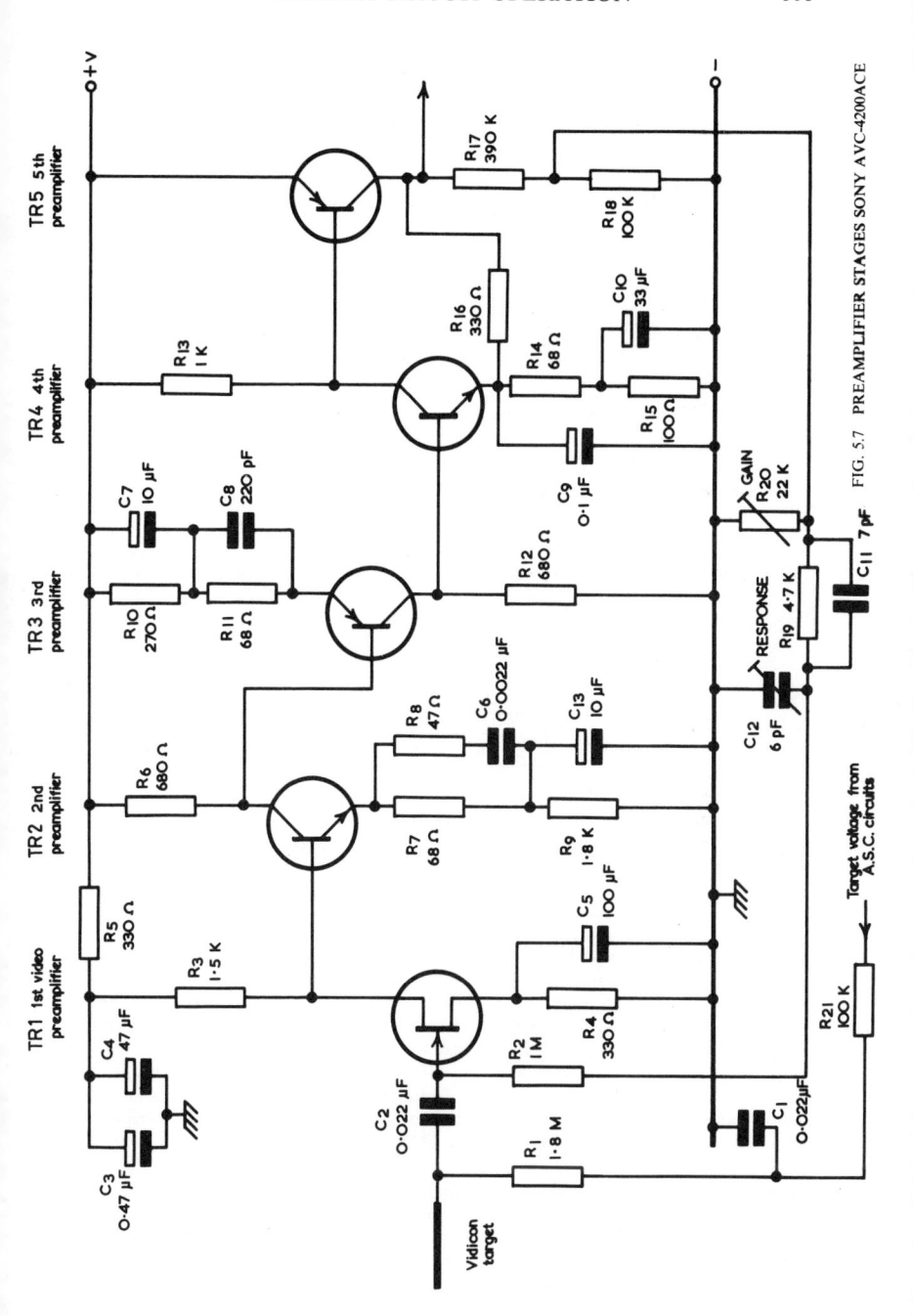

FIG. 5.7 PREAMPLIFIER STAGES SONY AVC-4200ACE

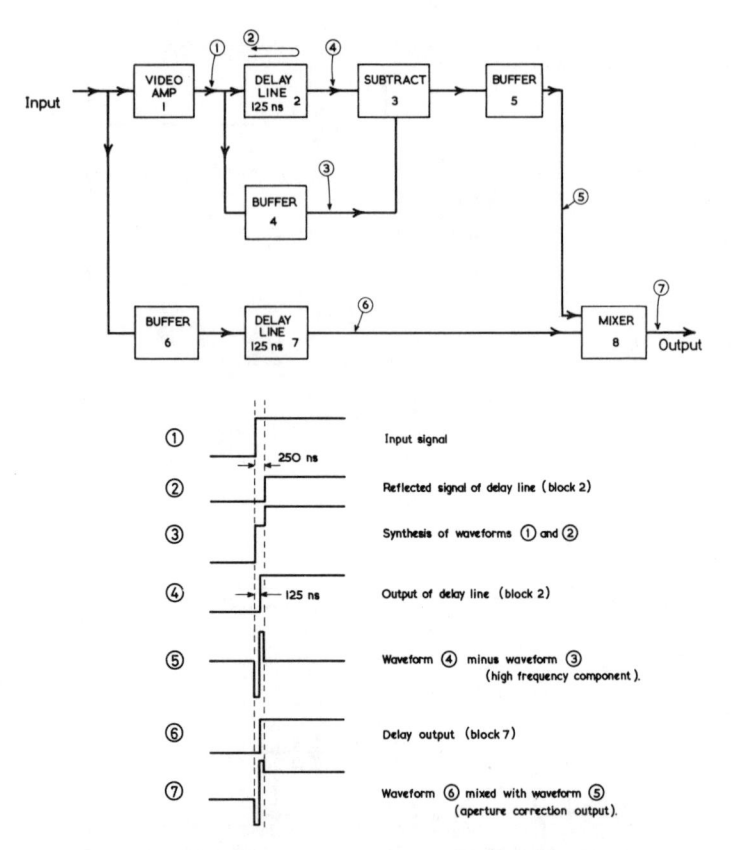

FIG. 5.8 PRINCIPLE OF APERTURE CORRECTION (SONY CAMERA)

block 2. The line, which gives a delay of 125 ns, is 'matched' at its input but 'mismatched' at its output. Because of the mismatch a sample of the wave arriving at the output of the line is reflected back to the input. Waveform 2 shows the reflected wave which arrives back at the input 250 ns later (having travelled 'up' and back 'down' the line). The waveform now present at the input of block 2 is the synthesis of waveforms 1 and 2 and is represented by waveform 3. This composite signal is supplied to the subtract circuit of block 3 *via* the buffer stage in block 4. Also fed to block 3 is the output of the delay line, waveform 4 delayed by 125 ns.

In the subtract circuit, waveform 3 is subtracted from waveform 4 the result of which is given by waveform 5. This waveform contains high-frequency components and is passed to the mixer stage *via* block 5. The other input to block 8 is waveform 6, a sample of the video input signal delayed by 125 ns in block 7 and supplied to the delay line *via* a buffer stage. In the mixer stage waveforms 5 and 6 are added together to produce waveform 7. This is the corrected output waveform from the aperture circuit. It will be seen that the effect of adding the high-frequency component of waveform 5 to the video signal (waveform 6) is to cause pre-shoot and overshoot whenever the video signal suddenly changes its level. As a result the change in level of the video waveform is accentuated. The level of the high-frequency component fed to the mixer stage can be adjusted so that just the right amount of correction is applied. Excessive correction may lead to instability at high frequencies.

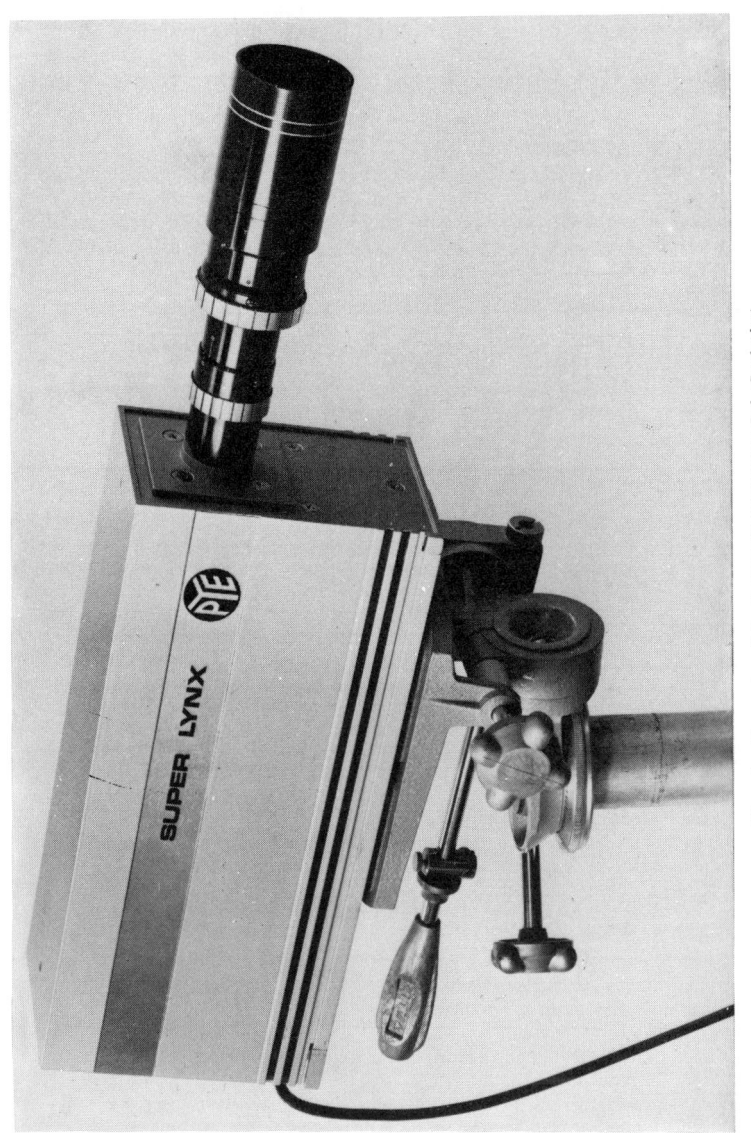

(Reproduced by courtesy of Pye Business Communications Ltd., Cambridge)

PYE SUPER LYNX CAMERA WITH MANUAL PAN AND TILT

SYNC. PULSE GENERATOR

Before considering how the sync. and blanking waveforms are added to the video signal we must first discuss the generation and timing of these waveforms. The Sync Pulse Generator (S.P.G.) generates the line and field drive pulses for the camera timebases, blanking waveforms (to be added to the composite signal and for suppressing the beam in the camera tube) and the line and field sync. pulses. As explained on page 63 the reference timing for all of these waveforms is provided by the master oscillator.

MASTER OSCILLATOR

In a 2:1 interlaced camera the master oscillator works at 31,250 Hz. A stable circuit must be used and it is normal practice to use either a crystal or a blocking oscillator. Figure 5.9 shows a crystal oscillator comprising TR_1, TR_2 and X_1. The frequency of oscillation is determined by the crystal X_1 which is a thin but accurate slice of quartz

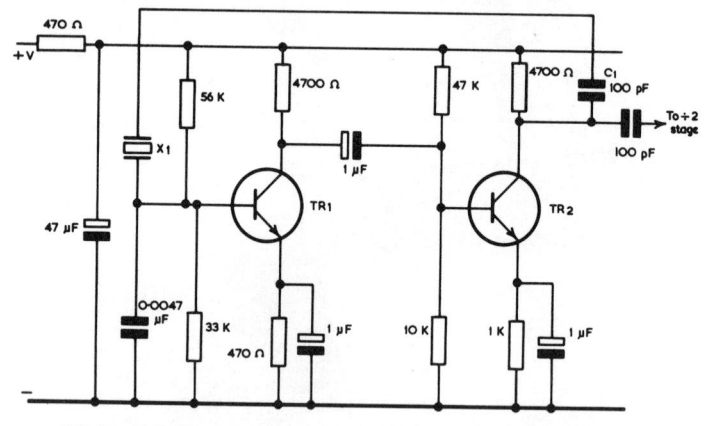

FIG. 5.9 MASTER OSCILLATOR USING CRYSTAL OPERATING AT 31,250 Hz

crystal with metal electrodes deposited on each side. When the assembly is supplied with an alternating e.m.f. a mechanical vibration is set up. The vibration gives rise to an oscillatory voltage which is of especially large amplitude when the applied e.m.f. is close to the mechanical resonance of the crystal. A crystal normally exhibits strong vibrations at two frequencies usually within 1% of each other. At the lower frequency the crystal operates in a series resonant mode and is of low impedance, whereas at the higher frequency the crystal operates in a parallel resonant mode where its impedance is high. A very high Q is to be expected from a crystal, often in excess of 20,000. Because of its high Q its frequency stability is excellent.

In figure 5.9 the crystal is operating in a series resonant mode. Positive feedback is supplied to the crystal from TR_2 collector via C_1. Provided the gain around the loop is greater than unity the circuit will oscillate. TR_1 and TR_2 provide the necessary gain to make up for losses in the circuit.

A master oscillator employing a blocking oscillator is given in figure 5.10. TR_1, T_1 and C_1 form the blocking oscillator. When TR_1 is conducting, C_1 is charged by base current with the polarity shown and the voltage across this capacitor causes TR_1 to go OFF. C_1 now discharges via the constant-current transistor TR_2, R_3 and R_4. As C_1 discharges, the base potential of TR_1 rises and when above that of its emitter it conducts. The rise of current in the collector winding of T_1 causes a voltage to be induced in the base winding, the polarity of which drives the transistor harder ON. At the same time, base current is flowing which causes C_1 to recharge once more. Eventually, the current in TR_1 will limit, at which point feedback between collector

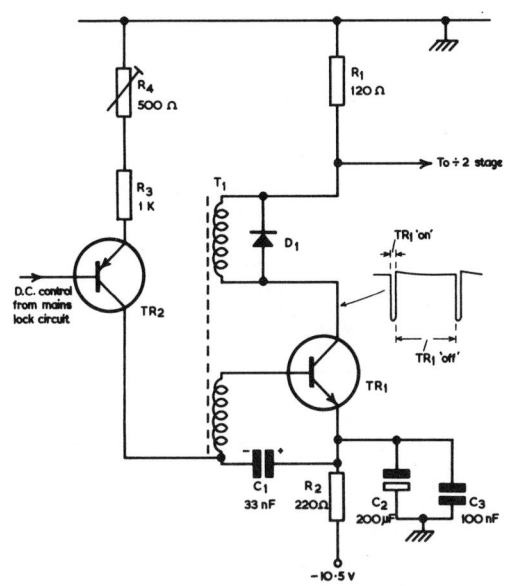

FIG. 5.10 MASTER OSCILLATOR USING BLOCKING OSCILLATOR CIRCUIT, OPERATING AT 31.250 Hz WITH CONSTANT CURRENT TRANSISTOR

and base windings will cease. TR_1 will now cut off due to the voltage across C_1. When the transistor goes OFF, any tendency for the collector voltage to reverse is damped by D_1.

With a constant discharge current in C_1, the instant at which TR_1 cuts on is more precisely determined, hence the timing of the output waveform is more accurately maintained. The constant current transistor TR_2 is fed from the 'mains lock' circuit (see page 117) which gives out a d.c. correction voltage. This voltage, which is applied to TR_2 base, varies the current in TR_2 and thus determines the time that TR_1 is cut off. Initially, the circuit is set up with the aid of R_4 which since it is in the discharge circuit of C_1 acts as a variable frequency control.

In a random interlace camera where there is no timing relationship between the horizontal and vertical timebases, the master oscillator (or horizontal oscillator) may operate directly at line frequency. A blocking oscillator or multivibrator circuit may be used but sometimes an L,C oscillator is adopted and one example is shown in figure 5.11. Here, a Colpitts oscillator is used with L_1, C_1, C_2, C_3 and C_4 forming the tank circuit of the oscillator. Feedback from the emitter of the transistor to across C_4 sustains the oscillations. Initial starting bias for the oscillator is provided by R_1 and R_2. Capacitor C_2 blocks the d.c. from L_1 which has a variable core to adjust the frequency of oscillation. Although the oscillations in the base circuit are sinusoidal, overdriving the transistor results in a pulse waveform appearing at the collector which is suitable for driving the horizontal scan output stage.

DIVIDE-BY-TWO STAGE

The output of the master oscillator in a 2:1 interlace camera must be divided by two in order to produce a 15,625 Hz drive to the horizontal output stage of the camera (see figure 5.1).

Bistable multivibrators are commonly used for frequency division in camera circuits and a typical circuit is shown in figure 5.12. As its name implies, a bistable oscillator has two stable states. The two transistors are cross-coupled by R_2 and R_4 so that positive feedback occurs. The two transistors cannot be conducting

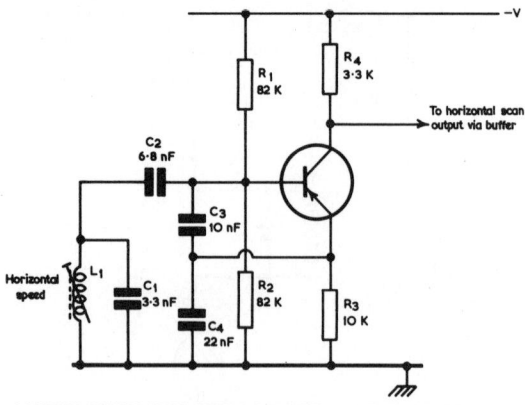

FIG 5.11 MASTER OSCILLATOR USING COLPITTS CIRCUIT OPERATING AT 15,625 Hz

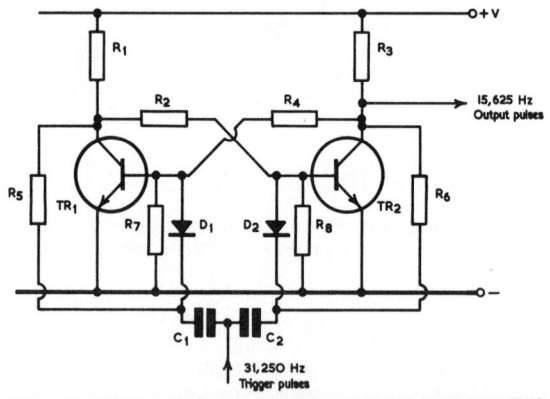

FIG. 5.12 USING A BISTABLE MULTIVIBRATOR TO DIVIDE BY TWO

simultaneously, except for a brief period during changeover. Thus, initially when the supply is connected, one transistor is conducting and the other is cut off.

Suppose that TR_1 is conducting and TR_2 is cut off. As a result TR_1 collector will be 'bottomed' and, due to the low collector voltage (say 0·2 V), the base voltage of TR_2 will be insufficient to turn TR_2 ON (this assumes that TR_1 and TR_2 are silicon type transistors which require a forward bias of the order of 0·6 V to 1·0 V). This is the condition illustrated in figure 5.13 just prior to instant t_1.

To change the circuit over, a negative trigger pulse is required to be fed to the base of TR_1 to switch TR_1 OFF. The trigger pulses are fed to the bistable *via* C_1 and C_2.

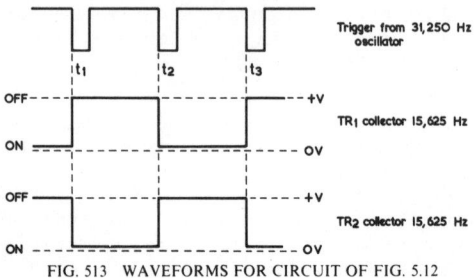

FIG. 513 WAVEFORMS FOR CIRCUIT OF FIG. 5.12

These pulses are derived from the output of the master oscillator. To ensure that each trigger pulse is fed to the appropriate transistor, 'steering diodes' D_1 and D_2 are employed. When TR_1 is ON, D_1 is biased ON but D_2 is OFF. Thus, the first trigger pulse to arrive is steered through D_1 to TR_1 base. Here it causes TR_1 to start to turn OFF. The resulting rise of voltage at TR_1 collector is passed via R_2 to TR_2 base causing TR_2 to conduct. As a result TR_2 collector voltage commences to fall which is passed via R_4 to TR_1 base causing TR_1 current to reduce even further. This action is repeated rapidly many times resulting in TR_1 quickly going OFF and TR_2 rapidly coming 'hard on' (instant t_1).

Thus the state of the circuit has been changed over with TR_1 cut off and TR_2 conducting. The circuit will remain in this new state until the next trigger pulse arrives. This pulse will be steered through D_2 which is now biased ON to TR_2 base (D_1 is biased OFF). TR_2 now starts to turn OFF and the rise in voltage at TR_2 collector is passed on to TR_1 base via R_4. This action causes TR_1 to turn ON and the fall in voltage at its collector is transmitted to TR_2 base where the effect is to further reduce the current in TR_2. Again, this action is rapidly repeated many times resulting in TR_1 coming 'hard on' and TR_2 going OFF (instant t_2). The next trigger pulse is steered through D_1 causing TR_1 to cut off and TR_2 to conduct (instant t_3) . . . and so on.

It will be noted that for every two trigger pulses, one positive or negative excursion of the waveform at TR_1 or TR_2 collector occurs, *i.e.* the circuit has divided-by-two. An output may be taken from either collector depending upon the polarity required.

As we shall see later, bistable oscillators are also used in the 625-divider; in modern cameras these circuits are fabricated using integrated circuit techniques.

LINE SCAN OUTPUT

To deflect the electron beam horizontally across the target layer of the camera tube a linear current waveform has to be supplied to the line deflector coils. This current waveform is supplied by the line scan output stage. Because the camera employs a low velocity scanning beam and the tube dimensions are small, the energy fed into the scan coils is small compared with energy supplied to the line deflector coils in a television monitor or receiver. There is no need for an e.h.t. overwinding as is found on the line output transformer in television receivers, because the voltages required by the camera tube are quite modest and so can be generated directly from the mains source. In some cameras, however, a small output transformer is used which may have several windings to generate voltages up to about, say, 300 V to supply the electrodes of the camera tube.

Figure 5.14 shows the basic circuit of a transistor line scan output stage. Essentially, the circuit consists of an inductor L_1 in parallel with C_s (the self-capacitance of the coil) and C_1, a bi-directional switch (TR_1 and D_1) and a d.c. supply.

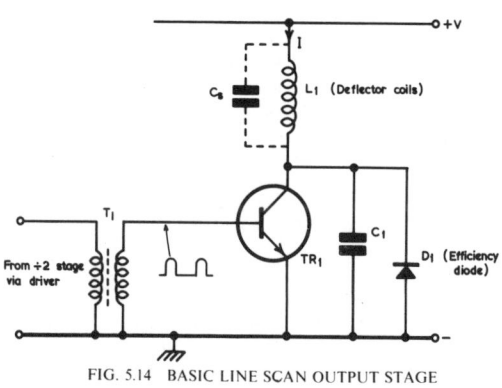

FIG. 5.14 BASIC LINE SCAN OUTPUT STAGE

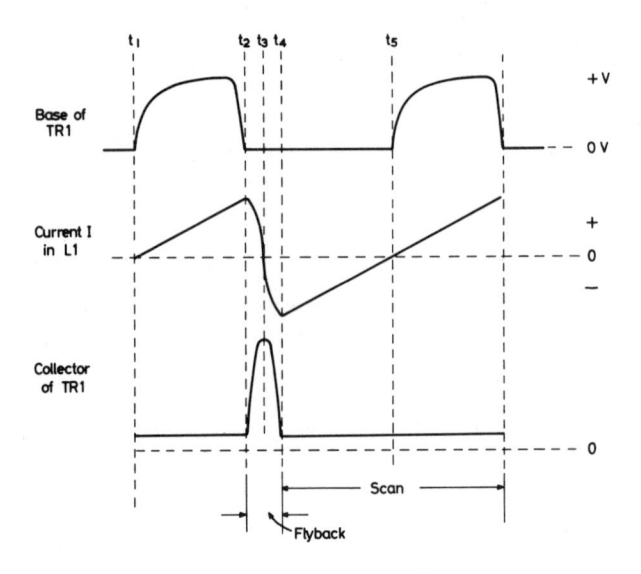

FIG. 5.15 WAVEFORMS ASSOCIATED WITH OPERATION OF CIRCUIT IN FIG. 5.16

At instant t_1 in figure 5.15 the transistor is caused to conduct by the positive-going waveform applied to its base. TR_1 acts like a switch connecting the d.c. supply across L_1. This allows the current I to flow through L_1 in the direction indicated and to grow in a linear manner, provided the circuit resistance is small. With a linear rising current in L_1 there is a constant voltage drop across it. Thus, during the interval t_1—t_2 the collector voltage of TR_1 is constant and low (as the transistor is bottomed). Throughout this period D_1 is OFF as it is reverse biased.

TR_1 is switched OFF at instant t_2 when the base voltage is reduced to zero. As there is now no voltage to maintain the current in L_1, the magnetic field around the inductor collapses inducing a large voltage into it. This voltage drives a diminishing current into C_1 and C_s charging them. At instant t_3, C_1 and C_s are fully charged and the current is zero. C_1 and C_s now discharge through L_1 and the current reverses direction (t_3—t_4). At instant t_4 the capacitors will be fully discharged. As there is no voltage to maintain the current in L_1, the magnetic field around L_1 collapses once more, inducing a voltage into the inductor. The direction of this induced e.m.f. is opposite to that during the interval t_2—t_4 and the collector voltage of TR_1 attempts to swing below zero potential. The voltage is prevented from falling below zero by D_1 which now conducts. Current now decreases in a linear manner as it flows through D_1 (assuming D_1 resistance to be small).

When the current reaches zero (instant t_5), TR_1 conducts once again and the cycle is repeated. Thus it will be seen that energy is put into the deflector coils during the interval t_1—t_2, recovered from the coils in the period t_2—t_4 and then used to produce the other part of the scan during t_4—t_5. Not all of the energy that is put into the deflector coils is recovered due to the inevitable resistance losses. The duration of the flyback period is governed by the natural resonant frequency of L_1, C_1 and C_s. Without C_1 this period would be shorter but its inclusion reduced the peak voltage at the collector of TR_1.

An example of a practical line scan output stage is shown in figure 5.16. Instead of

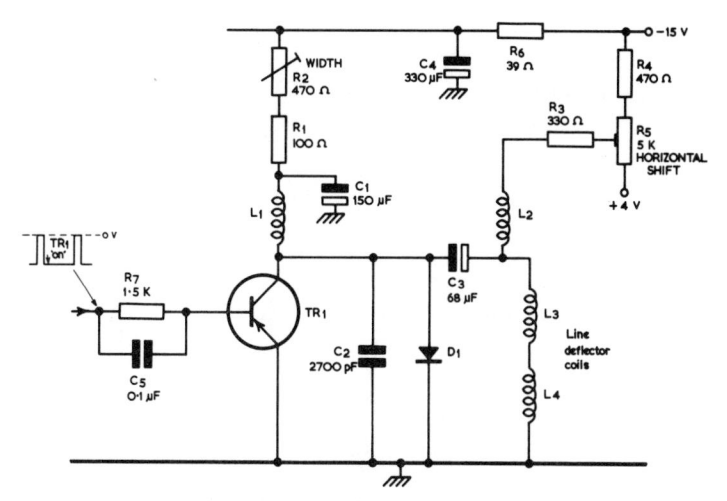

FIG. 5.16 LINE SCAN OUTPUT STAGE FOR SIMPLE CAMERA

placing the deflector coils in series with the collector circuit of the output transistor, they are shunt fed using choke-capacitance coupling (L_1 and C_3). The primary purpose of L_1 is to isolate line scan currents from the low voltage supply to TR_1. The inductance of the deflector coils is low in relation to the inductance of L_1, thus the major portion of TR_1 current flows in the deflector coils. C_3 blocks the d.c. component of TR_1 current from L_3, L_4 but passes on the a.c. component. D_1 damps the 'overswing' and serves as the efficiency diode as previously explained. The inductance of L_3, L_4 together with C_2 and the stray capacitance of the deflector coils determine the flyback time. Control of the horizontal scan amplitude is effected by R_2 which varies the current in TR_1. A reversible d.c. current is introduced into the deflector coils from R_5 for the purpose of horizontal shift. The inductance L_2 isolates the scanning current from the shift current source.

Another line scan output stage is shown in figure 5.17 but here an auto transformer is used for feeding the deflector coils. The output from the divide-by-two stage after suitable shaping is fed to the pulse transformer T_1 which supplies the switching waveform to the line output transistor TR_1.

When there is no positive pulse at the base of TR_1 the transistor is conducting and a constant voltage is applied across w_1. Thus a linear increasing current will flow in this winding and also in the deflector coils as they are tapped into the winding. During this interval D_1 will be conducting and C_2 will be charging. As soon as the positive pulse arrives at TR_1 base the transistor cuts off. The self-inductance of w_1, together with the stray capacitance of the winding, now gives rise to an oscillation which at first causes the stray capacitance to charge to a large negative value. A large negative pulse thus appears at TR_1 collector and D_1 eventually turns off. After the negative pulse, TR_1 collector voltage tends to become positive due to the oscillation but this is prevented by the fact that the diodes D_4, D_3 and D_2 become conductive.

The voltage across w_1 is again constant and a linearly decreasing current will again flow in the winding and the deflector coils. After a time the function of D_4, D_3 and D_2 is taken over by TR_1 when it becomes conductive at the end of the positive pulse applied to its base.

As a result of this action, a linear rising current flows in the deflector coils during the scan, the amplitude of which may be controlled by L_1.

As the negative-going flyback pulses present on the tap of w_2 are used in this camera to produce the line blanking waveform, the commencement of line flyback in the

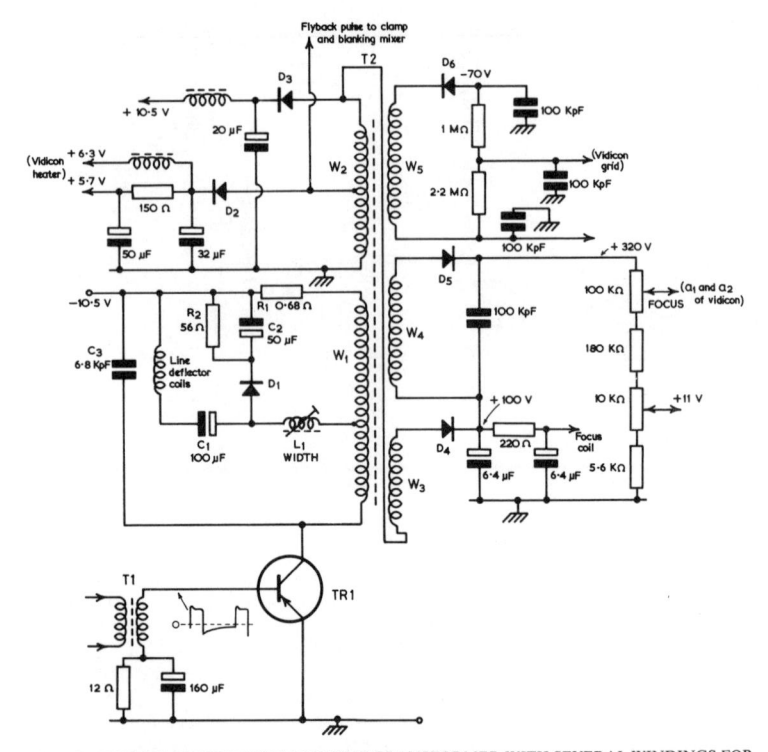

FIG. 5.17 LINE SCAN OUTPUT STAGE USING TRANSFORMER WITH SEVERAL WINDINGS FOR OBTAINING SUPPLY VOLTAGES FOR VARIOUS PARTS OF THE CAMERA

deflector coils must occur slightly later. This is ensured by D_1 which delays the start of flyback at the deflector coils. Due to the 'hole storage' of this diode it does not immediately cut off when the collector of TR_1 swings negative.

During scanning, constant voltages appear across windings w_2 and w_3 which cause the diodes D_2, D_3 and D_4 to conduct. After smoothing, $+6.3$ V and $+5.7$ V, $+10.5$ V and $+100$ V are respectively obtained from these diode circuits which are used in various parts of the camera. D_5 provides peak rectification of the positive pulses present across w_4. The d.c. voltage thus obtained is added to the $+100$ V output of D_4 and is used to supply the a_1 and a_2 electrodes of the vidicon. D_6 also provides peak rectification of the flyback pulses across w_5 to obtain a -70 V d.c. supply which is used for supplying the grid of the vidicon.

Other variations in the design of the line scan output stage are possible but they all tend to work on the same lines.

625-DIVIDER

To provide the field frequency timing, pulses at 50 Hz are required. In a 2:1 interlaced camera these pulses are obtained by frequency division. As was explained in Chapter 3, practical dividers can only divide by whole numbers. Hence the reason for operating the master oscillator at 31,250 Hz, since dividing 31,250 Hz by 625, pulses at 50 Hz may be obtained.

Bistable oscillators are often used in the 625-divider. We have seen that a bistable oscillator divides-by-two. Consider now a string of bistables connected in cascade as

CAMERA CIRCUIT OPERATION

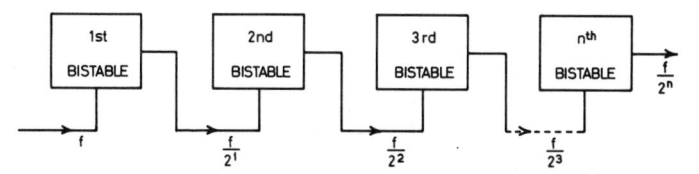

FIG.5.18 A STRING OF BISTABLES PRODUCING A DIVISION OF 2^n (where n is the number of bistables)

shown in figure 5.18 where the output of one bistable is used to trigger the following bistable. The first bistable divides the input (f) by a factor of two, the second bistable divides the input by a factor of four, the third bistable divides the input by a factor of eight and so on. Thus, in general a string of bistables produces a division of 2^n where n is the number of bistables. With the arrangement of figure 5.18 division by even numbers only is possible, thus it is now necessary to show how division by an odd number may be achieved.

Two bistables in cascade will produce a division of 4, but if the output of the second bistable is fed back to the first bistable a division of 3 is obtained. This is shown in figure 5.19 where the pulse output of the second bistable is used to reset the first bistable. In effect, the feedback pulse takes the place of one of the input pulses. Thus, in this case

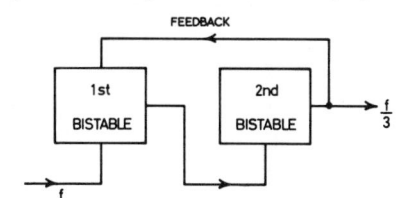

FIG. 5.19 TWO BISTABLES WITH FEEDBACK PRODUCING A DIVISION OF 3

only three input pulses are needed to produce one output pulse, *i.e.* a division of 3 is obtained. Similarly, with a string of 3 bistables and a feedback path from the output of the third to the first, a division of 7 may be achieved. Thus, in general for n bistables and one feedback path from the nth bistable to the first, the division is reduced from 2^n to (2^n-1).

Figure 5.20 shows how a division of 5 may be produced which is pertinent to the camera 625-Divider. In diagram (a) we have two bistables A and B with feedback (as in

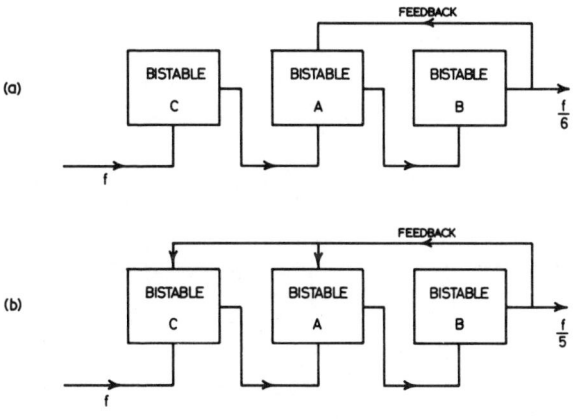

FIG. 5.20 OBTAINING A DIVISION OF 5

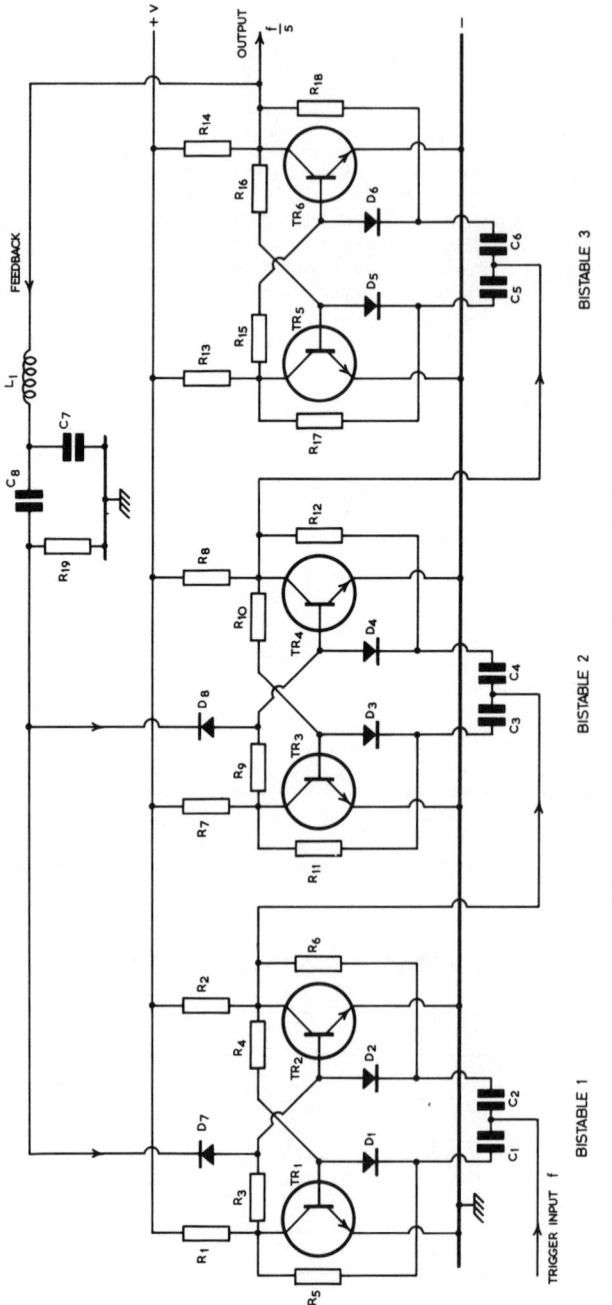

FIG. 5.21 PRACTICAL ARRANGEMENT FOR DIVIDING BY 5

figure 5.19) which will produce a division of 3 of an input to bistable A. If another bistable (C) is added this will increase the division by a factor of 2, *i.e.* producing a division of $2 \times 3 = 6$. By feeding back the pulse output of bistable B to bistable C as well as to bistable A as in diagram (b), the division of the arrangement in (a) is reduced by one. Thus the arrangement in (b) will divide $(6 - 1)$, *i.e.* 5. It is possible to achieve a division of 5 with the feedback arranged differently, *e.g.* with one feedback line from bistable A to bistable C and a separate feedback path from bistable B to bistable C. The actual arrangement used in practice is often influenced by the need to keep the number of components used to a minimum.

A practical circuit for producing a division of 5 will now be considered and one such arrangement is shown in figure 5.21 with associated waveforms given in figure 5.22. Three identical bistable oscillators are employed which are formed in one integrated circuit, except for the input capacitors C_1–C_6 and the components in the feedback path (L_1, C_7, C_8 and R_{19}). The feedback arrangement used follows the principle shown in diagram (b) of figure 5.20 with the feedback pulses fed to bistables 1 and 2 from the output of bistable 3 *via* the diodes D_7 and D_8.

In describing the action of the circuit we shall consider that, initially, the left-hand transistor in each bistable is conducting, *i.e.* TR_1, TR_3 and TR_5. Thus TR_2, TR_4 and TR_6 are OFF and the collector voltages of these transistors are high (practically equal to $+$ V) prior to the arrival of the first trigger pulse. In practice, at switch-on the transistors which are conducting is a matter of pure chance but after a few trigger pulses have been applied they will take up the assumed initial state.

Negative trigger pulses are fed to bistable 1 and pulse 1 causes TR_1 to switch OFF and TR_2 to come ON (the pulse being applied to TR_1 base *via* C_1 and D_1). The fall in voltage at TR_2 collector is passed *via* C_3 and D_3 to TR_3 base. This causes TR_3 to switch OFF and TR_4 to come ON. As a result, TR_4 collector voltage falls and this is passed *via* C_5 and D_5 to TR_5 base causing TR_5 to switch OFF and TR_6 to come ON. Thus, the first trigger pulse causes all bistables to change their state. Now, the fall in voltage at TR_6 collector is fed back to bistables 1 and 2 *via* the d.c. blocking capacitor C_8. The negative-going voltage which is fed back is delayed by L_1, C_7 and one may consider that these components form one section of an L.C. delay line. A delay is necessary to ensure that bistables 1 and 2 complete their switching action before the feedback is applied. The fall in voltage which is applied to the cathodes of D_7 and D_8 causes the diodes to conduct and as a result the bases of TR_2 and TR_4 are taken negative. This causes TR_2 and TR_4 to switch OFF, *i.e.* TR_1 and TR_3 switch ON. Thus, TR_2 and TR_4 collector voltages return to the levels that existed prior to the arrival of the first trigger pulse.

Pulse 2 is routed *via* C_1 and D_1 to TR_1 base switching TR_1 OFF and TR_2 ON. The fall in voltage at TR_2 collector is applied *via* C_3 and D_3 to TR_3 base, switching TR_3 OFF and

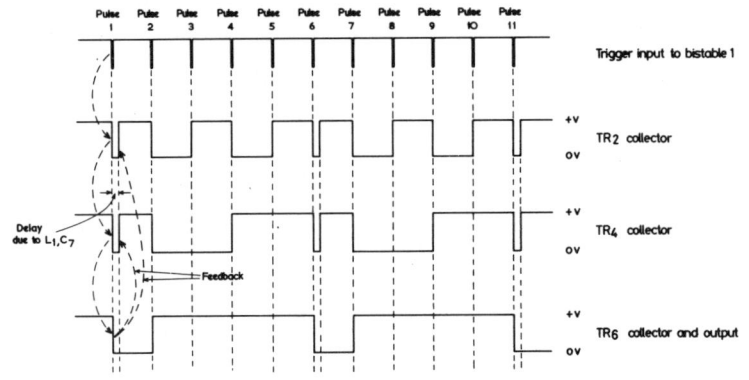

FIG. 5.22 TIMING OF THE WAVEFORMS FOR THE DIVIDE-BY-FIVE ARRANGEMENT OF FIG. 5.21

TR_4 ON. The fall in voltage at TR_4 collector is applied via C_6 and D_6 to TR_6 base causing TR_6 to switch OFF and TR_5 to come ON. The rise of voltage at TR_6 collector is fed back to D_7 and D_8 but has no effect as it will tend to bias back these diodes.

Pulse 3 is fed via C_2 and D_2 to TR_2 base switching TR_2 OFF which causes TR_1 to switch ON. As TR_2 goes OFF, its collector voltage rises but this will have no effect on bistable 2 as a negative trigger is required. As a result TR_4 collector voltage is stable at 0 V (corresponding to the ON condition), also TR_6 collector voltage is stable.

Trigger pulse 4 is fed via C_1 and D_1 to TR_1 base switching TR_1 OFF and TR_2 ON. The fall in voltage at TR_2 collector is passed via C_4 and D_4 to TR_4 base, switching TR_4 OFF and TR_3 ON. As TR_4 goes OFF its collector voltage rises but this has no effect on bistable 3 which requires a negative trigger at its input. In consequence bistable 3 remains stable.

The next trigger input, pulse 5, is fed via C_2 and D_2 to switch TR_2 OFF and TR_1 ON. As TR_2 goes OFF its collector voltage rises but this has no effect on bistable 2 which remains stable. In consequence bistable 3 remains stable.

Just prior to the arrival of pulse 6 the conditions of each bistable are the same as those which existed prior to the arrival of pulse 1. When pulse 6 arrives the cycle of events is repeated. Thus for every five trigger pulses fed to bistable 1, one negative-going pulse is obtained at the output of bistable 3, i.e. a division of 5 has been achieved.

To produce a 625-divider four divide-by-five stages are required as illustrated in figure 5.23. Each divide-by-five stage may have the circuit form of figure 5.21 and

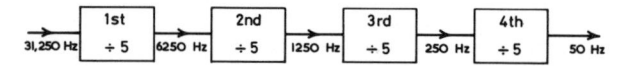

FIG. 5.23 FOUR DIVIDE-BY-FIVE STAGES IN CASCADE GIVING A DIVISION OF 5^4 (625-DIVIDER)

assuming the fabrication is in integrated circuit form, the output of one i.c. will feed the trigger input of the next i.c. Thus, with a trigger input at 31,250 Hz from the master oscillator to the first divide-by-five stage a final pulse output at 50 Hz may be achieved which can be used for the field frequency timing.

A frequency divider which operates on a different principle is shown in figure 5.24. This is really an energy storing arrangement and diagram (a) shows the basic circuit. The pulses to be divided which are negative-going and of fixed amplitude cause the voltage across the energy storing capacitor C_2 to jump in steps after each pulse. When the voltage across this capacitor reaches a predetermined level the discharge device operates which discharges C_2 and generates the output pulse.

On the application of the first pulse, D_1 conducts and C_1 charges with the polarity shown to the magnitude of the pulse voltage. When the pulse has ended, the voltage across C_1 will cause D_2 to conduct and C_2 will partly charge with the polarity shown. During this time D_1 is reversed biased and does not conduct. If the capacitance of C_2 is large compared with C_1 the voltage rise across C_2 will be small compared with the original voltage of C_1 (and the pulse magnitude). On the next pulse (2) D_1 will again conduct, replenishing the charge lost by C_1 and causing the voltage across C_1 to rise to the magnitude of the input pulse. At the end of the second pulse, D_2 conducts once more and C_1 partly discharges into C_2. However, because C_2 was partly charged at the end of the first pulse the rise in voltage will be less. Thus each pulse causes a jump in voltage across C_2, but each jump is smaller than the preceding one. When the voltage across C_2 reaches the trigger voltage level of the discharge device, C_2 is discharged and the process is then repeated. As shown in diagram (b) the trigger level is reached at the end of the fifth pulse. When the discharge device operates a pulse is obtained at its output. In this case, for every five input pulses one pulse is obtained at the output, resulting in a divide-by-five arrangement. The number of input pulses required to produce one output pulse is determined by the amplitude of the input pulses, the trigger

CAMERA CIRCUIT OPERATION

FIG. 5.24 STEP COUNTER (PUMP CIRCUIT)

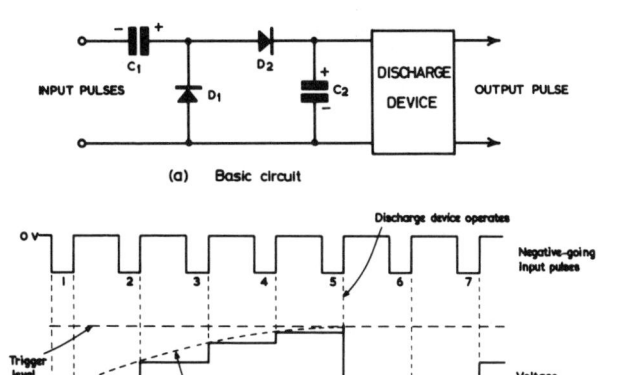

(a) Basic circuit

(b) Waveforms explaining operation

voltage level, the capacitance ratio of C_1 to C_2 and the initial voltage across C_2 (which was zero in the basic circuit).

Since the voltage jumps across C_2 get smaller and smaller, the step divider is not suitable for large division ratios as then the accuracy of the trigger is not reliable. The circuit may be modified to produce equal voltage jumps which improves the accuracy of division.

Various types of voltage operated switches may be used for the discharge device, e.g. bipolar transistor, s.c.r., unijunction transistor, etc.

Figure 5.25 shows the circuit of a step divider giving a division of five which is used in a Philips camera. The pulses to be divided are negative-going and are applied to C_1 from a preset resistor which sets their amplitude. On the negative-going edge of each pulse D_1 conducts causing C_1 and C_2 to charge. The voltage distribution here is inversely proportional to the capacitance of C_1 and C_2. At the end of the pulse on the positive-going edge, D_1 cuts off and, due to the rise of its emitter potential, TR_1 commences to conduct. C_1 now discharges through TR_1, and eventually TR_1 cuts off. Thus, after each pulse the voltage across D_1 is practically equal to the base-emitter drop of TR_1. On the negative-going edge of the next pulse D_1 again conducts, causing a voltage jump across C_2 which equals the preceding jump. As before on the positive-going edge of the pulse, D_1 is cut off, TR_1 conducts and C_1 discharges until TR_1 turns off which sets the delay voltage across D_1.

After five input pulses the voltage at C_2 (which is practically the emitter potential of TR_2) reaches a value such that TR_2 becomes conductive. The fall in voltage at TR_2 collector turns TR_3 ON which results in a rise of the voltage at TR_3 collector thereby turning TR_2 harder ON. Thus, TR_2 and TR_3 are brought hard ON (previously both were OFF) when the voltage across C_2 reaches the trigger level. C_2 now discharges through TR_2 and when the voltage across C_2 falls almost to zero TR_2 and TR_3 cut off and the cycle of events is then repeated. During the brief ON—OFF interval of TR_3 a positive-going pulse is developed at TR_3 collector. This pulse is fed to the next divide-by-five stage which operates with positive-going pulses.

With four similar divide-by-five stages a 625-divider may be constructed, but the component values and transistor types will be different to accommodate the different frequencies and polarities of the input pulses.

MAINS LOCK

As explained in Chapter 3 the field frequency is synchronized with the 50 Hz mains supply to eliminate the annoyance caused by drifting hum bars. In a 2:1 interlaced

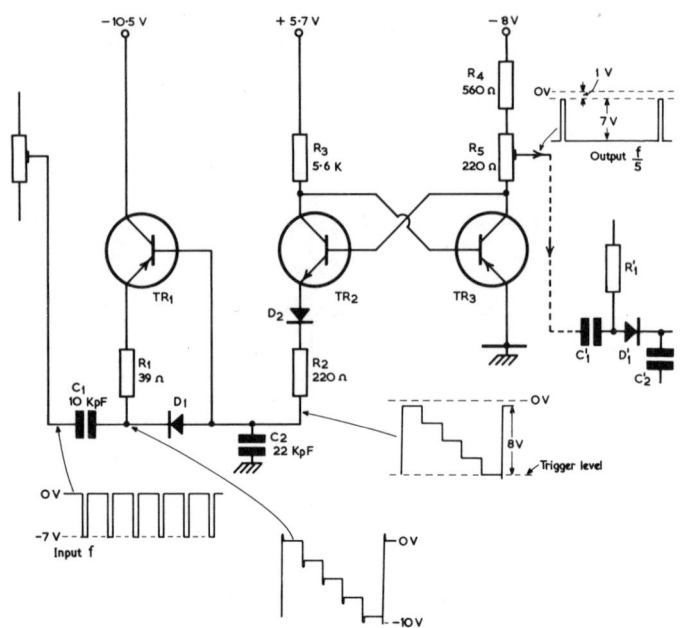

FIG. 5.25 STEP-DIVIDER GIVING DIVISION OF 5 (PHILLIPS 'MINICOMPACT' CAMERA)

camera an arrangement like that shown in figure 3.9 (also shown in figure 5.1) is usually adopted. The mains lock circuit is actually a phase comparator and one example is given in figure 5.26. This particular circuit feeds into the master oscillator circuit given in figure 5.10 (both from the same camera circuit).

One input to the phase comparator TR_4 is a sample of the mains voltage derived from a winding on the mains transformer and this is applied to TR_4 emitter. The other input is 50 Hz negative-going field timing pulses from the output of the 625-divider which are applied to TR_4 base. Each time a pulse is applied to TR_4 base, the transistor is turned hard ON and the collector-emitter voltage drop is very small. During the field pulse period, the collector voltage is practically equal to the emitter voltage. Consequently, during a field pulse the voltage on the lower plate of C_1 is equal to the instantaneous value of the sine wave applied to TR_4 emitter. This voltage, after smoothing by R_5, C_3, R_6 and C_4, is applied to the base of an emitter-follower TR_3.

When the phase between the field pulse frequency and the mains frequency varies, see diagrams (b) and (c) of figure 5.27, the charge on C_1 alters and the d.c. at TR_3 base changes. The d.c. from TR_3 emitter is fed to the master oscillator circuit via the constant current transistor (see figure 5.10). As a result, the frequency of the master oscillator will vary. Since the field pulse timing is derived from the master oscillator by means of the 625-divider, the field pulse frequency is also corrected. In this way the field pulse frequency is synchronized with the mains frequency. The frequency of the master oscillator is adjusted so that the field pulses coincide with the zero crossings of the 50 Hz sine-wave input [see diagram (a)].

The control time of the circuit is determined by the filter R_5, C_3, R_6 and C_4.

Other circuit arrangements are possible but they all tend to work on the same basic principle. With a random interlaced camera a phase comparator is not needed, and a simpler mains lock circuit can be used. Figure 5.28 shows the circuit arrangement and figure 5.29 the relevant waveforms for a simple random interlaced camera.

A sample of the mains frequency sine wave (taken from the vidicon heater supply) is applied to TR_1 base via R_1 and C_2. The standing bias on TR_1 is set so

CAMERA CIRCUIT OPERATION

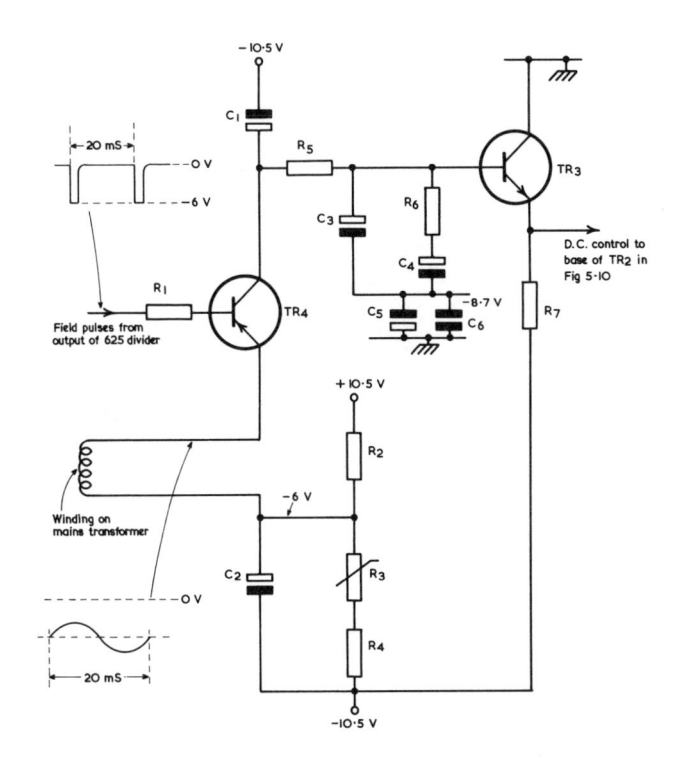

FIG. 5.26 MAINS LOCK CIRCUIT

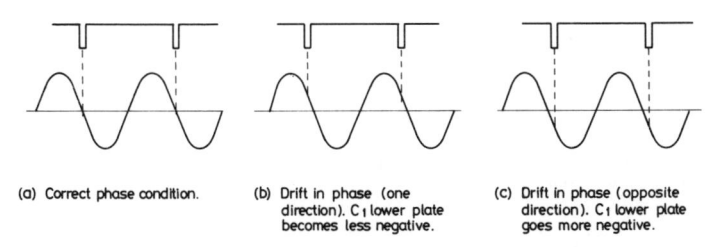

(a) Correct phase condition.

(b) Drift in phase (one direction). C_1 lower plate becomes less negative.

(c) Drift in phase (opposite direction). C_1 lower plate goes more negative.

FIG. 5.27 WAVEFORMS ASSOCIATED WITH MAINS LOCK CIRCUIT

that on the positive part of waveform *B* the transistor is cut off, but on the negative-going part the transistor is hard ON. Thus, TR_1 is overdriven which results in a positive-going pulse waveform at the collector with a suitable locking edge. After shaping, this pulse is used to provide the field blanking waveform. The pulse is also fed to TR_2 base *via* a differentiating circuit comprising C_4, R_7 and R_6 which produces a positive-going spike on the leading edge of the waveform. TR_2 is normally conducting but on the application of the positive spike to its base it cuts off for a short period. An amplified and limited negative-going pulse waveform of very short duration is developed at TR_2 collector. The pulse voltage from across R_{10}, which has been produced from the mains sine wave, is fed to the camera field timebase where it is used to trigger the field oscillator. Thus, the field is positively locked to the mains. In

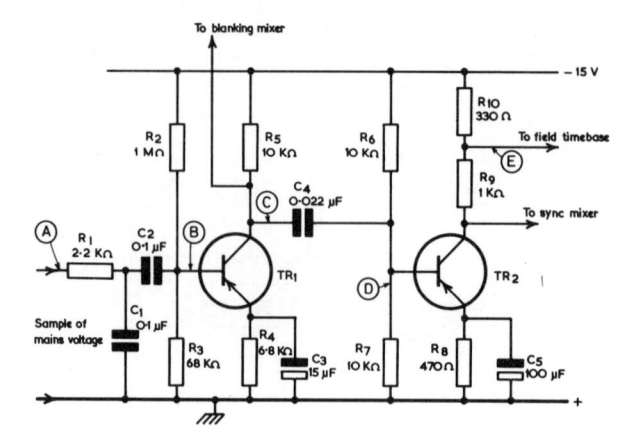

FIG. 5.28 MAINS LOCK IN A RANDOM INTERLACE CAMERA

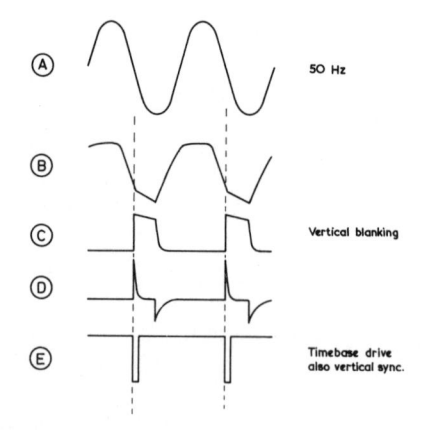

FIG. 5.29 WAVEFORMS ASSOCIATED WITH FIG. 5.28

addition the pulse voltage from across R_9 and R_{10} is supplied to the sync. mixer where it is mixed with suitable line pulses for insertion into the composite video waveform.

FIELD DEFLECTION CIRCUITS

To provide vertical deflection of the beam in the camera a linear sawtooth current at 50 Hz is required in the field deflector coils. A repetitive waveform at 50 Hz is obtained from the field oscillator which may be either free-running or driven. Normally, a sawtooth waveform is supplied by the oscillator but in some cases a pulse waveform is generated which is converted to sawtooth form by suitable circuitry. The sawtooth is fed to the field output stage which provides the sawtooth current drive to the deflector coils. It is most important that the scanning current is quite linear as any distortion present is reflected through the entire CCTV system (this, of course, applies to the horizontal scanning current as well). Because of this, various feedback circuits may be adopted to ensure good linearity of scan. Whilst vertical deflection circuits follow a similar pattern there are often detailed circuit differences which the designer incorporates to achieve the required current amplitude and linearity when working into a particular set of deflector coils from selected active devices. It is therefore

CAMERA CIRCUIT OPERATION

(Reproduced by courtesy of Pye Business Communications Ltd., Cambridge)

ALL-WEATHER CAMERA COVERING FACTORY ENTRANCE AND PARKING AREA FOR SECURITY SURVEILLANCE

120 INDUSTRIAL AND COMMERCIAL CCTV

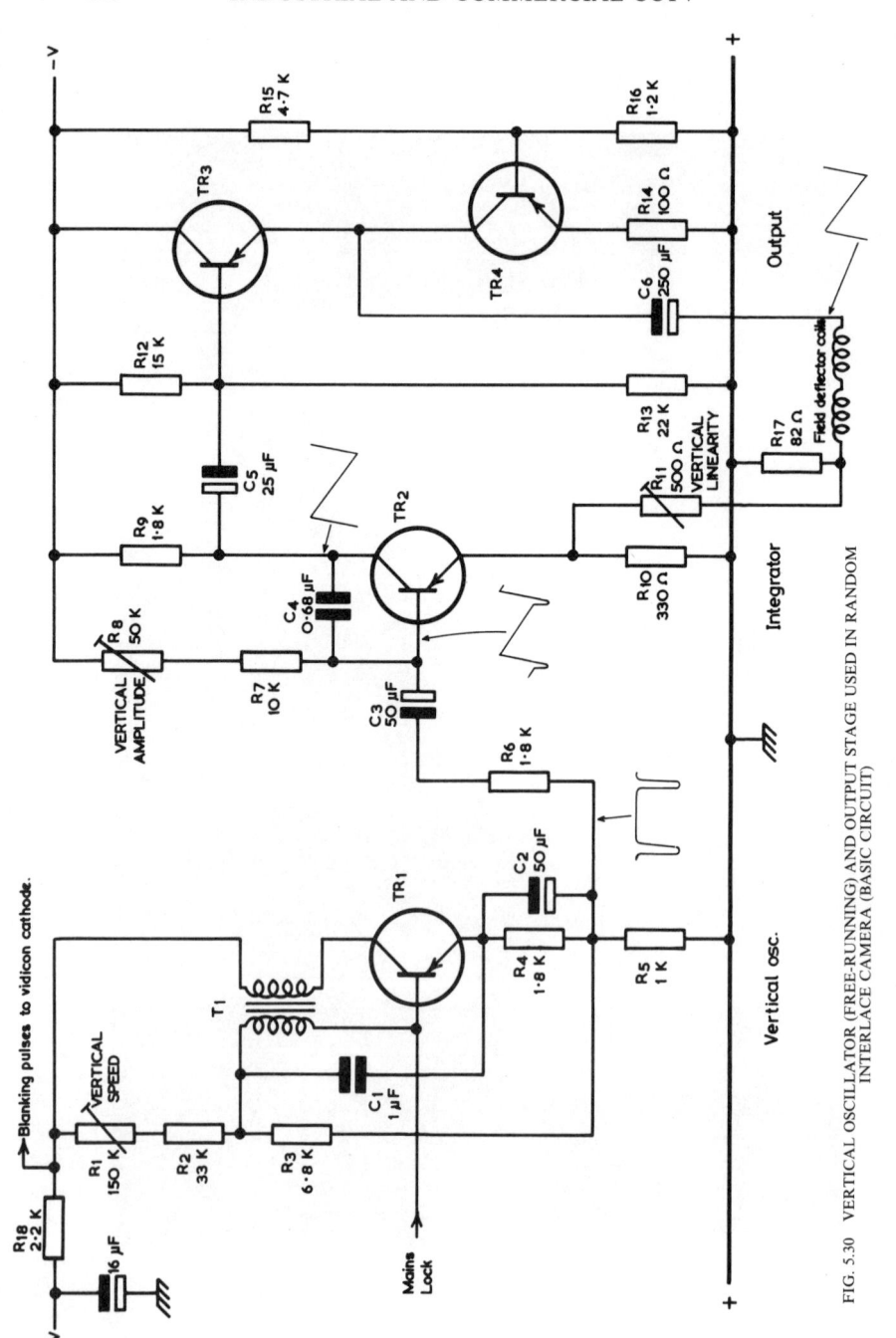

FIG. 5.30 VERTICAL OSCILLATOR (FREE-RUNNING) AND OUTPUT STAGE USED IN RANDOM INTERLACE CAMERA (BASIC CIRCUIT)

difficult to generalise on the circuits used, which is particularly true for the output stage.

An example of a free-running oscillator and output stage is given in figure 5.30. TR_1 stage is a blocking oscillator with feedback *via* the blocking transformer T_1. The frequency of the oscillations is determined mainly by the time-constant C_1, R_2 and R_1. Thus, variation of R_1 will alter the frequency of operation and hence serves as the vertical speed control. Locking to the mains frequency is by means of negative-going pulses, originating in the mains rectifier circuits, and applied to TR_1 base. During the intervals that TR_1 is conducting, negative-going pulses are developed across R_5 in the emitter circuit and these are fed to TR_2 base *via* R_6 and C_3. Also, during TR_1 ON periods, positive-going pulses are developed across R_{18} in the collector circuit. These pulses, after suitable shaping, are fed to the vidicon tube to cut off the electron beam during the field blanking period.

TR_2 stage is an integrator which works on the Miller principle. The feedback between collector and base *via* C_4 integrates the pulse input to the base to produce the sawtooth at the collector which is necessary for scanning. C_5 couples the sawtooth to the base of the output transistor TR_3. This transistor is connected as an emitter-follower and feeds the field deflector coils *via* C_6 which blocks the d.c. component. The transistor TR_4 serves as the d.c. path for the emitter circuit of TR_3, at the same time presenting a high impedance to the a.c. sawtooth waveform. The scanning current is passed through R_{17} which produces a feedback voltage to TR_2 emitter to improve the linearity of the scanning current. Vertical linearity control R_{11} adjusts the amount of feedback.

The circuit diagram of a driven field oscillator is shown in figure 5.31. This oscillator uses a simple CR charging circuit driven by pulses obtained from the output

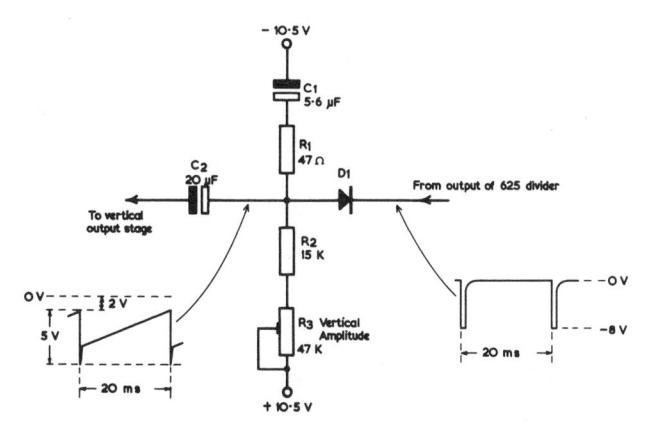

FIG. 5.31 DRIVEN FIELD OSCILLATOR

of the 625-divider. When there is no input pulse at D_1 cathode, C_1 is charged through R_1, R_2 and R_3 so that a positively increasing voltage is applied to the anode of D_1. This corresponds to the field scan period. During each negative field pulse applied to D_1 cathode, C_1 discharges through R_1, D_1 and the source circuit of the driving pulses. This corresponds to the field flyback period. Thus at the junction of R_1, R_2 a sawtooth waveform is developed which is fed *via* C_2 to the output stage where it is used to provide the drive waveform to the scanning coils. The charging current of C_1 can be adjusted by R_3 which serves as the vertical amplitude control. Note that if the drive to D_1 fails there will be no vertical scan, therefore some form of protection circuit is required for the vidicon to guard against scan failure. This of course applies to free-running oscillators as well (either line or field).

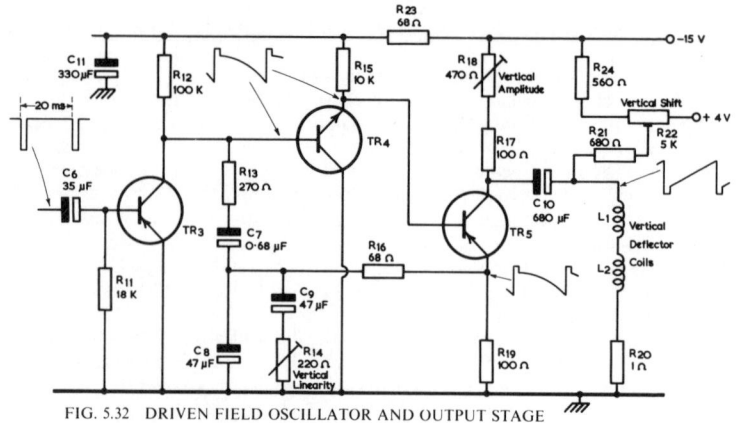

FIG. 5.32 DRIVEN FIELD OSCILLATOR AND OUTPUT STAGE

Another driven field oscillator circuit together with its output stage is shown in figure 5.32. The timebase is driven by field pulses derived from the output of the circuit in figure 5.28.

In the absence of negative drive pulses to TR_3 base, C_7 charges towards the supply line via R_{12}, R_{13}, R_{16} and R_{19} causing the voltage at TR_3 collector to travel exponentially in a negative direction. This constitutes the scan period. When a negative drive pulse is applied to TR_3 base via C_6, TR_3 conducts and C_7 rapidly discharges via R_{13}, R_{16}, R_{19} and TR_3 thereby forming the retrace period. At the end of the trigger pulse TR_3 cuts off and C_7 recharges once more. The waveform generated by this action is applied to the base of TR_4. An emitter-follower connection is used for this stage because its high input impedance reduces the loading on the input charging capacitors and its low output impedance is suitable for driving the output transistor. D.C. coupling is used from TR_4 emitter to the base of the output transistor TR_5. Feedback from across R_{19} in the emitter circuit of TR_5 to across C_8 linearises the scan. Adjustment over the amount of correction is provided by R_{14} which serves as the vertical linearity control.

The low impedance deflector coils are capacitively coupled to TR_5 via C_{10}. R_{18} adjusts the amplitude of the waveform fed to the coils. The deflection current may be inspected by placing a c.r.o. across the low value resistor R_{20} which lies in the earthy side of the deflector coils. A variable d.c. is fed into the deflector coils from R_{22} to provide vertical shift for the vidicon beam. D.C. shift current in the scan coils has a role different from that for alignment coils or magnets. Alignment coils compensate for the off-axis gun position in the camera tube, whereas the vertical and horizontal shifts centre the aligned scanning on the target image. R_{21} stands-off the deflection current from the shift circuit.

BLANKING AND SYNC. PULSE GENERATION

The pulse generation circuits described so far have been concerned with the production of line and field DRIVE pulses which ensure correct operation and timing of the camera timebases. In addition to the drive pulses, the camera pulse generating circuits must produce blanking and sync. waveforms with the correct timing and duration.

We will first consider the line frequency waveforms required, figure 5.33. In quite a number of industrial cameras, the source of the timing for the blanking and sync. pulses is the line flyback pulse taken from a suitable point on the line scan output stage (usually the transformer). The flyback pulse may be either positive- or negative-going and will have a duration which, within limits, is quite arbitrary (but not longer than the line blanking period) and depends upon the design of the line scan output stage.

The blanking waveform starts at the same instant as the flyback pulse but it has a

CAMERA CIRCUIT OPERATION

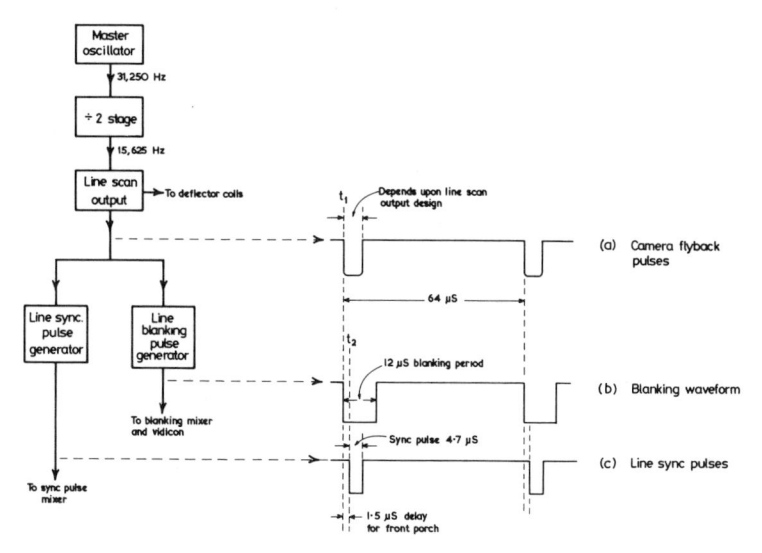

FIG. 5.33 TIMING OF LINE WAVEFORMS

longer duration. Thus, an electronic circuit is required which will produce a 12 μs output pulse with its leading edge commencing at instant t_1. Matters are more complicated for the line sync. pulse which must start at instant t_2, 1·5 μs later than the leading edge of the blanking pulse and has a duration of 4·7 μs. The brief delay at the start is necessary to produce the front porch of the composite line waveform (see page 66, Chapter 3). Thus in the case of the sync. pulse production a delay circuit is required in addition to a pulse-forming circuit.

In other cameras the master timing for these waveforms is not the flyback pulse but line frequency pulses derived from the output of the divide-by-two stage. In both cases it is seen that the camera line timebase commences its flyback a little earlier than the line timebase in the monitor. In normal circumstances this would not appear to be important because the video signal is suppressed for the line blanking period. Even if the camera and monitor flyback occurred at the same instant there is no guarantee that the commencement of scans would be coincident, as the time for flyback in the camera and monitor will most likely be different. However, it should be noted that in some camera circuits the commencement of the flyback of current in the LINE DEFLECTOR COILS is delayed for a brief period which ensures alignment with the start of flyback in the monitor.

The timing of the field waveforms is shown in figure 5.34. Commonly, the source of the timing is the field drive pulse derived from the output of the 625-Divider. The field blanking waveform commences at the same instant as the drive waveform but lasts for 1·28 ms which corresponds to 20-line periods. There is no need for a front porch before the commencement of a field sync. pulse, and this waveform may commence immediately and is commonly about 120 μs in duration in CCTV cameras. Similar circuit techniques may be employed in the field pulse production to those used in line pulse production (see later) but the complication of achieving a front porch delay is not required in the case of the field. After the field sync. and blanking waveforms have been generated they are fed to the sync. pulse mixer and the blanking mixer where they are mixed with the line sync. and blanking waveforms prior to insertion in the video signal chain. Also, both line and field blanking pulses are fed to the camera tube to suppress the scanning beam during the retrace periods.

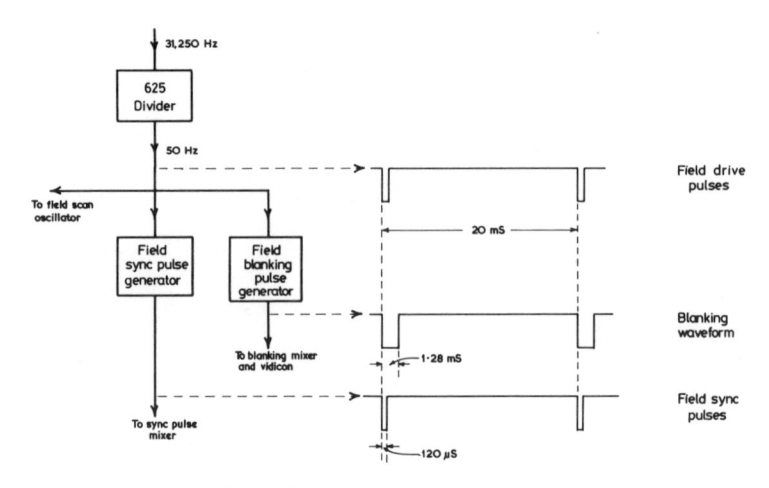

FIG. 5.34 TIMING OF FIELD WAVEFORMS

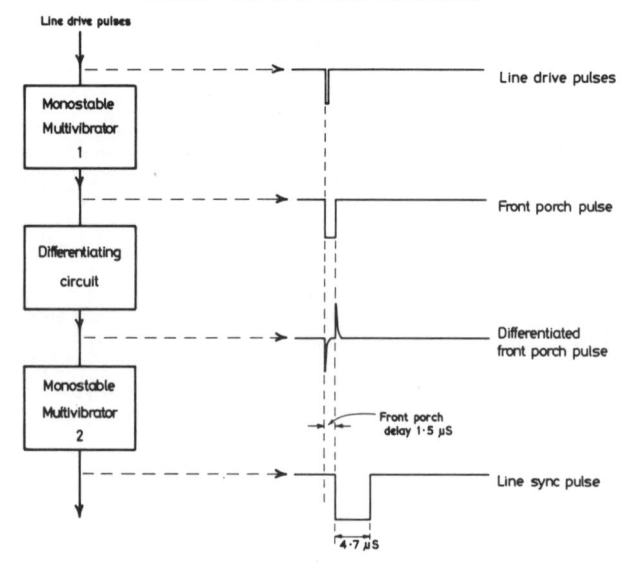

FIG. 5.35 OBTAINING THE LINE SYNC. PULSE WITH FRONT PORCH DELAY

SYNC. PULSES

One method of obtaining a line sync. pulse with the necessary delay for the front porch is illustrated in figure 5.35. Here, line drive pulses are used to trigger a monostable multivibrator which produces pulses at its output with a duration equal to the front porch period. The output pulse is then fed to a differentiating circuit which gives out negative- and positive-going spikes as shown. A second monostable oscillator is used to generate the line sync. pulses and this is triggered by the positive-going spikes at its input. In this way, line sync. pulses with a fixed time delay from the leading edge of the line drive pulses are created.

CAMERA CIRCUIT OPERATION

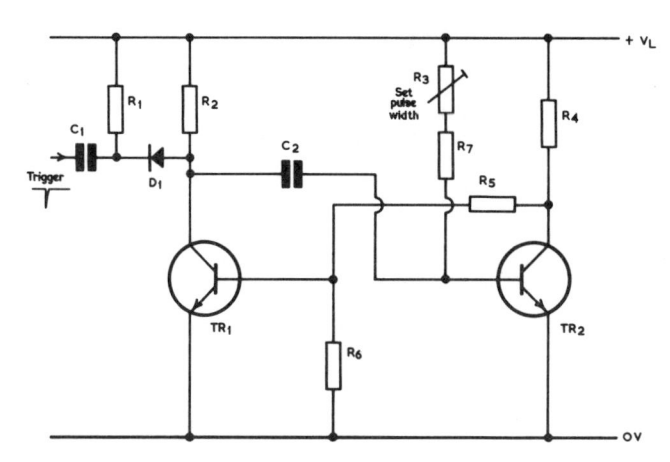

FIG. 5.36 MONOSTABLE MULTIVIBRATOR

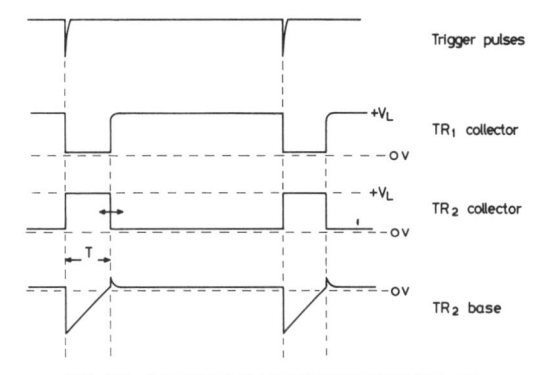

FIG. 5.37 WAVEFORMS ASSOCIATED WITH FIG. 5.36

The basic circuit of a monostable multivibrator is given in figure 5.36 and its associated waveforms in figure 5.37. A monostable oscillator has only one stable state. At switch-on the circuit takes up this particular state with one transistor ON and the other transistor OFF. On the application of a trigger pulse it will switch over and reverse the transistor conducting states. However, this is not a stable state and after a period of time determined by the circuit components it reverts back to its former stable state. It then remains in this state until another trigger pulse is applied.

A.C. coupling is used between TR_1 and TR_2 through C_2 but d.c. coupling is employed between TR_2 and TR_1 via R_5. At switch-on, TR_2 conducts hard as its base is fed from the positive line via R_3 and R_7. Thus, TR_2 collector potential is low and by suitable choice of R_5 and R_6 values the base potential of TR_1 can be arranged so that there is insufficient forward bias for conduction. Therefore TR_2 is ON and TR_1 is OFF; this is the stable state. It is assumed here that silicon transistors are used, but in circuits employing germanium types the lower end of R_6 may be taken to a negative potential to ensure that TR_1 is fully OFF. As TR_1 is OFF its collector potential is high and C_2 will be practically charged to the positive supply line voltage.

Trigger pulses are fed to TR_2 base via C_1 and the trigger diode D_1. When TR_1 is OFF both sides of the diode are at the positive supply voltage. Thus, on receipt of a negative trigger pulse at D_1 cathode, the diode conducts and a negative pulse is produced across R_2 which is fed to TR_2 base via C_2. The negative pulse applied to TR_2 base reduces the current in TR_2 and as a result its collector voltage rises. This rise is reflected at TR_1 base causing TR_1 to turn ON. In consequence TR_1 collector voltage falls and this is passed to TR_2 base via C_2 turning TR_2 further OFF. This action is repeated rapidly and many times resulting in TR_2 turning sharply OFF and TR_1 coming ON quickly. C_2 now discharges through R_7 and R_3 and the base potential of TR_2 rises exponentially. When TR_2 base voltage has risen sufficiently above zero, TR_2 conducts. The fall of potential at TR_2 collector commences an action which results in TR_1 rapidly switching OFF and TR_2 quickly turning ON. The circuit is now back in its original stable state and will remain so until the next trigger pulse is applied. Note that when TR_1 is conducting, D_1 is reversed biased and the fall in TR_1 collector potential is not fed back to the trigger pulse source.

It will be seen that at the collectors of the transistors, pulse waveforms are available having a duration T. This interval is essentially governed by the discharge of C_2 through R_7 and R_3. By varying the time-constant with the aid of R_3 the duration of the output pulse may be varied over quite wide limits when suitable component values are employed.

There are a number of advantages in replacing one of the collector-base couplings with emitter coupling. A basic circuit is given in figure 5.38 where it will be noted that the coupling from TR_2 collector to TR_1 base used in figure 5.36 has been replaced by coupling via a common-emitter resistor R_5.

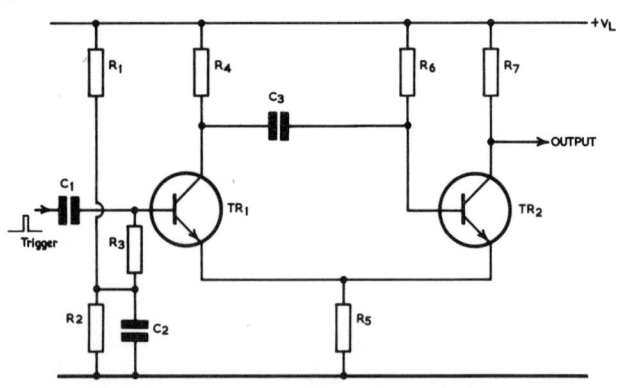

FIG. 5.38   EMITTER-COUPLED MONOSTABLE MULTIVIBRATOR

In the stable state TR_2 is conducting as its base is returned to the positive line via R_6. The current due to TR_2 flowing in R_5 makes the emitter of TR_1 positive and by suitable choice of values for R_1 and R_2, TR_1 can be arranged to be cut off. When a positive trigger pulse is fed through C_1 to TR_1 base the transistor conducts and its collector voltage falls. The fall in voltage is passed through C_3 to TR_2 base causing TR_2 to turn OFF. As there is now no emitter current in TR_2, the voltage across R_5 falls a little which maintains TR_1 in the ON state. C_3 now discharges through R_6 and when TR_2 base potential rises slightly above the voltage across R_5, TR_2 conducts once more. As a result of this the voltage across R_5 rises which turns TR_1 OFF. The circuit is now back in its original stable state.

As with the collector-coupled circuit, the pulse width at the output is settled by the time-constant C_3, R_6. This arrangement has the following advantages: (a) the output may be taken from TR_2 collector which is isolated from the coupling path between the transistors; (b) similarly, the trigger circuit is isolated from the feedback path as the trigger is applied direct to TR_1 base; and (c) a greater flexibility is allowed in choosing the operating points of the transistors which is of importance in high-speed switching applications where the recovery time of a saturated transistor is of significance.

By employing two monostable oscillators interconnected via a differentiating circuit the line sync. pulses may be produced using the method outlined in figure 5.35.

CAMERA CIRCUIT OPERATION

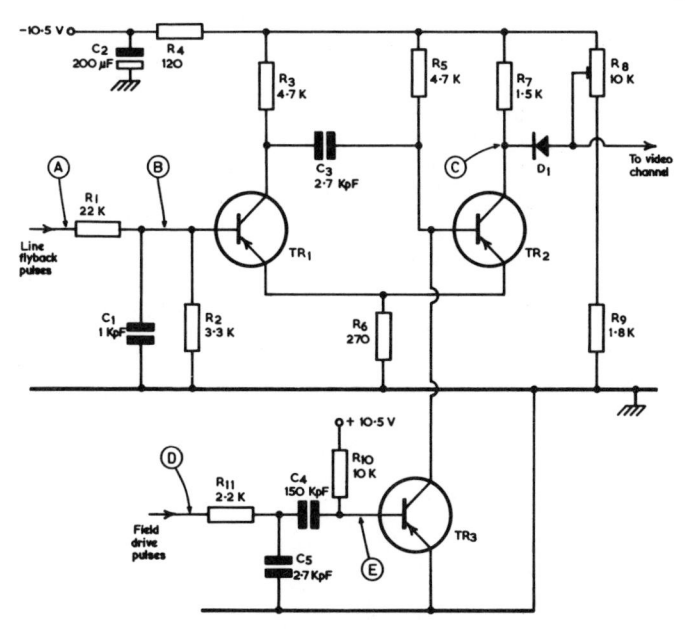

FIG. 5.39 SYNC. MIXER EMPLOYING AN EMITTER-COUPLED MONOSTABLE MULTIVIBRATOR

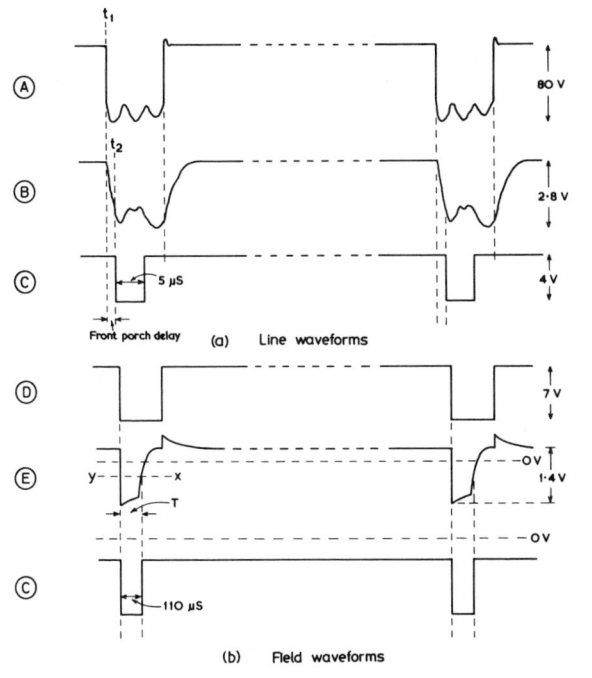

FIG. 5.40 WAVEFORMS ASSOCIATED WITH FIG. 5.39

An alternative method of generating the line sync. pulses is shown in figure 5.39. An emitter-coupled monostable multivibrator is used, comprising TR_1 and TR_2. In the stable state, TR_2 is conducting and TR_1 is cut off. Negative-going lines flyback pulses are applied to TR_1 base via an integrating circuit comprising R_1, C_1 and R_2. As shown in Fig. 5.40, diagram (a), the effect of the integrating circuit is to smooth out the input pulse to a certain extent but in particular to increase its rise-and-fall times. At a particular voltage level corresponding to instant t_2, the negative-going edge of TR_1 base waveform causes TR_1 to switch-on and TR_2 to go OFF. C_2 now discharges via R_5 and after a while TR_2 switches back ON and TR_1 goes OFF. The time that TR_2 is in the blocked condition depends upon the time-constant C_3, R_5. Consequently the values of these components determine the width of the line sync. pulse generated at TR_2 collector. As the line blanking waveform starts at instant t_1, a suitable delay for the front porch is achieved.

Field drive pulses from the 625-divider are fed to TR_3 base via R_{11} and C_4. These pulses are partially differentiated by C_4 and the input resistance of TR_3 base circuit. TR_3 is normally biased OFF but when the negative-going differentiated pulse arrives, the transistor cuts-on at a level corresponding to, say, x-y on waveform E. When TR_3 conducts the base voltage of TR_2 drops which cuts off TR_2 for an interval T. Thus, at TR_2 collector a negative-going pulse of suitable duration is produced which may be used as the field sync. pulse. It will be seen that TR_3 stage acts as a PULSE-NARROWER CIRCUIT and this technique is sometimes used when the drive pulse width is too long.

Therefore, at TR_2 collector there will be available line sync. pulses due to the basic action of the monostable TR_1 and TR_2 and field pulses due to the effects of TR_3 conduction on TR_2, i.e. MIXED SYNC. is present at the output. D_1, which is biased by R_8, is set so that the correct quantity of synchronizing pulses is passed to the video channel to be added to the video signal.

BLANKING PULSES

One method of obtaining blanking pulses with suitable duration is by shaping the line and field drive waveforms, e.g. by using pulse lengthening circuits. Alternatively, similar techniques to those used for generating the sync. pulses may be employed, i.e. using monostable oscillators to produce pulses of the required width. In some cameras integrated circuits are used for generating the blanking pulses (also the sync. pulses) and an example is shown in basic form in figure 5.41. Here, two integrated circuits are used, one for horizontal pulse production and the other for vertical pulses. Each i.c. contains three monostables. One monostable is employed in the production of the blanking pulses and the other two are used in generating a sync. pulse with a suitable front porch delay. A front porch delay on the vertical sync. can be used to allow the introduction of pre-field sync. equalizing. External presets (P_1—P_6) permit the various pulses to be set to the correct width.

A blanking mixer is essentially a combining circuit where the line and field blanking pulses are added together to provide MIXED BLANKING at the output. In figure 5.41, TR_1 is the blanking mixer which is fed with blanking waveforms from the integrated circuits. The pulses are mixed in the base-emitter circuit of TR_1 and then receive amplification and inversion. Amplified mixed blanking from across R_3 is fed out to the video channel via an emitter-follower TR_2.

Other techniques may be employed in the mixing of the blanking waveforms, e.g. using two transistors which supply a common load but as these are straightforward amplifiers they do not present much difficulty.

EXTERNAL DRIVE

Some cameras have provision for receiving drives, sync. and blanking from external sync. pulse generators. This enables several cameras to be mutually synchronized to a common s.p.g. source. Figure 5.42 shows the internal sync. pulse generating system of a modern camera intended for fine quality work.

In position 1 of S_1, the camera generates its own drive, blanking and sync.

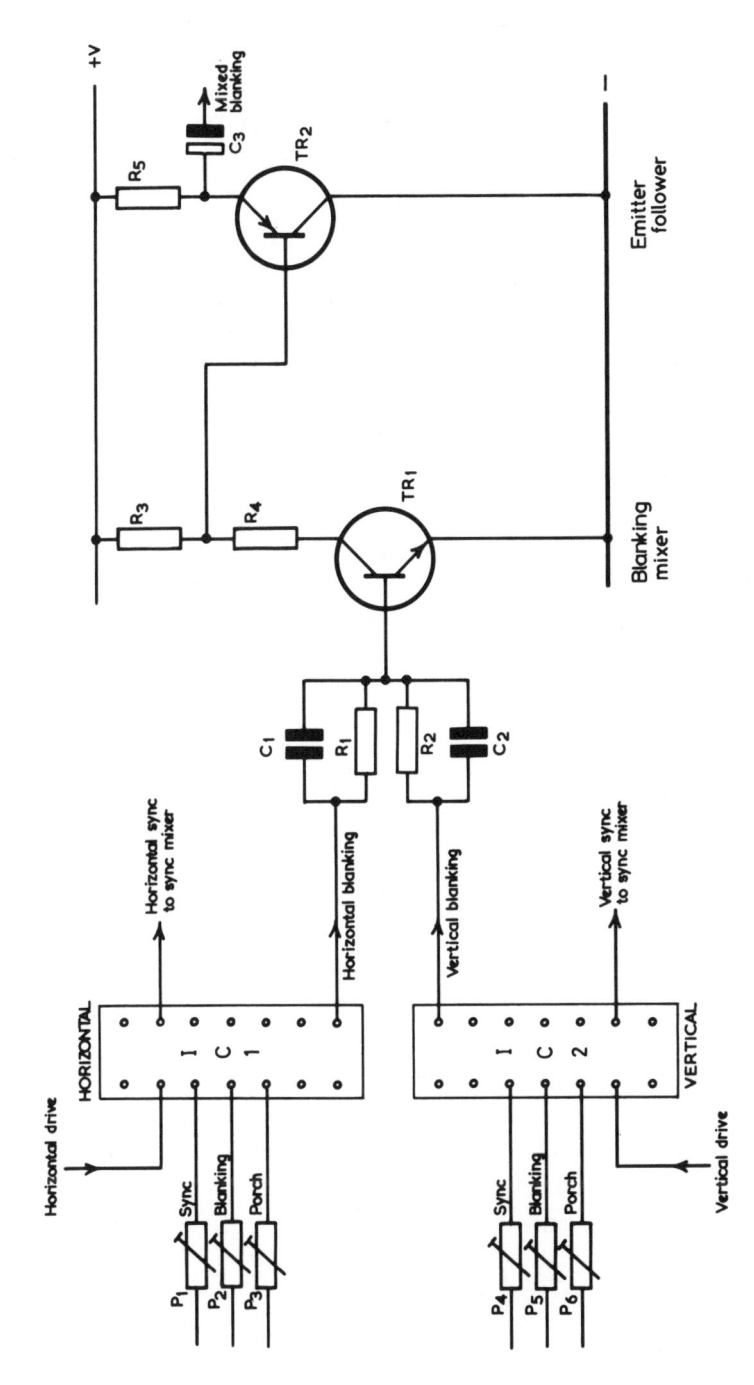

FIG. 5.41 BLANKING MIXER WITH INPUT FROM INTEGRATED CIRCUIT SYNC. AND BLANKING GENERATORS

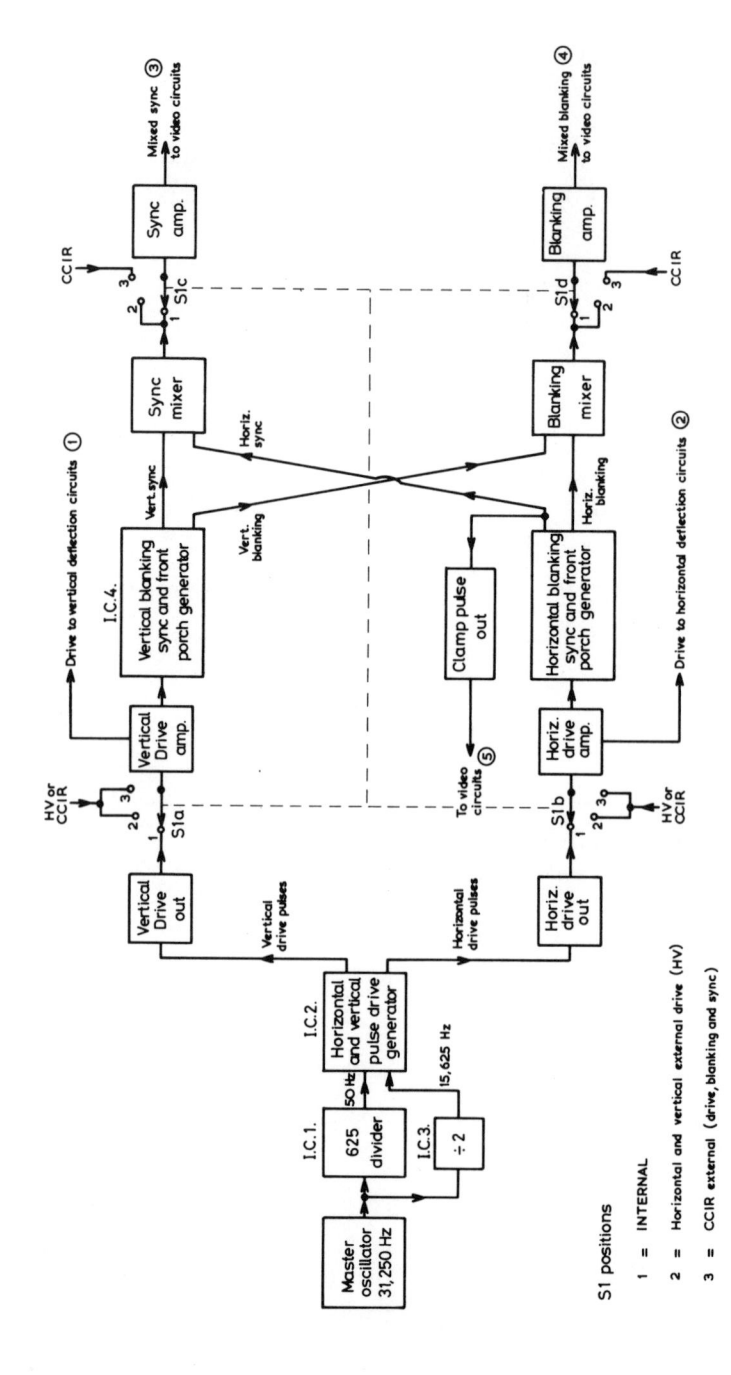

FIG. 5.42 MODERN CAMERA SYNC. GENERATOR WITH PROVISION FOR RECEIVING EXTERNAL DRIVE OR DRIVES, SYNC. AND BLANKING TO CCIR STANDARD

waveforms using the techniques previously outlined and a number of the operations are carried out with integrated circuits. The camera s.p.g. provides outputs of vertical and horizontal drive pulses, (1) and (2), to the camera deflection circuits, mixed sync. (3) and mixed blanking (4) to the video circuits and clamping pulses (5) to the video circuits.

When S_1 is set to position 2, external vertical and horizontal DRIVE pulses may be applied *via* S_1a and S_1b. This makes the generating blocks to the left of these switches redundant. Thus, drive to the deflection circuits, mixed sync. and blanking to the video channel and clamping pulses are locked to the external drives.

In position 3, external drives to S_1a and S_1b and external sync. and blanking to S_1c and S_1d respectively from a pulse generator to C.C.I.R. specification ensures a high standard of performance with regard to waveform timing and duration. Standards laid down by C.C.I.R. (Comité Consultatif International des Radiocommunications) are to encourage international standardization of 625-line television. The specifications for the sync. pulse train are shown in figure 3.17, page 70.

COMBINING BLANKING AND SYNC. WITH THE VIDEO SIGNAL

The composite video signal appearing at the camera output socket is usually formed in two stages. Figures 5.43 and 5.44 show the basic idea but it should be noted that whereas in these diagrams line and field are shown separately, in practice we are concerned with adding MIXED blanking and MIXED sync. to the video signal.

Diagram (a) of figure 5.43 shows two lines of video information. During the line blanking period there should be no video signal present as the camera scanning beam

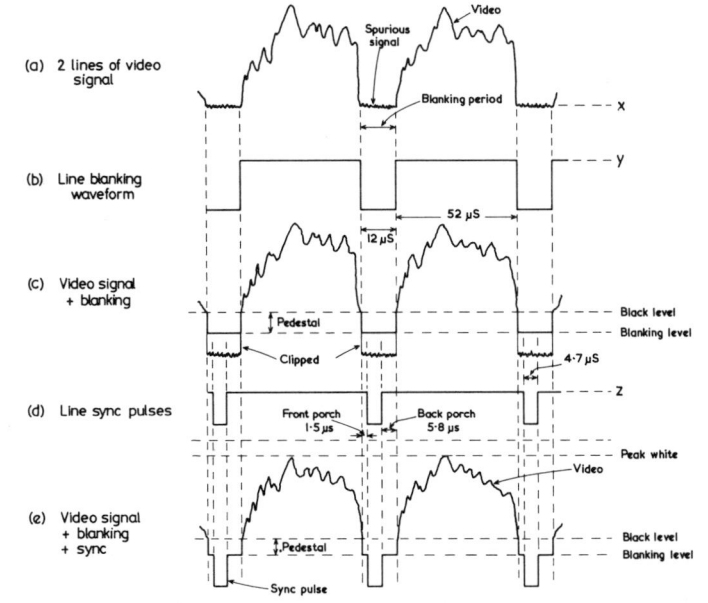

FIG. 5.43 ADDING LINE BLANKING AND SYNC. TO THE VIDEO SIGNAL

will be cut off. There may, however, be a small quantity of spurious signal and noise which can cause faulty synchronization and unwanted modulation of the monitor beam during its retrace period. If the blanking waveform of (b) is added to the video signal so that the levels x and y are coincident then waveform (c) will result. By clipping off the waveform at a suitable level (the blanking level) any spurious signal is lost. We now have a pedestal or 'set-up' on the waveform of the correct amplitude. Line

132 INDUSTRIAL AND COMMERCIAL CCTV

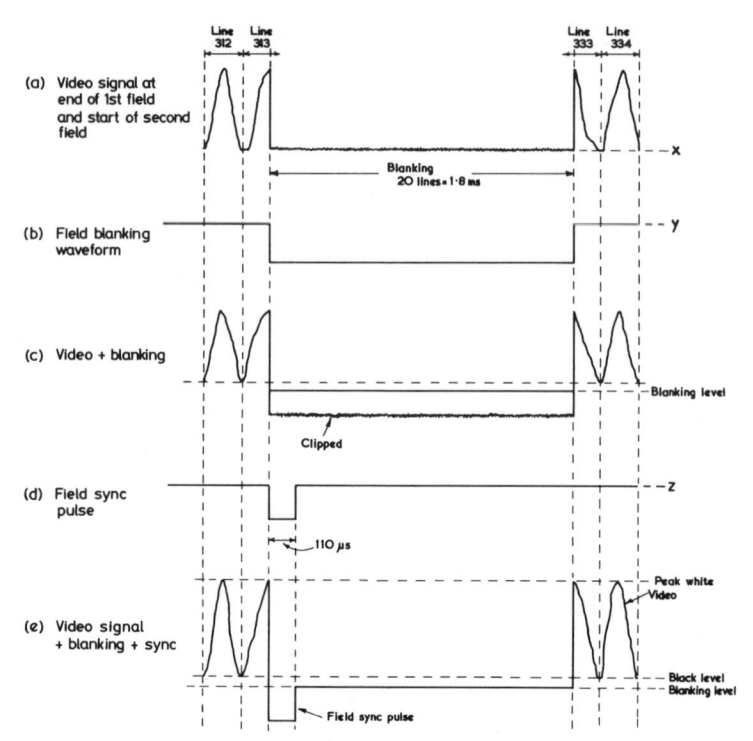

FIG. 5.44 ADDING FIELD BLANKING AND SYNC. TO THE VIDEO SIGNAL (ONE FIELD ONLY SHOWN)

synchronizing pulses with the necessary front porch delay [diagram (d)] are now added to waveform (c). By arranging alignment of the level z of the sync. pulses with the blanking level of diagram (c), the composite waveform of (e) will result. The amplitude of the added sync. pulses must be adjusted to give the correct picture-to-sync. ratio; standards levels for CCTV signals are given on page 67. The process for the field blanking and sync. is just the same. A front porch on the field sync. may be incorporated into which equalizing pulses are inserted. During the field blanking period there will, of course, be line sync. pulses inserted into the pulse train as indicated in figure 3.17 but these have been omitted from figure 5.44.

An example showing the electronics involved in carrying out the operations just described is given in figure 5.45. Video signals are applied via C_1 to the base of TR_1 which is the final amplifier of a four-stage preamplifier. Amplified and inverted video from across R_3 is fed to TR_2 base circuit where it is added to the line and field blanking pulses. The blanking pulses which are positive-going and about 60 V in amplitude are applied to the cathode of D_1, the anode of which is returned to a $+10.5$ V rail via R_8. This limits the amplitude of the blanking pulses to about 7·5 V which are then fed to TR_2 base via C_5 and R_7. The d.c. component of the pulses are blocked by C_5 and the blanking pulses are added to the video signal at TR_2 base.

Video, plus blanking, are then fed to the clamp transistor TR_3 via the emitter-follower TR_2 which is necessary to provide impedance matching to ensure correct operation of the clamp. The reasons for including a clamp were explained on page 99. Negative-going pulses at line rate are applied to TR_3 base and drive the transistor into

CAMERA CIRCUIT OPERATION 133

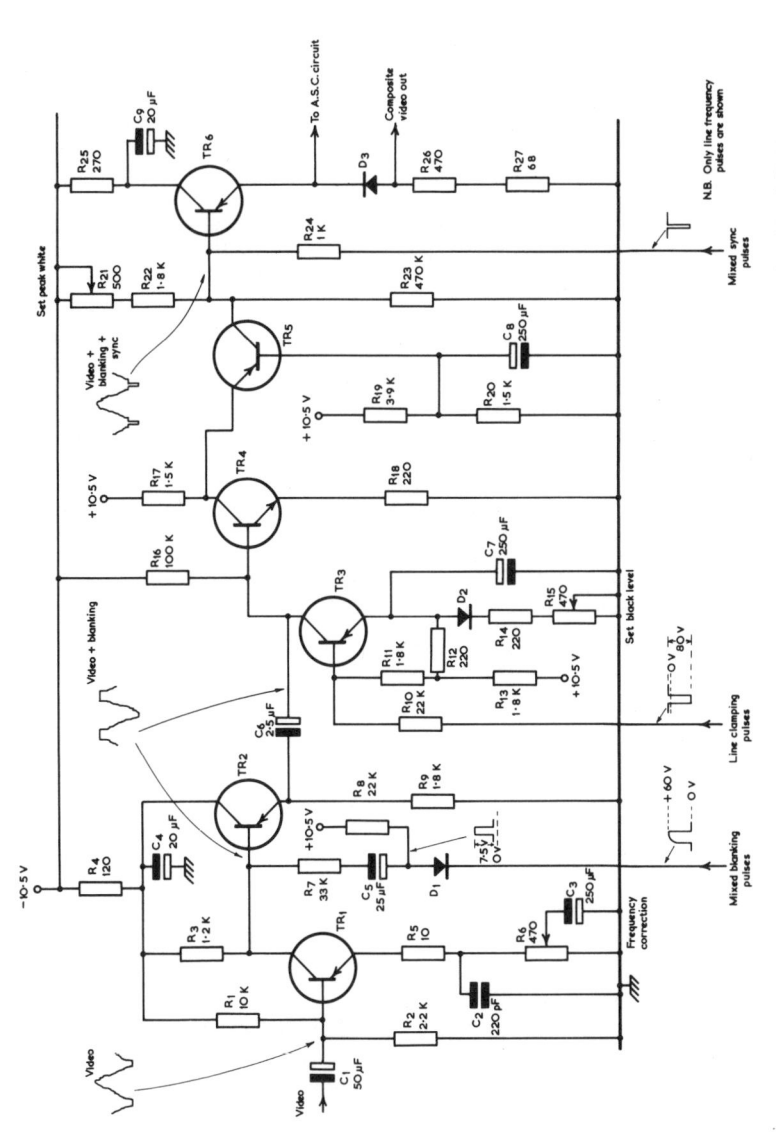

FIG. 5.45. COMBINING THE BLANKING AND SYNC. WAVEFORMS WITH THE CAMERA SIGNAL

full conduction. The peaks of the blanking pulses at TR_3 collector and TR_4 base will thus be clamped to the emitter potential of TR_3. This potential can be adjusted by R_{15}.

Clamped video and blanking is then applied to the common-emitter amplifier TR_4. The positive peaks of the blanking pulses drive TR_4 into conduction to such an extent that the fall in collector voltage causes TR_5 to cut off. Thus, the positive peaks of the waveform are clipped at a certain level. This level is determined by the setting of R_{15} in the clamp circuit so that this potentiometer is used as the black level control. After amplification in the common base amplifier TR_5 the signal is applied to the base of the output amplifier TR_6, connected as an emitter-follower. Negative-going synchronizing pulses are applied *via* R_{24} to the base of TR_6 and added to the drive from TR_5. The composite video signal is then fed to the video output socket from across R_{26}, R_{27}. A feed to the a.s.c. circuit is taken from the cathode of D_3. The base voltage of TR_6 is set by R_{21} which determines the emitter potential of this transistor and the cathode voltage of D_3. At a certain amplitude of the video signal D_3 cuts off and the video signal is clipped. Thus R_{21} serves to set the peak white level of the video signal.

AUTOMATIC SENSITIVITY CONTROL (A.S.C.)

A camera intended for use in industrial situations may be required to work under varying lighting conditions. For example, the light falling on the lens of a camera that is mounted outside will vary from summer to winter with local weather conditions and from day to night when there may be artificial lighting. Clearly, one setting of the manual target control will not produce the best pictures under such changing lighting conditions. A.S.C. eliminates the need for continual readjustments and allows the camera to be used by non-technical operators.

A.S.C. or 'automatic target control' is really a form of a.g.c. After the video output of the camera tube has been amplified it is fed to an auto-target circuit which detects either the peak level or the mean level of the video signal. It then uses this level, which is dependent upon the scene illumination, to automatically control the voltage fed to the target electrode of the vidicon tube. This then adjusts the sensitivity of the camera tube to suit the scene illumination.

Figure 5.46 shows one arrangement. The video input to the auto-target transistor TR_7 comes from the emitter of TR_6 in the previous circuit (figure 5.45). This is the

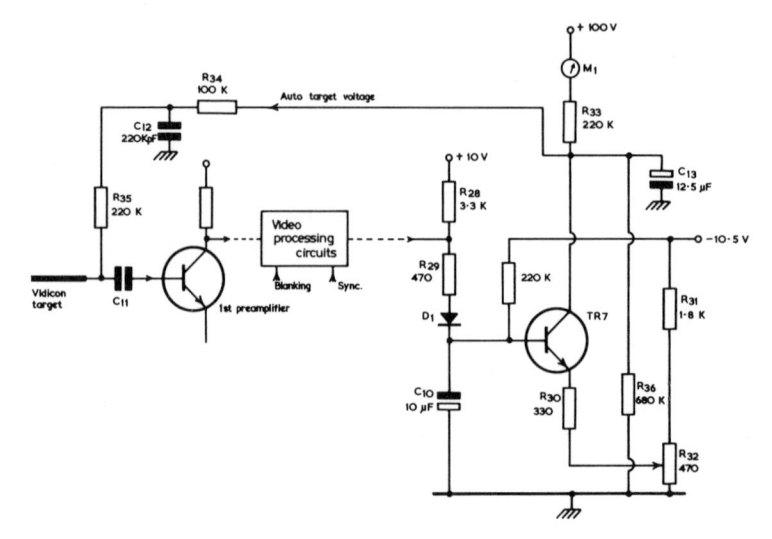

FIG. 5.46 EXAMPLE OF AUTO-TARGET CIRCUIT

unclipped output of TR_6 and is applied to the diode D_1 which acts as a peak detector. When the diode conducts, C_{10} is charged positively. The moment at which TR_7 will become conductive can be set by means of R_{32}. As soon as TR_7 starts conducting, the collector current will cause a larger voltage drop across R_{33} so that the target voltage drops. The resistance in series with D_1 (R_{29}) prevents detection on narrow signal peaks, *i.e.* highlights.

The target voltage in this arrangement is limited to about 70 V by the potential divider R_{33}, R_{36}. The control time of the circuit is determined by means of the capacitor C_{13}. Interference pulses are removed by the filter R_{34}, C_{12} and R_{35} in the target voltage line. The target voltage can be read from the meter M_1 (in terms of TR_7 collector current). This meter has a scale which is divided into two sections—red and green. When the pointer lies in the red region of the scale there is insufficient signal at the input to TR_7 to cause the transistor to conduct to any appreciable extent. A reading in the green region indicates that TR_7 is conducting hard, *i.e.* sufficient illumination is falling on the scene. Under these conditions the target voltage is lowered.

Thus, if the illumination of the scene falls the effect is that a reduced amplitude video signal is available at the camera tube output (see figure 5.47). As a result, C_{10} will charge positively by only a small amount causing the current in TR_7 to drop. Therefore the collector voltage rises increasing the voltage fed to the target electrode of the tube. This causes the sensitivity of the vidicon to increase resulting in a larger output signal which compensates for the lower scene illumination to some extent.

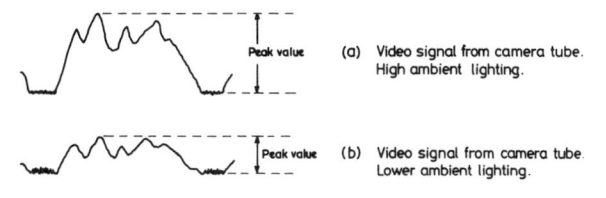

(a) Video signal from camera tube. High ambient lighting.

(b) Video signal from camera tube. Lower ambient lighting.

FIG. 5.47 DIAGRAM SHOWING THE EFFECT OF LIGHTING ON VIDEO SIGNAL AMPLITUDE

GAMMA CORRECTION

In Chapter 4, it was stated that no gamma correction is ordinarily required in a vidicon camera-monitor system with an overall gamma of about 1·3 which is suitable for direct viewing. However, for Telecine applications in broadcasting when *negative* film is used some gamma correction is desirable. The overall gamma of film plus the t.v. system may vary from about 0·9 to 2·0 depending upon operating conditions. Thus, if distortion of the tonal values is to be avoided, appropriate gamma correction is necessary. Additionally, gamma correction may be used to deliberately distort the tonal values to improve the contrast range when the camera is viewing scenes which have a limited tonal range, *i.e.* overall too dark or too light.

Gamma correcting circuits are usually only found in the more expensive cameras and normally employ biased diodes for tailoring the response of the video channel. Figure 5.48 shows one arrangement for reducing the gamma of the system. Clamped, non-composite video is applied to the base of the correcting amplifier TR_1. This transistor conducts harder on the peak whites of the signal as the input signal is negative-going at this point in the video channel. Amplified and inverted video is developed across the collector load R_1. Correction is carried out by the biased diode D_1. The cathode of the diode is returned to the emitter potential of TR_2 which can be set with the aid of R_4. As the input signal goes towards peak white, the collector potential of TR_1 is taken more positive and at some point D_1 will become conducting. When this occurs the effective collector load of TR_1 is reduced (D_1 is now effectively in parallel with R_1 for signals). As a result, the gain of TR_1 is lowered. In diagram (b) the level of the input signal at which D_1 conducts (A) is controlled by the setting of R_4 which determines the operating point of the gamma correction. The gamma corrected

(a) Basic circuit

(b) Transfer characteristic

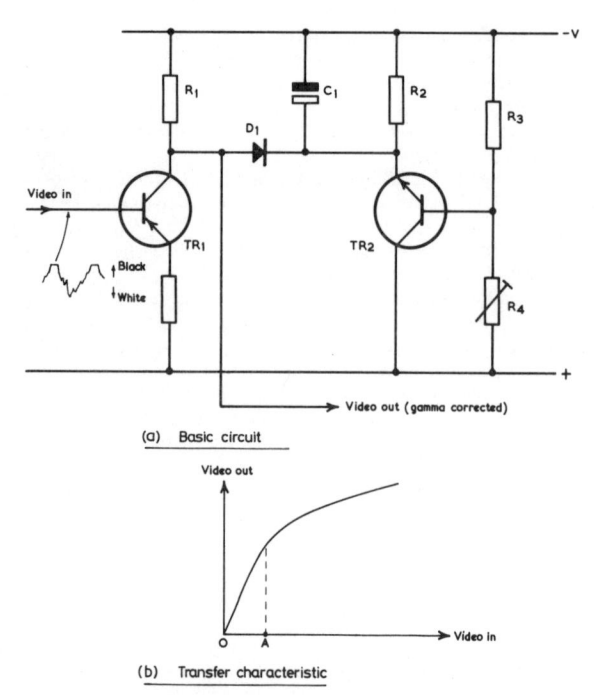

FIG. 5.48 ONE ARRANGEMENT FOR DECREASING THE GAMMA OF THE SYSTEM (WHITE CRUSH)

(a) Basic circuit

(b) Transfer characteristic

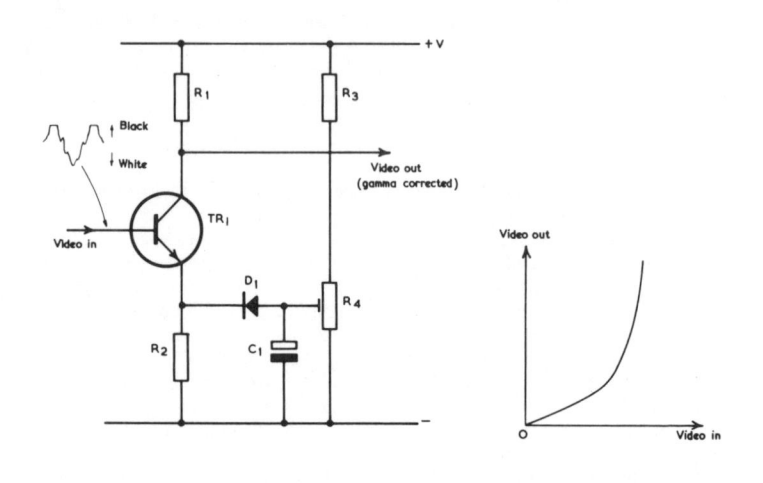

FIG. 5.49 AN ARRANGEMENT FOR INCREASING THE GAMMA OF THE SYSTEM (WHITE STRETCH)

(Reproduced by courtesy of Pye Business Communications Ltd., Cambridge)

CAMERA OBSERVING FLOW OF RAW MATERIAL (SUGAR BEET) THROUGH A NARROW SECTION OF A CONVEYOR SYSTEM—INDUSTRIAL PROCESS CONTROL

output from TR_1 collector is then fed on to the next video stage for subsequent processing.

To increase the gamma of the system an arrangement like that given in figure 5.49 may be used. TR_1 is the gamma-correcting amplifier providing amplified and corrected video from across R_1. As before clamped, non-composite video is fed in at the base but note here that as the signal goes towards peak white TR_1 conducts less. By adjusting R_4 it can be arranged that D_1 is non-conducting when the signal input is travelling towards black level (voltage across R_2 increasing). As the signal goes towards peak white D_1 conducts at some point due to the fall in voltage across R_2. When this happens, D_1 is effectively placed in shunt with R_2 to signals thereby increasing the gain of TR_1. The signal level at which D_1 becomes conductive and the gain increases is set by R_4 which thus determines the operating point of the correction.

FOCUS REGULATION

In some cameras the current in the focus coil is maintained steady and focus adjustment is carried out by varying the potentials of the electrostatic lens formed by the various vidicon anodes. This is the most common arrangement but in a few cameras the anode potentials are held steady, and focus adjustment is effected by varying the current in the focus coil. Regardless of the method used it is most important that the current in the focus coil does not vary, or the image will not remain in sharp focus.

The main causes of current variation are supply voltage changes and temperature variations. Stabilized power supplies may be employed to combat varying feed voltage to the focus coil. However, if the feed voltage is taken from the line scan output transformer by rectifying the flyback pulse voltage the stabilization may not be too good. In this case a regulator may be used to keep the focus coil current stable with changes in supply voltage. Changes in ambient temperature will cause the resistance of the focus coil to alter which produces corresponding changes in the coil current. If this results in noticeable de-focusing then some form of current stabilization should be used. Such an arrangement is given in figure 5.50.

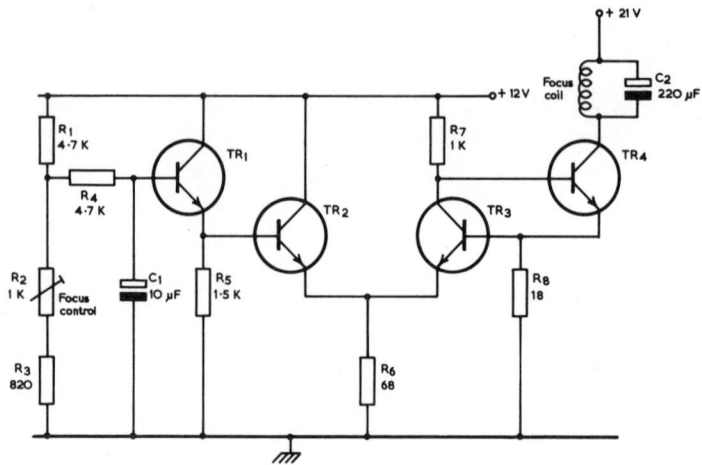

FIG. 5.50 ONE TYPE OF FOCUS REGULATOR

The regulator comprises four transistors including a differential amplifier TR_2 and TR_3. The focus coil is supplied from an unstabilized 21 V rail through a series control element TR_4, the emitter current of which flows through R_8. When a change of temperature occurs varying the resistance of the focus coil, the current in TR_4 and R_8 changes. As a result, the base voltage of TR_3 alters. This change in voltage is amplified and inverted by TR_3 and applied to TR_4 base from across R_7, thereby stabilizing the current in TR_4 and the focus coil. With an increase in ambient temperature, the coil resistance increases and less current flows in TR_4 and R_8. Therefore, TR_3 is biased back and its collector voltage rises. TR_4 is turned harder ON, *i.e.* its resistance is reduced which compensates for the rise in coil resistance and stabilizes the current. The circuit will also detect variations in the 21 V rail because if the supply rail voltage alters so does the current in R_8.

Focus adjustment is carried out by R_2 which sets the base voltage of TR_1. In its turn this sets the base voltage of TR_2, the current in TR_3 and hence the current in TR_4, and the focus coil. The use of a differential amplifier improves the temperature stability.

Simpler arrangements employing zener diodes or thermistors are to be found in the cheaper type camera.

MODULATED R.F. OUTPUT

A modulated r.f. output is available from some cameras thus enabling them to be used with a domestic t.v. receiver. One example of the electronics involved is shown in figure 5.51.

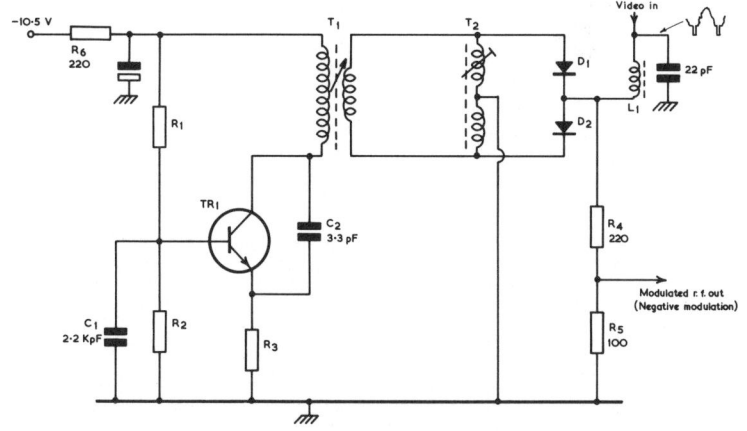

FIG. 5.51 R.F. OSCILLATOR AND MODULATOR

TR_1 is the oscillator with the primary of T_1 and stray capacitance forming the tuned collector load. Feedback, to sustain the oscillations, is *via* C_2 connected between collector and emitter. Starting bias is provided by the base resistors R_1 and R_2; C_1 grounds the base. By altering the core of T_1, the oscillations can be set to any channel within Band 1.

The modulator consists, in principle, of a bridge circuit comprising the diodes D_1 and D_2 and the tapped transformer T_2. The oscillatory voltage is developed across T_2 from the coupling winding on T_1. Modulating video signal is applied to D_1 and D_2 *via* the r.f. choke L_1. Balancing of the bridge is by means of T_2 core. The effect of the video signal is to throw the bridge out of balance causing an amplitude modulated h.f. current to flow in R_4 and R_5. A portion of the modulated h.f. signal (8–15 mV r.m.s. depending upon the channel) from across R_5 is fed to the h.f. output socket. The complete circuit is suitably screened as are the oscillator and modulator sections from one another.

VIEWFINDER

Better quality (and more expensive) cameras may incorporate a built-in electronic viewfinder which permits easy set-up and operation of the camera. Essentially, an electronic viewfinder is a miniature monitor which is fitted at the back of the camera. It

140   INDUSTRIAL AND COMMERCIAL CCTV

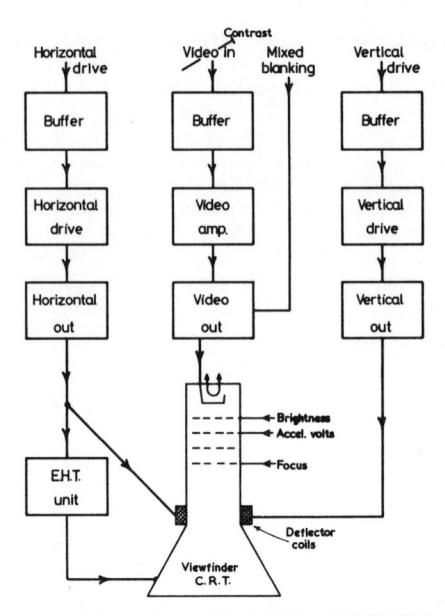

FIG. 5.52 BLOCK SCHEMATIC OF ELECTRONIC VIEWFINDER

displays a picture of the imaged scene on the face of a small c.r.t. screen (about 4 ins).

Figure 5.52 shows the block diagram of a representative arrangement. The c.r.t. employs magnetic deflection, electrostatic focusing and is provided with an e.h.t. of about 8 kV which is derived from rectified line flyback pulses. A raster is produced by horizontal and vertical deflection circuits which supply the deflector coils with linear sawtooth currents at line and field rate. Synchronism with the camera timebases is achieved by driving the viewfinder deflection circuits with horizontal and vertical drive pulses obtained from the camera s.p.g. A suitable portion of the video signal taken from the camera video circuits is amplified, mixed with line and field blanking pulses and finally fed to the cathode of the c.r.t. to intensity-modulate the raster. The level of the video signal is adjusted by a contrast control so that the picture on the viewfinder screen can be set to give the best results.

Independent brightness and focusing controls are provided which together with the contrast control are mounted on a control panel at the rear of the camera close to the screen. This is a most convenient method of mounting for the camera operator who can adjust the optics and position of the camera by considering the viewfinder picture until exactly the right shot is obtained.

CHAPTER 6

THE MONITOR DISPLAY TUBE

THE device universally used for reproducing television images is called a display tube or picture tube (in America it is referred to as a 'kinescope'). All display tubes in current use are cathode-ray tubes (c.r.t.). The c.r.t. converts the video signals back into light information, reversing the process which takes place in the camera tube.

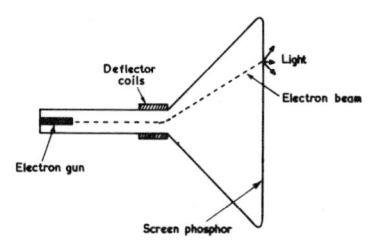

FIG. 6.1 BASIC IDEA OF DISPLAY TUBE

In a c.r.t. a beam of electrons is directed at high velocity towards a glass faceplate, the inside of which is coated with a layer of electroluminescent material which emits light on being struck by the electrons. This layer is called the 'screen phosphor' and the colour of the light emitted depends upon the chemical composition of the layer. To produce white light, a mixture of zinc sulphide (emitting blue light) and zinc cadmium sulphide (emitting yellow light) is used together with a silver activator. The particular proportions used in the mixture determine the type of white light that is emitted. The emission of light is due both to fluorescence and phosphorescence. Fluorescence occurs when the screen layer is excited by the electrons whilst phosphorescence occurs after the excitation has ceased, *i.e.* the afterglow. In television picture reproduction the afterglow is useful as it aids the persistence of vision. It would, however, degrade the picture quality if it exceeded the picture period (1/25th sec). A typical afterglow for a monitor c.r.t. is of the order of 0·1 ms.

The construction and operation of the c.r.t. may be considered under three headings:

(a) The electron gun assembly which produces the beam of electrons.

(b) A focusing system to produce a fine electron beam.

(c) The viewing area or screen which emits light on being excited by the arriving electrons.

In addition a deflecting system is required to move the beam vertically and horizontally over the screen face to produce the raster.

(a) The Electron Gun

The function of the electron gun assembly is to produce a finely pointed beam of electrons to strike the c.r.t. screen. The electrons must have high velocity and the intensity of the beam must be controllable so that luminance output from the screen can be varied. The gun assembly has the same basic principles as a thermionic valve, with the gun mounted in a glass envelope of high vacuum.

Figure 6.2 (a) shows a complete 4-anode assembly (a heptode tube) which is the most common gun assembly, but guns employing five electrodes (pentode) or seven electrodes (hexode) may be used. The production of the beam may be explained by considering diagram (b). An indirectly heated cathode is used but the oxide emitting area is of different form to an ordinary valve. In a c.r.t. only a small beam current is required and the emitting area is made smaller. Preferably, the emission should come from a point source to keep the beam dimensions small. One method of construction

142   INDUSTRIAL AND COMMERCIAL CCTV

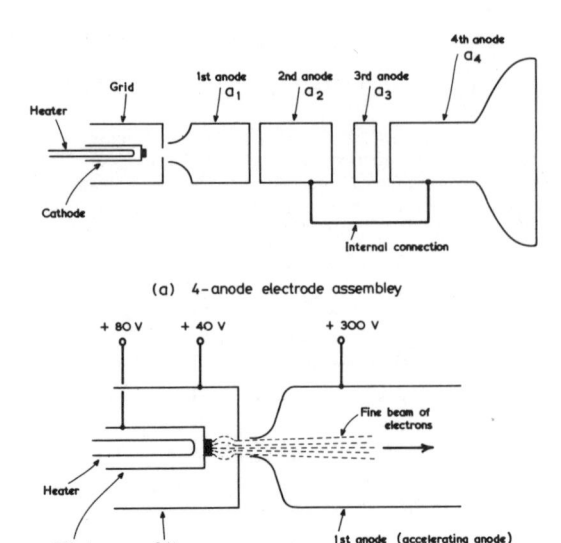

(a) 4-anode electrode assembley

(b) Producing a fine beam of electrons

FIG. 6.2   THE ELECTRON GUN

uses a small cylindrical tube which serves as the cathode but with the oxide emitter placed at one end as shown, and not all over the cylinder surface as in a valve. A heater is passed down the centre of the tube and when fed with a current raises the oxide to a sufficiently high temperature for emission to take place.

To get the electrons moving towards the c.r.t. screen the first anode is placed at a potential of some 200-400 V positive with respect to the cathode. The resulting electric field set up between the first anode and the cathode provides the initial acceleration for the electrons on their way to the screen.

Surrounding the cathode is the control grid which is constructed differently to the control grid in an ordinary valve. It consists of a cylinder which is open at one end but with a fine aperture at the other end. The grid cylinder is normally held at a potential which is negative with respect to the cathode. The electric field set up between grid and cathode is in such a direction as to return the electrons back towards the cathode. However, those electrons of high energy overcome the retarding field and converge on the aperture at the end of the grid cylinder. On entering the aperture the electrons come under the influence of the first anode potential and begin to accelerate. The resulting beam leaving the aperture is very narrow.

The intensity of the electron beam, and hence the light output from the screen, may be controlled by adjustment of the p.d. between grid and cathode (the grid bias). As the grid is made more negative to the cathode, fewer electrons have sufficient energy to reach the grid aperture. This results in a beam of lower intensity, and a reduction in light output from the screen. Conversely a reduction in grid bias permits greater numbers of electrons to reach the aperture with a resulting increase in beam intensity and increase in light output. If the bias is made sufficiently large, none of the electrons will have sufficient energy to overcome the intense retarding field. In consequence, there is no beam current and no light output from the screen. The value of bias which produces beam cut off is called the 'cut-off voltage.'

When viewing the screen, light of high intensity is interpreted as white; light of medium intensity as grey; and no light as black. The actual colour of an unactivated screen appears a greyish colour, thus 'black' is the sensation of viewing the natural screen colour in contrast to brighter energized areas.

Although the beam leaving the aperture in the grid cylinder is very narrow, the electrons tend to diverge due to the natural repulsion they have for one another. It is therefore necessary to focus the beam in some way so that on arrival at the screen a sharply converging beam is obtained.

(b) The Focusing Lens

In all modern c.r.t.s. the beam is focused using an electrostatic lens and figure 6.3 shows the basic action. The final anode potential is fed to a_2 and a_4 (these electrodes are

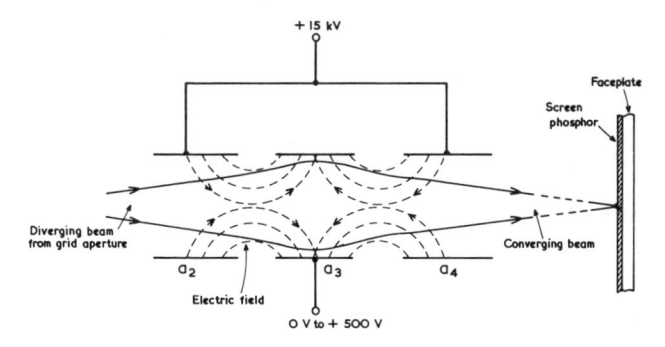

FIG. 6.3 AN ELECTROSTATIC LENS

commoned by an internal connection) whilst a much lower potential is supplied to the third anode. Electrostatic fields are set up between the three anodes as shown in the diagram. Electrons entering such a field are subjected to forces urging them to travel in paths exactly opposite to the direction of the lines of force. When an electron enters a field at an angle, its direction will therefore be changed. This principle is used in the electrostatic lens. Due to the shape of the anodes and the resulting electric field patterns, the electrons are brought to a focus at the screen of the c.r.t. When within the electric fields the electrons may travel in curved paths but on leaving the fields they travel in straight line paths.

To ensure that the point of focus coincides with the c.r.t. screen, the contour of the electric field is altered by varying the potential fed to a_3. This potential is supplied from the 'focus' potentiometer, and a_3 is usually referred to as the 'focusing anode'.

(c) The Viewing Area

As was mentioned earlier, the screen is made of an electroluminescent material which fluoresces on being bombarded by the electron stream. As the electrons approach a_2 they are accelerating under the influence of the high a_2 potential. On entering the focusing field they at first decelerate and then accelerate once more on their way to the screen. Once past a_4 the electrons coast along (but at high speed) as the screen is at the same potential as a_4. The velocity attained by the electrons is given by the expression

$$v = \sqrt{2\frac{e}{m}V} \text{ metres/sec, where } \frac{e}{m} \text{ is}$$

the ratio of electric charge to electron mass $= 1.759 \times 10^{11}$ coulombs/kg. With a final anode voltage of 15 kV the velocity is approximately 72×10^6 metres/sec. The kinetic energy acquired by the electrons during motion is given up to the screen coating on

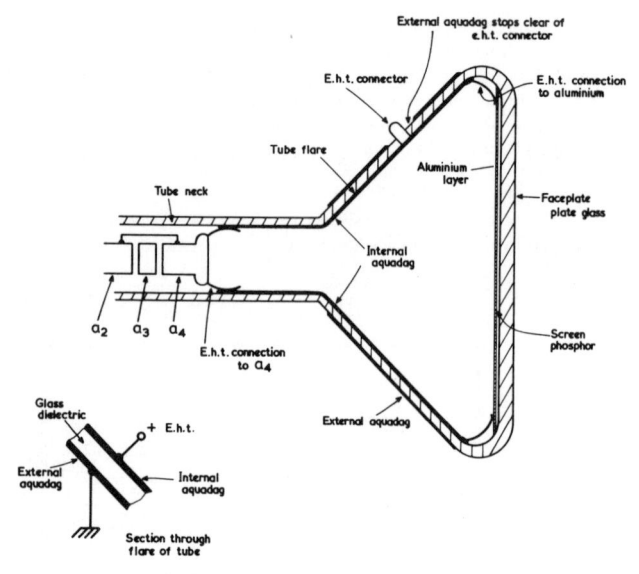

FIG. 6.4 DETAILS OF THE SCREEN SECTION OF THE TUBE

impact causing it to fluoresce.

The screen phosphor is deposited on the faceplate of the tube which is made from heavy reinforced plate glass. The faceplate protects the viewer from flying glass in the unlikely event of the tube imploding. In a 20" tube, the total pressure on the faceplate is in the region of about 1·5 tons and extreme care should be exercised in handling a c.r.t. A chance blow may fracture the glass envelope and a serious implosion could result.

On the back of the screen phosphor is deposited a very thin coating of aluminium and this serves two purposes:

(i) Prevention of Ion Burn

Negative ions generated in the region of the electron gun travel towards the screen along with the electron beam. The mass of an ion may be several thousands times greater than that of an electron and if allowed to strike the screen phosphor may result in permanent damage to the screen coating. When an ion collision has occurred, the light output of the phosphor at that point may be severely reduced or even nil; this results in darkened areas on the picture referred to as 'ion burns'.

When a coating of aluminium is used, the kinetic energy of the ions is given up to the aluminium on colliding with the aluminium atoms. Electrons, however, due to their smaller mass pass through the atomic spaces of the aluminium and strike the screen coating. Electron-aluminium collisions will occur but with e.h.t. values over 10 kV the loss is insignificant.

(ii) Increased Light Output

In a c.r.t. without an aluminium coating about 50% of the light emitted from the screen phosphor is directed away from the viewer, back down the c.r.t. With an alluminised screen, the coating is deposited in such a way that it forms a highly reflective backing for the screen phosphor. This results in the rearward light being reflected towards the viewer giving a greater light output from the tube.

AQUADAG COATING

The inside and outside of the tube flare are coated with a layer of graphite (applied in the form of a colloidal solution). The outer coating is connected to chassis and the inner coating is connected to the e.h.t. supply and *via* spring clips to the second and fourth anodes. These two coatings, separated by the thick glass of the tube, form the e.h.t. reservoir capacitor (see figure 6.4). Typical values for the capacitance between a_2, a_4 and the external coating are 1750–2500 pF. It is important to ensure that the tube capacitance is discharged before handling the c.r.t.. otherwise a very nasty 'kick' may be received; but what is probably more important, the tube may be dropped causing the c.r.t. to implode.

BEAM CURRENT RETURN PATH

Electrons which strike the screen phosphor must return to the e.h.t. supply to complete the beam current circuit. In non-alluminised tubes this was achieved by secondary emission. Each arriving electron caused secondary emission and the secondary emitted electrons were attracted to the internal graphite coating leaving the screen coating positively charged. In this way the potential of the screen phosphor increased until it reached the final anode potential. In an alluminised c.r.t., however, the aluminium coating is connected to the e.h.t. supply thereby providing the electrons with a return path.

OBTAINING THE RASTER

Now that a sharply focused beam has been obtained it is necessary to deflect the beam in order to obtain a raster. All display tubes use magnetic deflection, achieved by feeding the scanning currents into two sets of coils disposed at right angles to one another and fitted close to the tube neck as in figure 6.5.

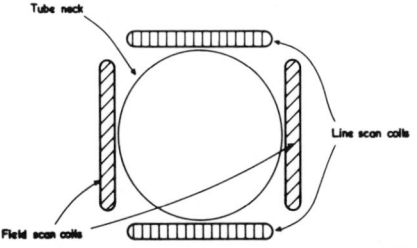

FIG. 6.5 DISPOSITION OF VERTICAL AND HORIZONTAL SCAN COILS

When an electron beam enters a field of constant magnetic flux it experiences a force acting in a direction at right angles to both the direction of the field and the direction of motion of the beam, figure 6.6. This force deflects the beam away from its original path with the result that the beam emerges along a path at an angle to its original path. Whilst in the deflecting field, the electron travels in a curved path, which is part of the circumference of a circle of radius r.. On leaving the deflecting field, the

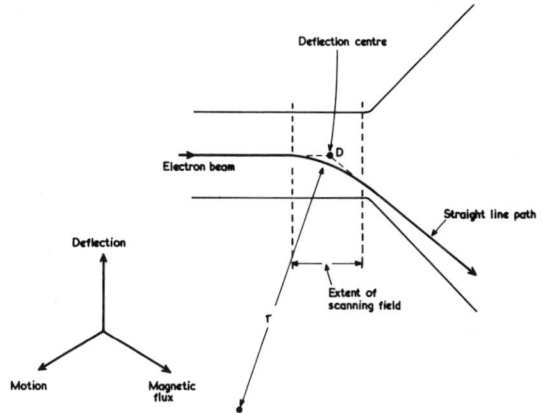

FIG. 6.6 DEFLECTING THE BEAM (MAGNETIC DEFLECTION)

beam path is tangential to the deflection curvature. The magnitude of the angle through which the beam is deflected upon the strength of the magnetic field, the time spent by the electrons in the field and the mass of the electron. With a 15 kV final anode potential and with the field extending over a distance of, say, 5 cm the time spent in the field is only about 0·6 nano-secs. Although the mass of the electron is very small, strong deflecting fields are required with m.m.f.s. of the order of 400–500 Ampere-turns. Because of the much greater mass of negative ions, they are deflected less than the electrons and thus in a non-alluminised tube they cause a burn near the tube centre.

If the deflecting field is uniform and all electrons have the same velocity, the electrons appear to have been deflected from a point D called the 'deflection centre'. With a non-uniform field the electrons will be deflected through different angles and an elliptical spot will be produced on the screen. This defect is called 'astigmatism'.

Figures 6.7 and 6.8 show the magnetic fields set up by the vertical and horizontal scanning currents fed to the coils. The direction of the resulting force acting on the beam may be found by applying Fleming's Left-Hand Rule but note that electron motion is opposite to conventional current flow. In both diagrams the electron beam is assumed to be moving out of the paper. Quite clearly, in order to produce deflection

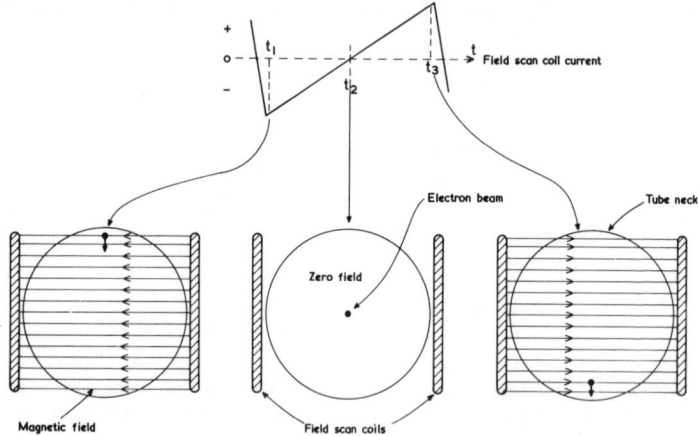

FIG. 6.7 VERTICAL DEFLECTION OF THE BEAM

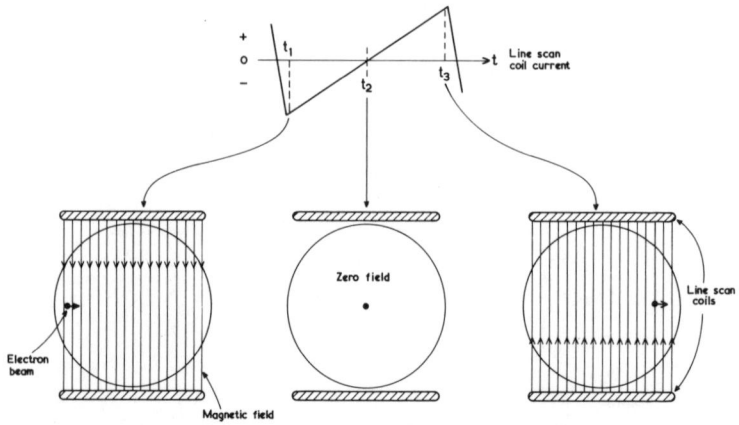

FIG. 6.8 HORIZONTAL DEFLECTION OF THE BEAM

THE MONITOR DISPLAY TUBE

left or right and up or down from the screen centre, the deflection currents must be a.c. type waveforms.

As the only useful flux produced by the scanning coils is that which passes through the tube neck, the coils are bent round the neck as in diagram (a) of figure 6.9 which shows one set of coils only as an example. At one time the coils were arranged in a

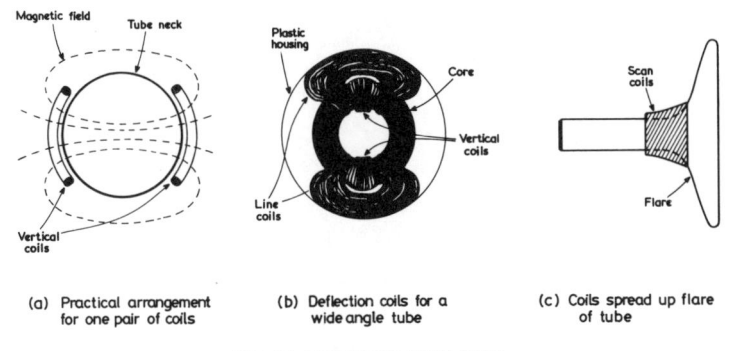

(a) Practical arrangement for one pair of coils

(b) Deflection coils for a wide angle tube

(c) Coils spread up flare of tube

FIG. 6.9 PRACTICAL SCAN COILS

castellated ferroxcube core to reduce the reluctance of the magnetic circuit and decrease the losses, thereby enabling a stronger magnetic field to be obtained for a given scanning current. With modern wide-angle tubes the castellated core is not used, the coils tend to be flatter and spread round the tube neck more as in diagram (b). The field coils are wound differently in this diagram than to those shown in (a). A ring core is used with the field coils wound over them. The two coils are connected so that the magnetic fields they set up oppose each other in the core. As a result a strong horizontal internal field is set up across the air-gap of the tube neck. Also, to reduce the possibility of 'corner-cutting', *i.e.* darkening in the corners due to the beam striking the neck of the tube at the extremities of deflection, the coils are taken up the tube flare as in diagram (c). The line coils are placed nearest the tube neck and extend further up the flare since the line coils must produce the greatest amount of deflection (width-to-height ratio = 4:3). Note that if the deflector coil assembly is not pushed home against the flare of the tube then corner cutting may occur. Also, if the coils are rotated away from their correct orientation, the raster will not be square with the sides of the tube.

The two halves of the scan coils may be arranged in series or in parallel with one another. The arrangement used depends upon the aim of the designer and to a large extent by the type of drive amplifier employed. Parallel connected coils provide a high current low impedance configuration whereas series connected coils provide a high impedance low current arrangement. Field deflector coils used in transistorised equipment generally have an inductance in the range 20–100 mH and a d.c. resistance of up to about 30 ohms. Line coils have a smaller inductance in the range 50–200μH and a d.c. resistance which is usually less than one ohm. These values depend on the line supply voltage, c.r.t. voltages and deflection angles required. With a field deflection coil of inductance 100mH and resistance 20 Ω, the ratio of

$$\frac{\text{reactance}}{\text{resistance}}$$

at field frequency is approximately 3:2 but with an inductance of 20 mH the ratio is 0·3:1. Thus, with field coils the inductive reactance or resistance may be predominant depending on the particular design. With line coils having an inductance of 50 μH and a d.c. resistance of 0·5 Ω, the ratio is approximately 10:1 (at line frequency) and with

148   INDUSTRIAL AND COMMERCIAL CCTV

an inductance of 200 μH (same resistance) approx 40:1. Thus the line coils are usually predominantly inductive.

Let us now consider the voltage waveform across the scanning coils when a linear sawtooth current flows in them. The line coils will be considered first and it will be assumed that the current passing through them is 1 A peak-to-peak and that the inductive and resistive components are as given in figure 6.10.

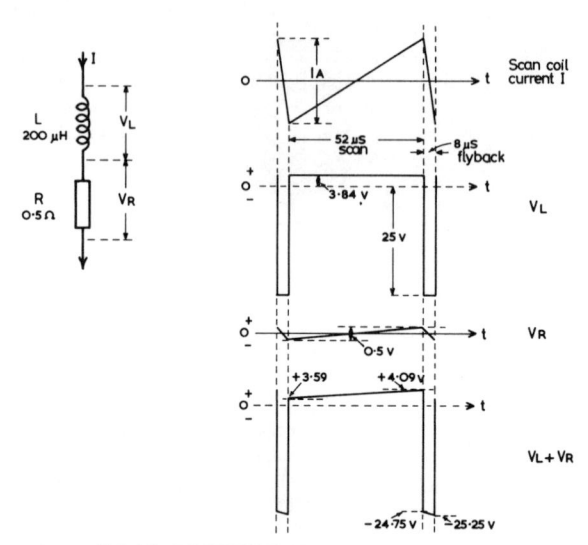

FIG. 6.10 WAVEFORMS ACROSS LINE SCAN COILS

The voltage across an inductor is given by

$$E = -L\frac{di}{dt} \text{ where } L \text{ is the inductance and}$$

$\frac{di}{dt}$ is the rate of change of current.

During the scan period (52μs) $E = \frac{200}{10^6} \times \frac{10^6}{52}$ volts

$$\simeq 3\cdot 84 \text{ volts}$$

Since $\frac{di}{dt}$ is constant over this period, the induced voltage will be constant at 3·84 V.

For the flyback period (say 8μs) $E = \frac{200}{10^6} \times \frac{10^6}{8}$ volts

$$= 25 \text{ volts.}$$

Again as $\frac{di}{dt}$ is constant the voltage across L will also be constant at 25 volts.

Now the peak-to-peak voltage across $R = 0\cdot 5 \times 1$ volts
$$= 0\cdot 5 \text{ volts and the voltage waveform will be sawtooth.}$$

THE MONITOR DISPLAY TUBE

It should of course be appreciated that we cannot measure the voltage across R as the resistance is distributed throughout the inductor and not 'lumped' as we have shown. The actual waveform across the line scan coils will therefore be $V_L + V_R$ as shown in the diagram. It will be seen that the ratio of

$$\frac{\text{flyback voltage}}{\text{scan voltage}}$$

is approximately 6·5:1. The voltage across the coils may greater than calculated during flyback as the flyback may not be linear. Over some parts of the flyback,

$$\frac{di}{dt}$$

will be greater. Also, due to the self-capacitance of the coils, the flyback pulse will be half sine and not rectangular as indicated.

Corresponding waveforms for the field coils are given in figure 6.11. The voltage values indicated are calculated for a coil inductance of 100 mH and d.c. resistance of 20

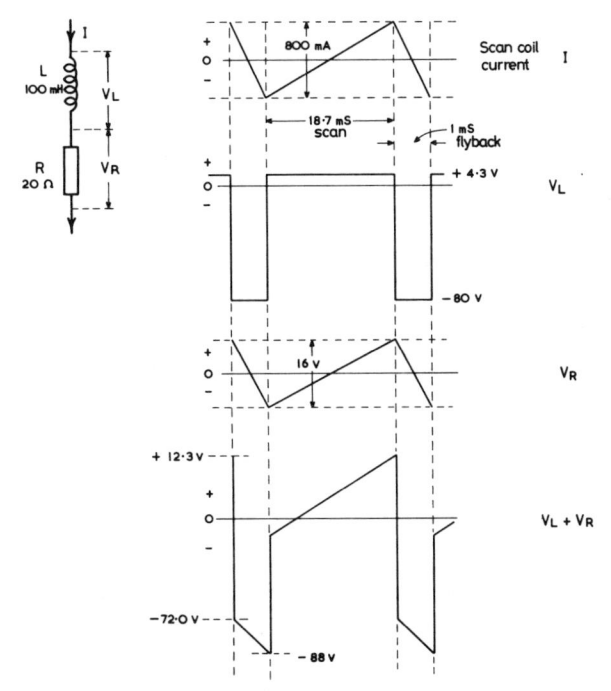

FIG. 6.11 WAVEFORMS ACROSS FIELD SCAN COILS

Ω with a peak-to-peak current of 800 mA. A scanning time of 18·7 ms has been used and a flyback time of 1 ms which is about the maximum time for the average monitor. The resultant waveform $V_L + V_R$ shows the predominating influence of the resistive component, whereas for the line coils it is the inductive component that determines the actual voltage waveshape.

150   INDUSTRIAL AND COMMERCIAL CCTV

TUBE DIMENSIONS

The important dimensions of a television tube which are referred to in technical literature are shown in figure 6.12. The screen size is measured diagonally and is given

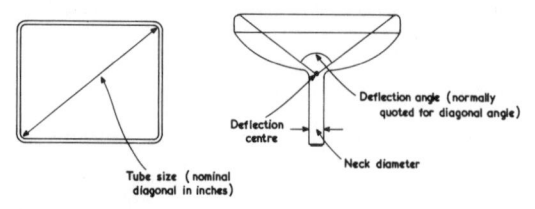

FIG. 6.12   SIGNIFICANT DIMENSIONS OF TELEVISION TUBE

in inches (or millimetres). Common screen sizes in monitors are 5″, 9″, 11″, 12″, 15″, 17″, 20″ and 24″. The screen size determines the maximum viewable distance. The maximum distance may be calculated from $d = 722AD^*$, where d is the distance from the screen to the observer, D is the diagonal measurement of the tube and A is the ratio of

$$\frac{\text{object height}}{\text{screen height}}$$

with all measurements in the same units. A table setting out the maximum distances for different tube sizes and for varying values of A are given in figure 6.13.

	1/400	1/200	1/100	1/50	1/25	1/10
5″	9″	1′ 6″	3′	6′	12′	30′ 1″
9″	1′ 4″	2′ 8″	5′ 5″	10′ 10″	21′ 8″	54′ 2″
11″	1′ 8″	3′ 4″	6′ 7″	13′ 3″	26′ 6″	66′ 2″
12″	1′ 10″	3′ 7″	7′ 3″	14′ 5″	28′ 11″	72′ 2″
15″	2′ 3″	4′ 6″	9′	18′ 1″	36′ 1″	90′ 3″
17″	2′ 7″	5′ 1″	10′ 3″	20′ 5″	40′ 11″	102′ 3″
20″	3′	6′	12′	24′	48′ 2″	120′ 4″
24″	3′ 7″	7′ 3″	14′ 5″	28′ 11″	57′ 9″	144′ 5″

D = Screen size in inches measured diagonally

$A = \dfrac{\text{Object height}}{\text{Picture height}}$

FIG. 6.13   TABLE OF MAXIMUM VIEWING DISTANCES (d) FOR VARIOUS SCREEN SIZES

In addition to the table being used for determining the maximum viewing distances it may also be used as a guide in selecting a suitable screen size for a particular application. For example, suppose that an instrument panel containing several meters which a process control operator is observing just fills the field of view of the camera. It may be vitally important that the operator can clearly discern the numerals on the dials of the instruments. Let us say that the height of the control panel is 20″ and that the smallest numeral is $\frac{1}{5}$″ in height (these dimensions can be taken from the instrument panel by direct measurement). Thus A in this case is 1/100. Now, suppose that at times the operator has to move away from the monitor screen to make adjustment on a control panel situated some 10 feet away from the monitor and that he still requires to clearly see the displayed numerals. From the table it will be noted that a suitable minimum screen size would be 17″.

The values of maximum viewing distances given the column for $A = 1/400$, which approximately represent the smallest detail of a 625-line system, show how close an

*See Appendix C, page 241.

operator must be to the monitor screen to take full advantage of the high definition a good system is capable of offering.

The deflection angle, *e.g.* 50°, 70°, 90°, 110° etc. of a c.r.t. is normally quoted for the diagonal angle. With, say, a 17″ tube using a deflection angle of 70° a certain amount of power is fed to the scanning system. If the tube size is increased to 20″ and the deflection angle kept the same, the distance between the deflection centre and the screen must be increased. If the e.h.t. is kept the same there is no need to supply the deflector coils with extra power as the deflection angle is unaltered. In practice the e.h.t. would be increased since the spot brightness is spread over a greater area hence the picture would appear dimmer with the larger screen size (although the total light output is the same as the smaller screen). An increase in e.h.t. increases the speed of the electrons, hence they spend less time in the deflecting field. Thus a larger power must be supplied to the scanning coils to maintain the same deflection angle and hence picture size. As the deflection angle increases a greater amount of power must be supplied to the scanning coils. However, with careful design it is possible to achieve large deflection angles without a great increase in scan power by reducing the neck diameter and by minimising the losses in the scan coils themselves. A decrease in the diameter of the tube neck allows the scan coils to be brought closer to the electron beam and a more intense field is produced for a given scanning current. With a small neck tube the length of the scan coils must be kept short to prevent the beam from striking the neck of the tube and causing darkening of the picture in the corners. As a short coil requires a stronger magnetic field, the coil is taken up the flare of the tube thereby eliminating the corner cutting problem. The design of neck flaring also helps in this respect.

The use of large deflection angles and eletrostatic focusing has considerably reduced the overall length of the display tube and therefore the size of the monitor cabinet. This has advantages where space is limited and in commercial applications a 'slim' monitor fits in better with contemporary furnishings.

'S' CORRECTION

When a tube with a round screen area is used as in diagram (a) of figure 6.14, where the distance between any point on the tube face to the deflection centre is the radius for

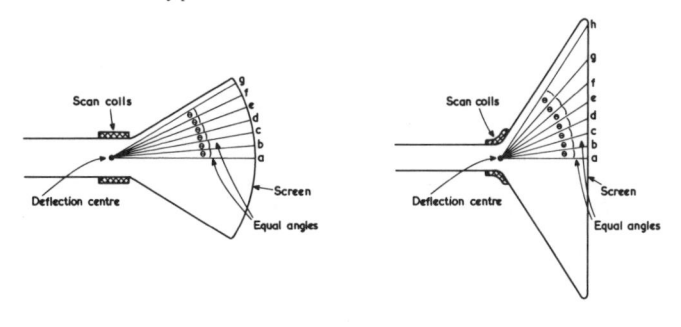

(a) Small deflection angle (round tube)
Linear display

(b) Large deflection angle (flat tube face)
Non-linear display

FIG. 6.14 DISTORTION OF PICTURE IMAGE ON FLAT SCREEN WITH LARGE DEFLECTION ANGLE

the circumference of the tube face, a linear picture will result when the scan coils are fed with linear currents. This is because a linear current produces equal increments θ in the deflection angle; thus the distances a–b, b–c, c–d, etc. on the screen are equal. The modern tendency is to use tubes with 'flatter' faces, thus the distance between the deflection centre and tube face increases with the deflection angle as in diagram (b) where the effect is exaggerated for clarity. Therefore, with equal increments in the deflection angle the distances a–b, b–c, and c–d, etc. are no longer equal and the display will be non-linear The effect becomes more pronounced as the size of the deflection angle is increased, so it is necessary to provide some form of correction with wide-angle tubes.

Correction is achieved by modifying the scanning current waveform. Figure 6.15 shows the waveform required. Clearly, the deflection angle increments must be

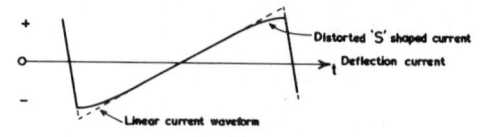

FIG. 6.15 'S' CORRECTING WAVEFORM

progressively reduced as the deflection angle increases. Thus the rate of change of current is progressively reduced towards the start and end of the scan as shown, producing an 'S' shaped waveform. To achieve this result, a capacitor of suitable value is placed in series with the scan coils. 'S' correction operates by resonance between the correcting capacitor and the scan coil inductance, the circuit being shocked into oscillation by the flyback. By arranging that the resonance is between one-third and one-half the scanning frequency, a suitable portion of the sine curve may be used. 'S' correction may be used in the vertical and horizontal scanning circuits but sometimes is only applied to the horizontal scanning current since a larger deflection angle is required for the horizontal deflection.

CENTRALISING THE PICTURE

Some means must be employed to centralise the picture on the screen due to manufacturing tolerances in the alignment of the gun assembly in the tube neck and production of the scan coil assembly. Picture centering is achieved by applying weak magnetic fields from permanent magnets fitted outside the tube neck. Commonly, the magnets are made in the form of magnetised annular rings as shown in diagram (a) of figure 6.16. The rings are magnetised in such a way so as to set up a magnetic field across the neck of the tube. Two rings are used which can be rotated independently of one another as in diagram (b). The resultant field of the two rings can be moved radially by moving the two rings together in the same direction and the strength of the field may be adjusted by moving the rings in opposite directions. Movement of the rings thus allows the electron beam to be deflected in any radial direction by a controlled amount causing an appropriate shift in the picture on the screen.

The shift magnets are mounted immediately behind the scan coil assembly and thus correct the beam position prior to its being deflected by the scanning fields.

RASTER CORRECTION MAGNETS

Shape distortion of the picture image may occur due to non-uniformity of the scanning field. One form of shape distortion arises when the magnetic field density falls off away from the deflection axis as illustrated in diagram (a) of figure 6.17. Here the flux density is greater along a line A–B (zero vertical deflection) than along a line such as C–D (vertical deflection now applied). Hence the line trace amplitude (x) at the screen centre will be larger than the trace amplitude (y) away from the screen centre. If both sets of scan coils produce flux distributions similar to that of diagram (a) the raster will exhibit outward curvature called 'barrel distortion' as in diagram (b).

Inward curvature of the raster is common with 'flat' screens, particularly when a large deflection angle is used. The reason for this may be explained by reference to diagram (a) of figure 6.18. Consider that only vertical deflection is applied to the beam so that it traces out the vertical line a–b. If the beam is now moved to the left and right of centre as it is scanning vertically, it will trace out vertical lines e-f and c-d at the screen edges. These lines are longer than at the screen centre because with a flat screen

THE MONITOR DISPLAY TUBE

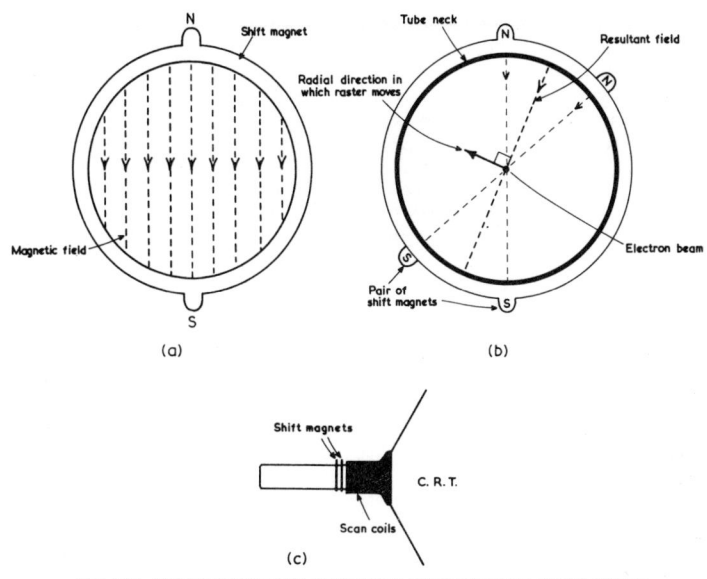

FIG. 6.16 PICTURE SHIFT MAGNETS FOR CENTRALISING THE PICTURE

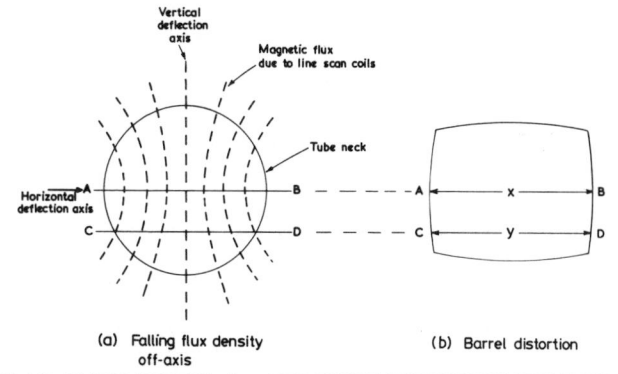

FIG. 6.17 BARREL DISTORTION CAUSED BY NON-UNIFORM DEFLECTING FIELD

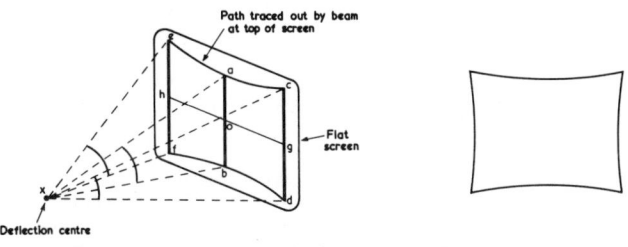

FIG. 6.18 PIN-CUSHION DISTORTION CAUSED BY EMPLOYMENT OF FLAT C.R.T. SCREEN

the distance between the deflection centre and the screen is greater at the edges than in the centre. Similarly, by considering the beam scanning a horizontal line only and then moving the beam up and down from centre, it may be shown that the vertical sides of the picture will also exhibit inward curvature. This form of picture shape distortion is called 'pin-cushion distortion,' diagram (b).

Pin-cushion distortion may offset barrel distortion (if any) produced by the deflecting field. However, either form of distortion, or their combined result, may be corrected by passing specially shaped currents through the deflector coils; this technique is used in colour monitors. In monochrome monitors correction is achieved by employing small permanent magnets mounted on the scan coil assembly as in diagram (a) of figure 6.19.

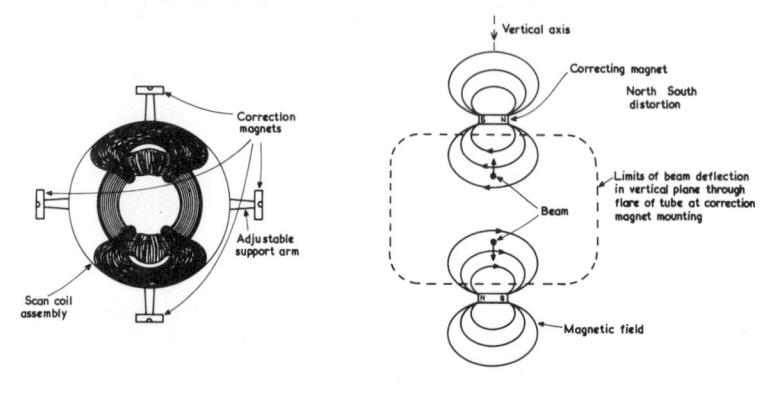

(a) Correction magnets mounted on scan coil assembly

(b) Effect of magnets on beam

FIG. 6.19 RASTER CORRECTION MAGNETS

Diagram (b) shows the effect of a pair of magnets on the electron beam assuming that the raster is exhibiting pin-cushion distortion. As the beam is traversing the screen at the top the effect of the field due to the upper magnet is to exert a force on the beam deflecting it upwards. At the bottom of the screen the field due to the lower magnet produces a force deflecting the beam downwards. The deflection forces due to the magnets are greatest around the vertical axis of the tube but diminish towards the screen centre as is required. The two magnets will thus correct for distortion in the north-south direction. If both magnets are rotated through 180°, they will correct for barrel distortion.

As the magnets are usually mounted on soft metal flanges, they can be positioned to give the correct strength of field or twisted to reverse the direction of the field.

MODULATING THE RASTER

To produce an image on the face of the display tube, the light output from the screen phosphor must vary from instant to instant as the beam scans the screen. This is done by varying the intensity of the beam with the video signal from the camera after it has received suitable amplification in the monitor.

The video signal is applied to the c.r.t. (either grid or cathode) so that the bias between grid and cathode is varied as in figure 6.20. Under NO SIGNAL conditions, the brightness control is set so that 'cut-off' voltage is applied to the gun resulting in zero beam current. The video signal, commencing at blanking level progressively reduces the bias causing the beam current to increase. Thus a high beam current corresponds to 'peak white', medium beam current to 'grey' and zero beam current to 'black'. The relationship between the grid drive voltage V_D and the light output L is not linear but

THE MONITOR DISPLAY TUBE

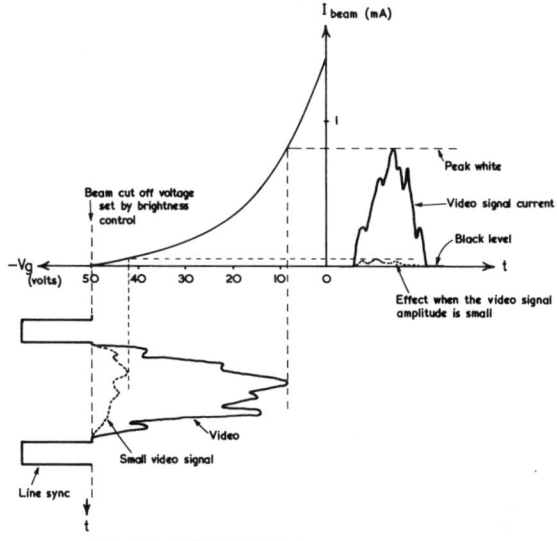

FIG. 6.20 MODULATING THE BEAM CURRENT

follows the law $L = kV_d^\gamma$ where γ is the gamma of the tube (see page 86). But this is normally corrected for at the camera.

If the contrast control is set too low, resulting in a small signal being applied to the c.r.t., there will be insufficient beam current on the peaks of the video signal and the picture will appear pale or 'thin'. The CONTRAST control thus sets the WHITE level of the picture.

The effect of maladjustment of the brightness control is shown in figure 6.21. In diagram (a) the brightness setting is too high causing a reduction in the standing bias. In consequence there is some beam current under NO SIGNAL conditions. This beam current level corresponds to black level as far as the video signal is concerned. Thus 'black' on the video signal is reproduced as a grey on the screen and the picture will be overall too light or lacking in blacks.

If the brightness control is set too low as in diagram (b), the part of the video signal excursions beyond cut-off will not produce any beam current variations. Thus, grey on the video signal is now reproduced as black and the picture will be overall too dark. The setting of the BRIGHTNESS control thus determines the BLACK level of the picture.

CATHODE AND GRID MODULATION

The beam may be intensity modulated either by applying the video signal to the grid or the cathode of the c.r.t. With grid modulation, the video signal must be positive-going so as to reduce the bias on excursions towards peak white; see diagram (b) of figure 6.22. For cathode modulation, a video signal of opposite phase must be used, *i.e.* negative-going on the video content. Note that a negative-going signal applied to the cathode of a c.r.t. (or valve) is equivalent to a positive-going signal on the grid. Hence, with cathode modulation, peak white on the signal will still reduce the bias of the tube as for grid modulation. The magnitude of the brightness control voltage in both methods must be such that the standing voltage existing between grid and cathode is sufficient to bias the gun correctly.

Whether grid or cathode modulation is used depends upon the aims of the designer. There are a number of points to be considered in assessing the relative merits of the two methods. One advantage of cathode modulation is that it is more sensitive than grid modulation. With grid modulation, the beam current is only influenced by the drive

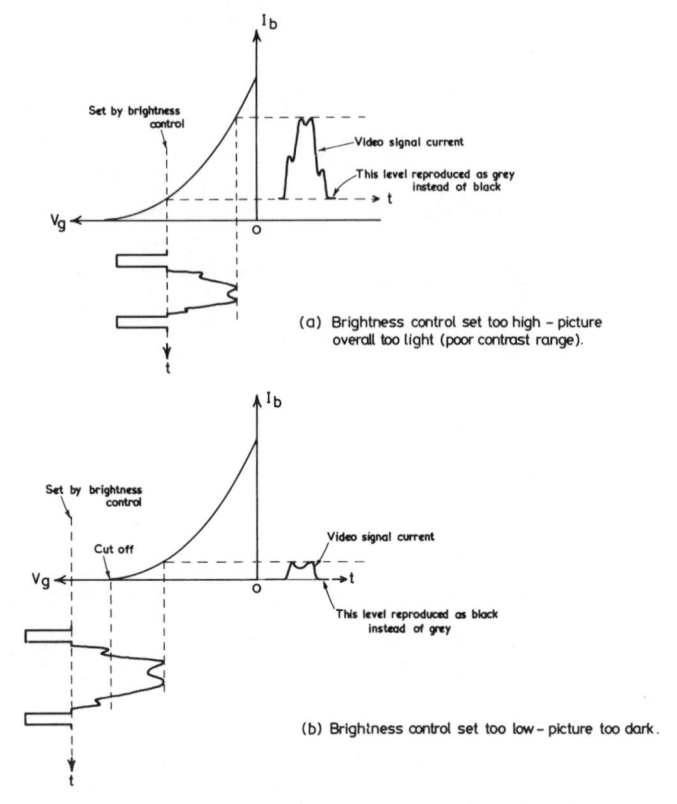

FIG. 6.21 INCORRECT SETTINGS OF BRIGHTNESS CONTROL

(a) Brightness control set too high – picture overall too light (poor contrast range).

(b) Brightness control set too low – picture too dark.

(a) Cathode modulation

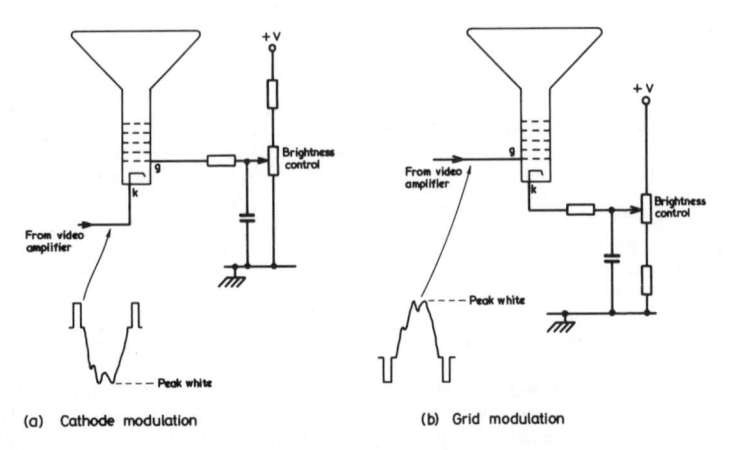

(b) Grid modulation

FIG. 6.22 GRID AND CATHODE MODULATION OF THE C.R.T.

signal applied at the grid with respect to a constant cathode potential. However, with cathode modulation, the first anode of the c.r.t. has an important influence. An increase in drive signal at the cathode relative to a steady grid potential makes the cathode more negative with respect to the first anode potential. Now the anode-to-cathode voltage has a significant effect on the beam current; thus, in addition to the increase in beam current due to the reduction in grid-to-cathode voltage there is a further increase as a result of the larger anode-to-cathode voltage.

A comparison between grid and cathode sensitivities is shown in figure 6.23. The effect of the increase in the a_1-to-cathode voltage is to increase the slope of the Ib/Vg

(a) Grid modulation (b) Cathode modulation

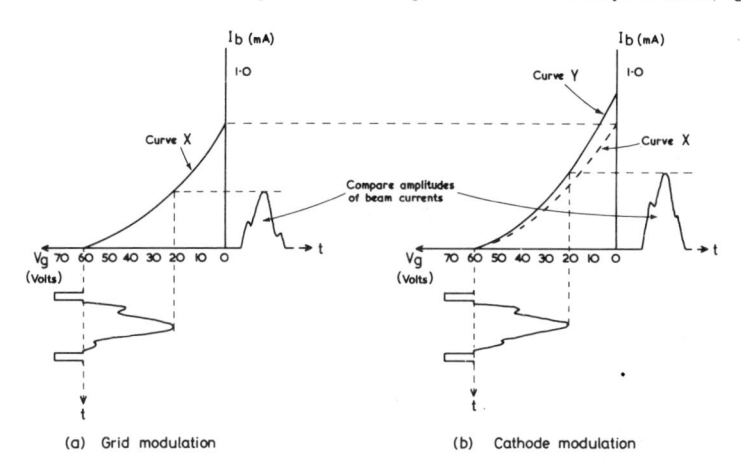

FIG. 6.23 DIAGRAMS SHOWING COMPARISON BETWEEN CATHODE AND GRID MODULATION SENSITIVITIES

curve when cathode modulation is used [curve Y of (b)]. Note that the slope of the curve progressively increases with the amplitude of the applied video signal excursions as the anode-to-cathode voltage increases. With equal drive signal amplitudes, cathode modulation is about 30% more sensitive than grid modulation, *i.e.* for a certain value of beam current less drive is required on the cathode than on the grid. This is a particular advantage when transistors are used in the video output amplifier.

ESSENTIAL SUPPLY CIRCUITS OF THE DISPLAY TUBE

Figure 6.24 shows a typical basic arrangement for the c.r.t. when cathode modulation is employed. Also included are the circuits for controlling the brightness and focus of the image.

All of the c.r.t. electrode connections are brought out to the base of the tube except for the second and fourth anodes which are internally connected to the e.h.t. connector on the flare of the tube. Spark-gaps G_1 and G_2 are included to prevent damage to the external monitor circuits in the event of a 'flash-over' within the tube. The spark-gaps are precision gaps and are usually formed in the copper-track between each electrode and chassis on the printed circuit boards associated with the supply circuits. Stand-off resistors such as R_1 and R_7 help to keep the discharge currents out of the supply circuits where they could cause permanent damage to the semiconductor devices and other components. Flash-over within the c.r.t. is unpredictable and its cause is rather obscure. It appears to be due to the formation of large voltages between the final anodes and neighbouring electrodes as a result of charge accumulation. During flash-over large currents flow between the electrodes but only last for brief periods, less than 1 μs or so.

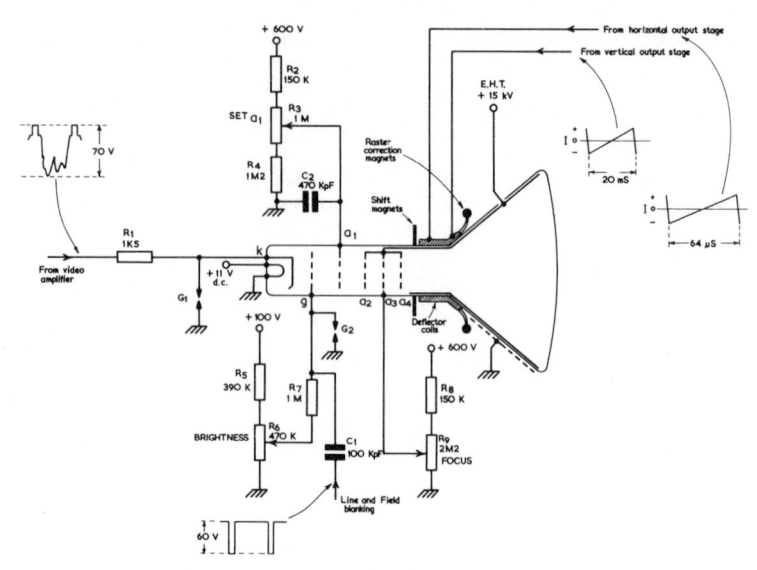

FIG. 6.24 THE DISPAY TUBE AND ITS ASSOCIATED SUPPLIES (TYPICAL ARRANGEMENT)

In some monitors the first anode potential is fixed but in others it may be adjustable so that the black level of the c.r.t. may be carefully set. This may be done after setting the brightness control to give a certain voltage difference between grid and cathode. R_3 is then adjusted so that the screen just lights up, *i.e.* the voltage set between grid and cathode corresponds to beam cut-off.

Line and field blanking pulses are fed to the grid of the c.r.t. to cut off the beam during the vertical and horizontal retrace intervals. The circuits associated with these pulses are dealt with in the next chapter.

CHAPTER 7

THE VIDEO MONITOR

A CCTV video monitor is very similar in design to a domestic t.v. receiver, except that in a video distribution system there is no need for a u.h.f. tuner, i.f. stages or vision detector which are so essential to the domestic receiver. Also, for many industrial and other usages there is no sound signal to deal with, so the audio stages encountered in the domestic set are not required. Some monitors do, however, incorporate audio stages for relaying the signal from a studio or other sound source.

Most video monitors are produced for working in different types of environment and are usually housed in metal cabinets for robustness and electrical screening. Unlike the domestic receiver, the video monitor may be in continuous use when it forms part of a CCTV system monitoring an industrial process (it may be kept switched on for 3 months or more). Thus the monitor may have to withstand the rigours of constant low or high temperature, *e.g.* 0°C to 50°C ambient and varying mains supply voltages, without noticeable deterioration of performance. Video monitors are available in a number of different screen sizes, *e.g.* 5″, 9″, 11″, 12″, 15″, 17″, 20″ or 24″ and may be free-standing or designed for panel or multi-rack mounting.

A block diagram of a typical monitor arrangement is given in figure 7.1. This schematic may be used in conjunction with the descriptions of basic circuits that follow. Some of the basic circuits in the monitor are quite similar to those in the camera and will thus be dealt with in less detail than those which are pertinent to the monitor only.

POWER SUPPLIES

Video monitors in current production feature solid-state electronics throughout (apart from the c.r.t.). Thus, only low voltage d.c. supplies are required up to about 30 V for feeding the various discrete transistor stages or integrated circuits. Higher voltages are required for the output video amplifier (around 100 V) and for the various electrodes of the c.r.t. (up to about 18 kV on the final anode) but these are normally generated by the line output stage.

A power supply, typical of modern design, is shown in figure 7.2. Here a conventional full-wave rectifier is employed comprising T_1 tapped secondary and the diodes D_1 and D_2. The mains input to T_1 is fused by F_1 (a slow-blow fuse) and filtered by L_1, L_2 and C_1 which assist in keeping r.f. spikes out of the d.c. supply. D_1 and D_2 are protected against voltage spikes by the r.f. by-pass capacitors C_2 and C_3. Particular attention may have to be given to the filtering of the mains in industrial environments where heavy electrical machinery is being switched ON and OFF because the large voltage spikes generated can have disastrous effects on the semiconductor devices.

The output of the rectifier circuit is fed to C_4 (reservoir capacitor) and then to the series regulating transistor TR_1. This transistor may be considered as a variable resistive element in series with the d.c. output of the regulator. If the d.c. output changes as a result of a mains input or load variation, the resistance of TR_1 is automatically altered thus keeping the d.c. voltage output substantially constant. Mains or load variations are sampled by the potential divider R_3, R_4 and the zener diode D_3/R_1 combination connected across the 14 V line. TR_3 amplifies the 'error' signal applied between its base and emitter and the output from the collector is used to drive the current amplifier TR_2. This transistor feeds the base of the series regulator to provide the required output correction.

C_5 aids the regulator in filtering the 14 V line and provides a low impedance feed to the line output stage. As the source of the energy generated by the line output stage comes from the 14 V rail, the potentiometer R_4 is set to give the required value of e.h.t. voltage (11·5 kV for a 11″ tube). A 12 V output is provided for feeding the video preamplifiers and this line is filtered by R_5, C_6. The pilot lamp mounted on the front

160   INDUSTRIAL AND COMMERCIAL CCTV

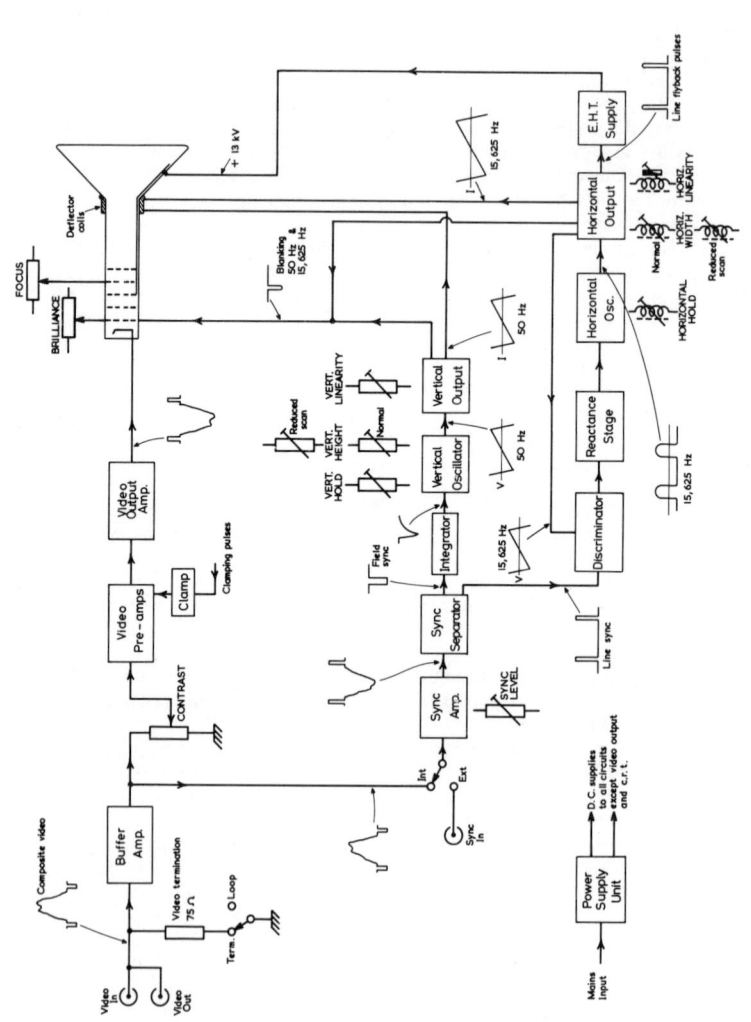

FIG. 7.1   BLOCK SCHEMATIC OF TYPICAL MONITOR ARRANGEMENT

THE VIDEO MONITOR

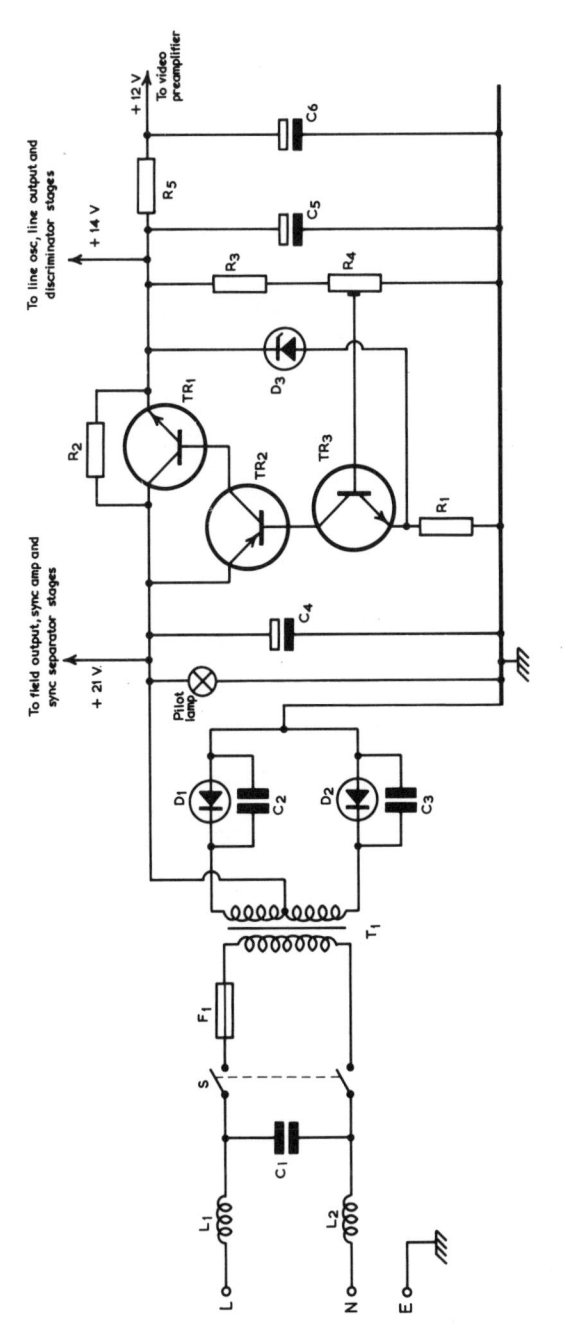

FIG. 7.2 POWER SUPPLY

162   INDUSTRIAL AND COMMERCIAL CCTV

panel of the monitor and connected across the 21 V supply line gives ON/OFF indication to the operator.

Regulated power supplies are now quite common for feeding solid-state circuits where stability of supply voltage is important. In recent years complete regulators have been fabricated in integrated circuit form with the series regulating transistor, error amplifier and voltage reference all included on a single chip. Figure 7.3 shows the circuit of a power supply featuring a full-wave bridge rectifier feeding an integrated circuit regulator. C_1 acts as the reservoir capacitor and C_2 the smoother. The electrostatic screen placed between the windings of the mains transformer helps to remove r.f. spikes from the d.c. supplies to the monitor.

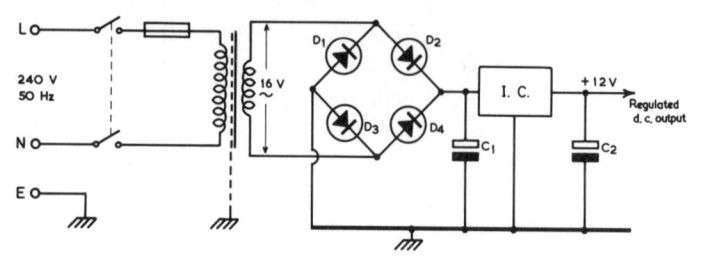

FIG. 7.3 POWER SUPPLY USING I.C. REGULATOR

VIDEO STAGES

These stages provide the necessary gain in the video signal path so that the video drive to the c.r.t. is sufficient to take the tube into full conduction on peak white. This may be of the order of 50–60 volts of video signal at the c.r.t. cathode. Thus, an overall voltage gain of between 70–85 is required when receiving a standard video signal level of 0·7 V. The requisite gain must be provided over the full bandwidth of, say, 0–10 MHz within a level variation of ± 3 dB or better.

LOOPING-THROUGH

The composite video signal input from the camera is normally fed to an emitter-follower amplifier *via* the input socket as shown in figure 7.4. This first stage in the video amplifying chain provides a high input impedance and permits 'looping-

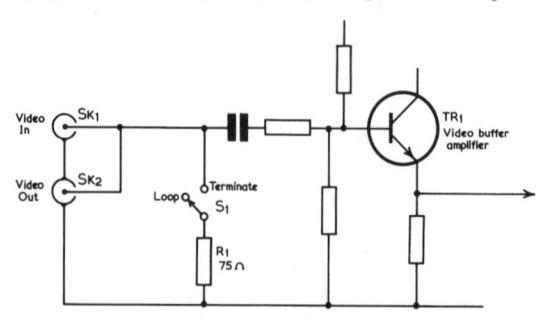

FIG. 7.4 LOOP-THROUGH INPUT PROVIDED IN SOME MONITORS

through' of the video signal. If the camera is to supply only one monitor, S_1 is set to the 'terminate' position. In this position the feed cable is terminated by R_1 which provides the correct match to the feedline (usually coaxial cable of 75 Ω characteristic impedance).

When it is necessary to supply several monitors or, say, two monitors and a video tape recorder from a common camera signal line, the looping facility may be used.

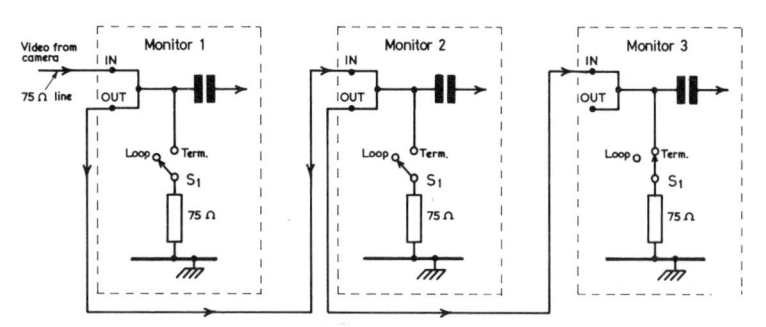

FIG. 7.5 LOOPING-THROUGH ALLOWING SEVERAL MONITORS TO BE DRIVEN FROM COMMON VIDEO SIGNAL LINE

Figure 7.5 shows the correct switch positions for S_1 when three monitors are supplied from a common line. Monitors 1 and 2 are set to 'loop', whilst monitor 3 at the end of the cable run is set to 'terminate'. The video signal now 'sees' the correct termination at the end of the feed line and all of the energy of the distributed signal is absorbed by the termination. There will be a small mismatch at the input circuit of monitors 1 and 2 as the input impedance of the emitter-follower stage is finite. Also, there is a break in the continuity of the coaxial line between the sockets on each monitor. It would be inadmissable in this arrangement to have, say, monitors 2 and 3 both terminated or monitors 1 and 3 set to loop and monitor 2 terminated. In both cases there would be a mismatch resulting in reflections travelling back down the line. This would cause standing waves to be set up resulting in larger signal amplitudes than normal, and sometimes ghosting. Because of the loss that results from the small mismatch at each of the looped inputs this method of distribution is limited in practice to about five monitors, depending upon the amount of picture deterioration that can be tolerated.

In situations where a large number of monitors are to be fed with a common camera signal a video distribution amplifier (V.D.A.) may be used. This has a single video input and commonly up to about five video outputs with each output capable of supplying five monitors, figure 7.6.

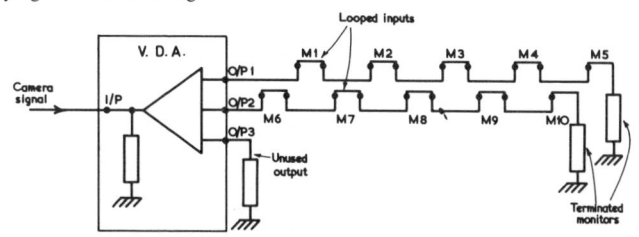

FIG. 7.6 VIDEO DISTRIBUTION AMPLIFIER WITH THREE OUTPUT LINES FOR SUPPLYING UP TO 15 MONITORS FROM COMMON CAMERA SIGNAL LINE. UNUSED OUTPUTS SHOULD BE TERMINATED AS SHOWN

AMPLIFICATION

A circuit of a video channel amplifier is shown in figure 7.7. The composite video signal with negative-going sync. is fed to the base of the buffer stage TR_1 via C_1 which blocks the d.c. component of the signal. TR_1 is connected as an emitter-follower thereby providing a high input impedance and low output impedance. R_2 and R_3 set the forward bias for TR_1, with values chosen to deal with a 1·5 V composite signal input and to maintain a high input impedance. TR_1 drives the contrast control P_1 in its emitter circuit and also supplies a composite signal to the sync. processing stages via R_5. Adjustment of P_1 slider sets the level of the signal passed on to the following stages and hence the contrast of the picture on the monitor screen. The signal, after passing

FIG. 7.7 TYPICAL VIDEO AMPLIFYING CHANNEL

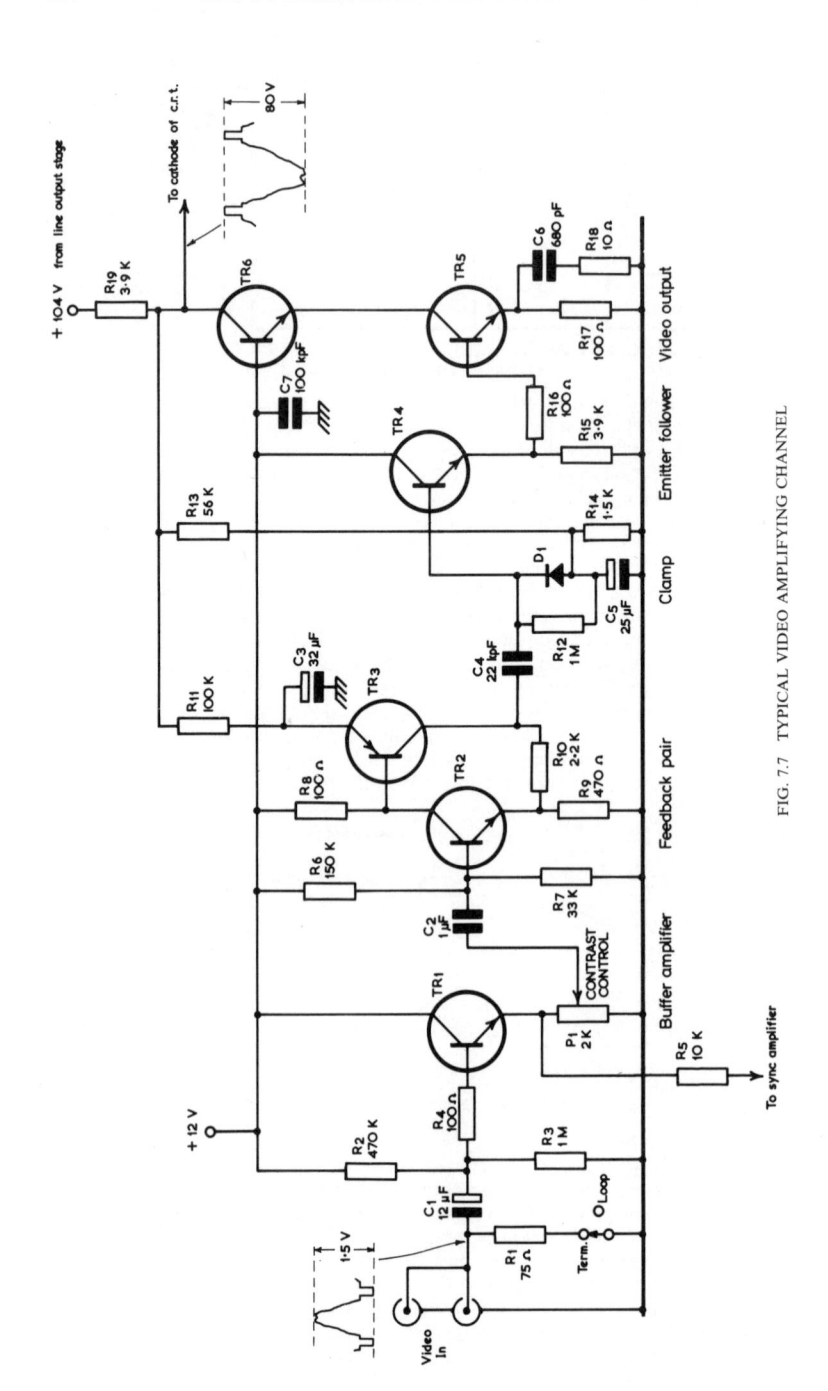

through C_2, is applied to the base of TR_2. Together with TR_3, TR_2 forms a feedback pair. The output from TR_2 emitter (from across R_9 and R_{10}) is fed to the clamp circuit but the signal from across R_8 is fed to the base of TR_3 which supplies an amplified signal to R_{10} and R_9. TR_3 acts as a variable gain controlled stage by the d.c. feedback to its emitter via R_{11} from the collector of TR_6.

TR_2 and TR_3 deliver a low impedance drive signal to the clamp diode D_1 via C_4. The diode conducts on the negative-going sync. pulse tips clamping the base of TR_4 to a d.c. level determined by the potential divider R_{13}, R_{14}. This action restores the d.c. component lost in the a.c. couplings of earlier stages and ensures a steady black level.

TR_4 is an emitter-follower which feeds the output amplifier TR_5 and TR_6 via R_{16}. The video output pair are connected in cascode with TR_5 operating in common emitter and TR_6 in common base (C_7 grounds the base to signal). TR_5 provides current drive to the emitter of TR_6 which produces the final amplification of the video signal. The drive for the c.r.t. is taken from across the collector load R_{19}. A comparatively high supply rail voltage is required by the output pair to produce the necessary drive signal amplitude. This voltage is generated in the line output stage (see later). Frequency and phase compensation of the amplifier is accomplished by negative feedback in the emitter circuit of TR_5 due to the effects of R_{17}, C_6 and R_{18}.

The use of a cascode video output stage reduced the risk of transistor failure as the comparatively high d.c. rail voltage is shared between the two transistors. Recent improvements in transistor technology allow a single transistor to be used without the risk of collector-emitter breakdown, figure 7.8. Here, the video output transistor TR_2 is driven at its base from an emitter-follower TR_1 as is normal practice. R_3 and L_2 form

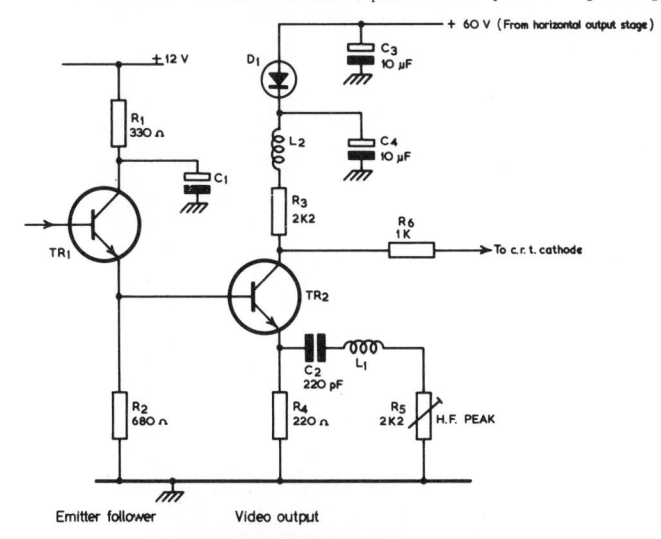

FIG. 7.8 VIDEO OUTPUT AMPLIFIER USING SINGLE TRANSISTOR

the collector load of TR_2 with L_2 (shunt peaking coil) providing h.f. lift. Variable amplitude boost in the 4–6 MHz band is provided by adjustment of R_5 in the emitter circuit. This is achieved by n.f.b. due to the series resonant circuit formed by C_2 and L_1 which sets the frequency band and R_5 which adjusts the circuit Q. As this particular monitor is intended for use as a camera viewfinder (screen size approximately 10 cm by 8 cm), the variable boost facility may be adjusted to give fine picture detail overshoot which is helpful in obtaining accurate camera optical focus despite the small screen size.

The signal output at TR_2 collector is directly coupled to the cathode of the c.r.t. via

COTRON 15″ VIDEO MONITOR (PM SERIES) FOR USE IN SECURITY SURVEILLANCE, EDUCATIONAL, BROADCAST AND DATA DISPLAY APPLICATIONS

(Photograph courtesy of Cotron Electronics Ltd., Coventry)

R_6. Diode D_1 and C_4 ensure that no spot burn can occur on the monitor screen after switching off. When the monitor is switched off, the voltage across C_4 maintains the c.r.t. cathode at high potential thereby biasing off the c.r.t. until its heaters have cooled down. D_1 becomes reverse biased at switch off (when the 60 V rail drops) thus isolating C_4 from the 60 V supply line (which would otherwise discharge C_4).

BEAM CURRENT LIMITER

Because the video output amplifier is d.c. coupled to the c.r.t., excessive beam current may flow in the c.r.t. when the d.c. at the collector of the video amplifier falls to a low value. To prevent overdissipation of the c.r.t. some manufacturers incorporate a beam current limiting circuit and one example is shown in figure 7.9.

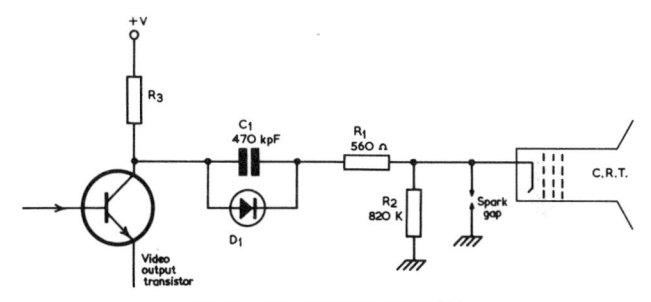

FIG. 7.9 BEAM CURRENT LIMITER

The limiter circuit consists of C_1, R_2 and D_1. Under normal signal drive conditions D_1 is conducting and the c.r.t. cathode is d.c. coupled to the collector of the video amplifier. When the average beam current exceeds a certain level, say, 100 μA, sufficient voltage drop occurs across R_2 to reverse bias D_1. This breaks the d.c. path and the signal becomes a.c. coupled by C_1. The operating point of the c.r.t. is then shifted to a safe level.

The spark-gap and R_1 prevent damage to the video circuit components in the event of a flash-over within the c.r.t.

SYNC. PULSE SEPARATION

The purpose of the sync. separator stage is (a) to remove the line and field sync. pulses from the composite video signal; and (b) to process the line and field sync. pulses to make them suitable for synchronizing the horizontal and vertical oscillators, see figure 7.10. As far as possible the operation of the sync. separator should be immune from changes in the signal level and from the effects of noise pulses.

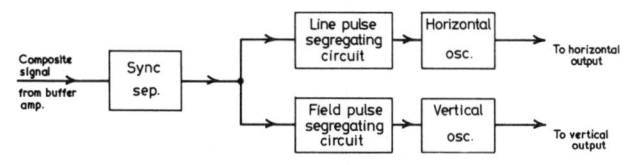

FIG. 7.10 BASIC PROCESSES

A basic circuit of a sync. separator stage is given in figure 7.11. Separation of the sync. pulses from the video content is merely a question of biasing the transistor so that it is cut off during video but conducting during the sync. pulses. The composite video input signal to the circuit may or may not have a d.c. component present, depending upon the point at which the signal is taken out from the video channel and the method of coupling between stages. We will assume that no d.c. component is present which is usually the case. The sync. pulses will normally be negative-going if the input waveform

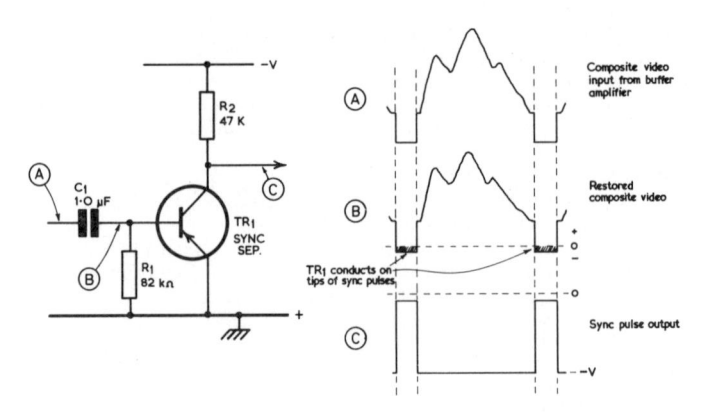

FIG. 7.11 BASIC CIRCUIT OF SYNC. SEPARATOR AND WAVEFORMS

is taken from the emitter of the buffer stage in the video channel. In such circumstances a PNP transistor is required as shown (assuming a sync. amplifier is not employed).

To separate sync. from video a fixed reference level is required so that, say, voltages above this level may be blocked but those below the reference level may be passed. This is done by using a d.c. restorer circuit. In figure 7.11 C_1, R_1 and the base-emitter junction of TR_1 form a negative d.c. restorer with the tips of the sync. pulses restored to a level which is negative of zero and of magnitude below zero equal to the forward voltage drop of the base-emitter junction. Thus, the transistor is made to conduct during the sync. pulses but cut off during the video content. Therefore, amplified sync. pulses are developed across R_2 in the collector circuit. By suitable choice of R_2 value the transistor is caused to 'bottom' during the period of the sync. pulses thereby ensuring clean output pulses, *i.e.* free from the effects of noise. In practical circuits some fixed bias may be provided for the transistor in addition to the self-adjusting bias resulting from the action of the d.c. restorer. This ensures that the transistor is driven well into 'bottoming' and gives noise-free pulses at its output.

Figure 7.12 shows a practical circuit using a PNP transistor with some fixed bias provided by the base potential divider R_1, R_2 and R_3. The basic d.c. restorer action is provided by R_1, C_1 and the base-emitter junction of the transistor. Clean output pulses of about 10 V amplitude are obtained with a standard composite signal input.

In some monitors, the sync. separator stage is preceded by a stage of sync. amplification and selection, see figure 7.13. Here, composite video or composite sync.

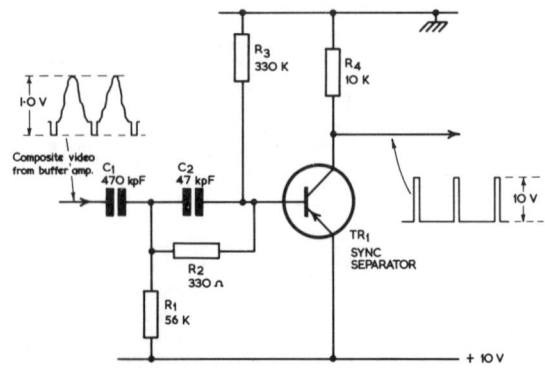

FIG. 7.12 PRACTICAL SYNC. SEPARATOR STAGE

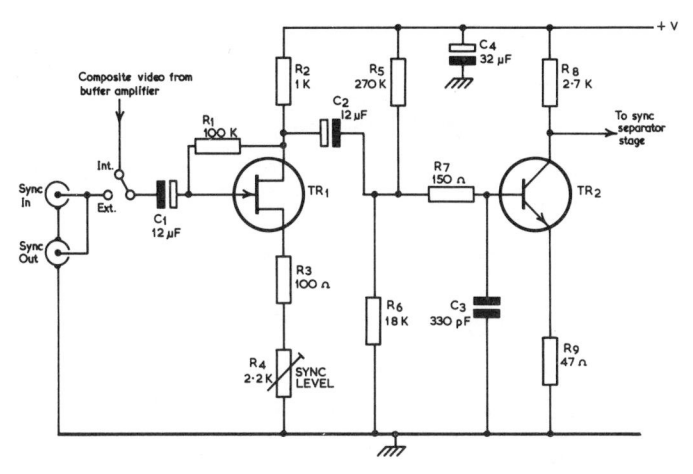

FIG. 7.13 SYNC. AMPLIFIER STAGES WITH LOOPED EXTERNAL SYNC. INPUT

(supplied to the monitor on a separate feed line) may be selected by S_1. As will be noticed, the sync. may be looped to other monitors, allowing several monitors to be supplied from a common sync. feed line.

Composite sync. or composite video is amplified in the f.e.t. stage TR_1. The sync. level control R_4 adjusts the gain of TR_1 by varying the gate bias and the amount of feedback in the gate-to-source circuit. This control, which may be accessable from the front panel of the monitor, is adjusted on extreme input levels to avoid video contamination in the following sync. processing circuitry. The amplified composite video signal (S_1 set to INT) from across R_2 is passed to the base of TR_2 via C_2. R_7 and C_3 form a low-pass filter limiting the bandwidth of the signal passed to TR_2 to improve the noise immunity. TR_2 provides further amplification and re-inversion of the signal prior to its passing to the sync. separator stage.

SEGREGATION OF HORIZONTAL AND VERTICAL SYNC.

(i) Vertical Sync.

All sync. pulses are of the same amplitude in the composite signal arrival from the camera; the difference between line and field sync. is one of time duration only. It is this time difference that enables the pulses to be segregated from one another. Essentially, the time difference is converted into an amplitude difference which makes it easy to ensure complete separation.

For extracting the vertical (field) pulses from the sync. pulse train use is commonly made of an integrating circuit, see diagram (a) of figure 7.14. This CR combination has a time-constant [t(secs) $= C$(farads) $\times R$(ohms)] which is generally about 30 μs. The value is not critical and variations will be met with in practice.

The composite sync. waveform after the video information has been removed in the sync. separator stage is applied to the integrator circuit C_1, R_1. We will assume that the sync. waveform is positive-going and 10 V in amplitude as in diagram (b). Consider the effect on the first line pulse shown, pulse a. During the period of the pulse a current will flow in R_1 charging C_1. If the time-constant of C_1, R_1 is 30 μs the voltage across the capacitor will only rise a little as the pulse is only 4·7 μs in duration. Between pulse a and pulse b (a time interval of 59·3 μs) the capacitor will discharge through R_1. When pulse b arrives the capacitor will attempt to charge once again but as each equalizing pulse is only 2·3 μs the charge obtained will be smaller than for a line pulse. Between pulse b and pulse c the capacitor will discharge once more. After a period of five equalizing pulses, the voltage across C_1 tends to stabilize at a low voltage of about 1 V

170 INDUSTRIAL AND COMMERCIAL CCTV

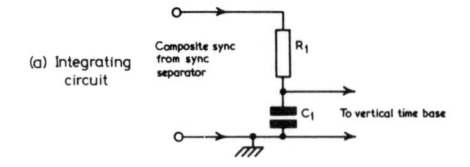

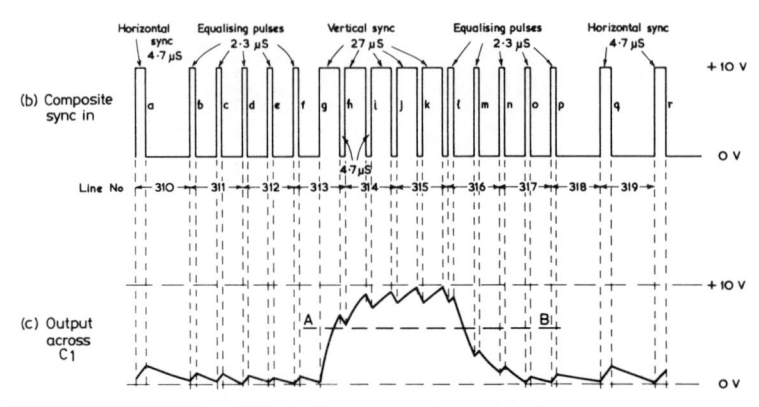

FIG. 7.14 SHOWING HOW AN INTEGRATOR CIRCUIT PRODUCES AN OUTPUT LOCKING PULSE WHEN FED WITH SYNC. PULSE TRAIN (AT END OF EVEN FIELD)

as the time between pulses of 29·7 μs is approximately thirteen times the equalizing pulse width.

When the vertical pulse g arrives, C_1 will charge to a much higher voltage (approximately 6 V) as the pulse width is almost equal to the time-constant of the circuit. Between vertical pulses the capacitor will discharge but only by a small amount as the interval is only 4·7 μs. Subsequent vertical pulses h, i, j and k will cause the voltage to build up as shown in diagram (c). Following the vertical sync. pulse period the intervals between the equalizing pulses will allow the voltage across C_1 to fall rapidly as it discharges. Each equalizing pulse has the effect of delaying the fall of voltage as C_1 charges a little during each pulse.

Thus the effect of the integrating circuit on the sync. pulse train is to produce a large voltage output during the vertical sync. interval but only a small voltage output due to the effects of the horizontal and equalizing pulses. The actual voltage across the integrating capacitor* after the application of each pulse may be found from

$$V_C = V_1 + V_D \left(1 - e^{\frac{-t_1}{CR}}\right)$$

where
V_1 = voltage across capacitor prior to start of charge
V_D = the VOLTAGE DIFFERENCE between the pulse amplitude and V_1
t_1 = pulse duration (secs)
CR = integrating circuit time-constant (secs)

*See Appendix A, page 225.

THE VIDEO MONITOR

Also, by using

$$V_C = V_2 \left(e^{\frac{-t_2}{CR}} \right),$$

the voltage across the capacitor at the end of the discharge period may be arrived at,

where V_2 = voltage across capacitor prior to start of discharge
         t_2 = interval between pulses.

The output from the integrator may be fed to a limiter to remove the effects of the line pulses by clipping off the waveform below a level such as $A–B$ in diagram (c) prior to the pulse being applied to the vertical oscillator. The purpose of the equalizing pulses can be seen by reference to figure 7.15 which shows the integrator output on odd and even fields. At the start of the equalizing pulse period there is a discrepancy between the

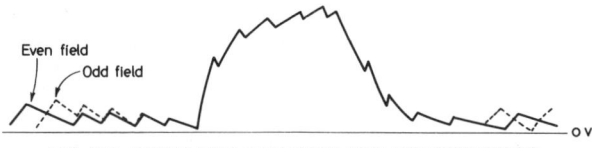

FIG. 7.15   INTEGRATOR OUTPUT ON ODD AND EVEN FIELDS

two output waveforms due to the 'half-line difference'. After a period corresponding to two and a half lines of equalizing pulses the difference disappears and the pulses become coincident.

In many CCTV cameras only a single field synchronizing pulse is generated as was mentioned in Chapter 3. As shown in figure 7.16, the field pulse commonly lasts for about 100–160 μs. The diagram shows the output of an integrating network employing

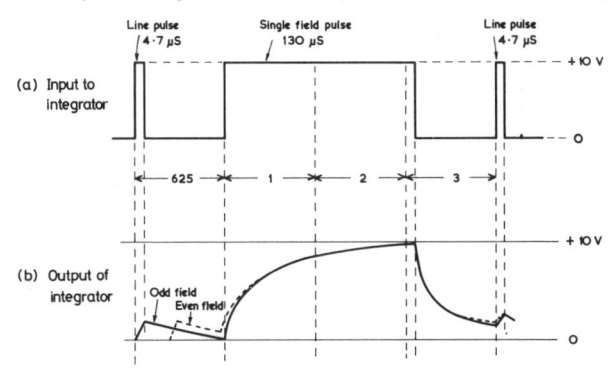

FIG. 7.16   INTEGRATOR WAVEFORM WHEN SINGLE FIELD PULSE IS USED
(TIME-CONSTANT = 30 μs)

a time-constant of 30 μs for both odd and even fields. Because of the absence of equalizing pulses there is now a discrepancy in the output pulse of the integrator on different fields. This small timing error can give rise to errors in interlace, but the error depends upon the level chosen on the leading edge of the pulse to trigger the field timebase oscillator and on the integrating network used in a particular monitor.

A practical integrating network is shown in figure 7.17. Here a diode D_1 is included in the circuit to improve the interlace as this monitor is designed to work without equalizing pulses. The composite sync. waveform from the sync. separator emitter-follower TR_1 is applied to an integrating network R_2, C_1 which produces a voltage

172   INDUSTRIAL AND COMMERCIAL CCTV

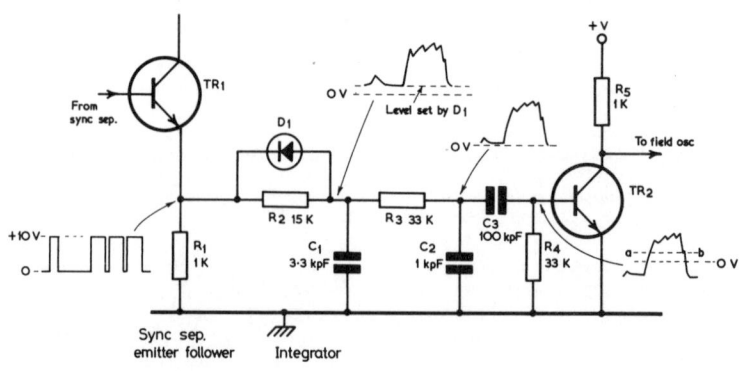

FIG. 7.17   PRACTICAL INTEGRATOR NETWORK INCLUDING AN INTERLACE DIODE

waveform across C_1 as shown. During the period of a pulse when C_1 is charging, D_1 is reverse biased. However, during the interim period between line pulses, D_1 becomes forward biased by the voltage across C_1. Thus, D_1 conducts and C_1 rapidly discharges via D_1 and R_1 to a voltage level set by D_1 (germanium diode). This action ensures a similar starting voltage across C_1 prior to the arrival of the field pulse(s) on odd and even fields. R_3, C_2 provides further integration of the waveform thus reducing the effects of the line pulses. The waveform across C_2 is passed via C_3 and R_4 to the base of TR_2. C_3 blocks the d.c. component and at a suitable level up the pulse such as a-b, TR_2 conducts. Thus negative-going vertical locking pulses are produced at TR_2 collector for feeding to the vertical oscillator.

A different arrangement for cancelling out the effects of the line pulses is shown in figure 7.18. Positive-going composite sync. is fed to TR_1 via C_1 and the integrating network comprising D_1, R_2, C_2 and R_3. Under NO-SIGNAL conditions D_1 is conducting,

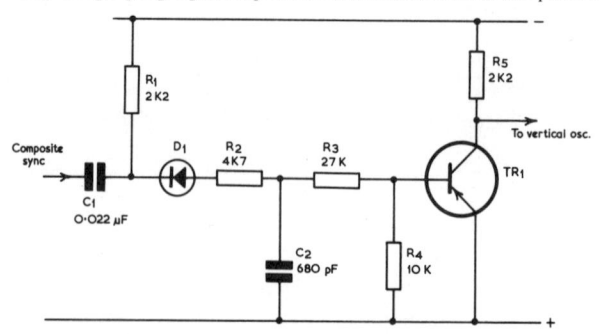

FIG. 7.18   A DIFFERENT ARRANGEMENT FOR REMOVING THE EFFECTS OF THE LINE PULSES

C_2 is charged and TR_1 held in saturation by the forward bias from across R_4. When a horizontal sync. pulse arrives at D_1 cathode, the diode cuts off and C_2 discharges via the base-emitter circuit of TR_1. However, C_2 provides the base current of TR_1 long enough to prevent the transistor from switching off during the period of a line pulse. When a field pulse is applied to D_1, the diode is held off long enough for C_2 to discharge and TR_1 switches OFF. The action of switching OFF TR_1 triggers the vertical oscillator.

(ii) **Horizontal Sync.**

Monitors of modern design invariably use a phase locked loop or 'flywheel' sync. for the horizontal timebase in common with present-day domestic t.v. receivers. At one time 'direct' sync. was used in domestic receivers where each line sync. pulse was used to synchronize the line timebase directly. This method suffers from the disadvantage that if a line sync. pulse is missing or mutilated by noise or interference, the synchronism is upset. Also, with 'direct' sync. the pulses must have a constant timing relationship to avoid horizontal displacement of the picture.

With flywheel sync. the system does not respond to each individual pulse but rather to a succession of line pulses. Thus, if a pulse is missing, mutilated by noise or incorrectly timed, the synchronizm is unaffected. This is because the system has momentum like a flywheel and keeps the horizontal oscillator synchronized at the correct frequency even if the line sync. pulses are missing for a few lines.

The ability of flywheel sync. to combat the effects of interference is useful in industrial CCTV where there is heavy electrical interference. Also, since flywheel sync. is able to counteract the effects of line pulse timing deviations, this method of synchronization is essential in a monitor which is to be coupled with a helical-scan v.t.r. or cheap camera.

BASIC IDEAS

Flywheel sync. systems are based on the phase-lock loop principle which is shown in block schematic form in figure 7.19. In the phase detector, synchronizing pulses of fundamental frequency f_s are compared in frequency and phase with a reference signal

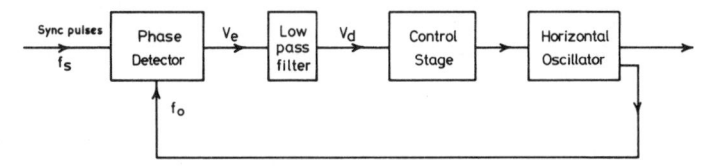

FIG. 7.19 PRINCIPLE OF PHASE-LOCK LOOP

from the horizontal timebase oscillator of fundamental frequency f_o. If there is a frequency or phase difference between the two inputs, the phase detector generates a correcting signal V_e which is fed to the control stage *via* a low pass filter. The filter is an *RC* network with values chosen so that only low frequencies and d.c. are passed on to the control stage. The control stage, which is usually of the variable reactance type, uses the output of the filter to correct the frequency or phase of the horizontal oscillator in such a direction as to reduce the initial error.

With no sync. input to the phase detector, the error voltage V_d is equal to zero and the horizontal oscillator operates at its free-running frequency. When sync. is applied the phase detector compares the frequency and phase of the sync. pulses with that of the oscillator and generates a correcting signal V_e which is related to the frequency and phase difference of the two signals. V_e is then filtered and the resulting output V_d is applied to the control stage which alters the frequency of the oscillator. The voltage V_d forces the oscillation to vary in a direction to reduce the frequency difference between f_s and f_o. If f_o is close to the input sync. f_s, V_d causes the oscillator to lock with the sync. except for a finite phase difference. This net phase difference is necessary to generate a correcting voltage V_d to shift the oscillation from its free-running frequency f_o to the sync. frequency f_s.

LOCK AND CAPTURE

Suppose that f_o is not at the desired frequency f_s when the sync. pulses are applied. The phase detector mixes the two inputs to produce sum-and-difference frequencies. The low pass filter will remove $f_s + f_o$ and if $f_s - f_o$ falls outside the band edge of the filter it, too, will be removed and no information will be sent to the control stage, *i.e.* the horizontal oscillator will remain at its free-running frequency. As f_o approaches f_s $f_s - f_o$ decreases and when it falls within the bandpass of the filter it will be passed to the control stage. The effect of the 'beat frequency' applied to the control stage will be to cause frequency modulation of the horizontal oscillator. This produces a small d.c. output from the phase detector which pulls the oscillator into synchronism.

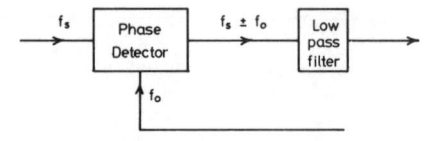

CAPTURE (PULL-IN) RANGE
This is the frequency range centred about the oscillator free-running frequency over which the loop can obtain lock with the input sync. It depends primarily upon the band edge of the low pass filter.

LOCK (HOLD-IN) RANGE
Once in lock, the oscillator will track f_s. There is now d.c. output from the phase detector. The lock range is the frequency range over which the oscillator will track f_s centred on the free-running frequency of the oscillator. It is limited by the range of error voltage that can be generated and the corresponding frequency deviation produced in the oscillator. It is essentially a d.c. parameter and is not dependent upon the low pass filter characteristics. The capture range is never greater than the lock range.

Typically the capture range may be ± 300 Hz and the lock range ± 900 Hz centred on the nominal line frequency of 15,625 Hz.

PHASE DETECTOR AND LOW PASS FILTER
A common arrangement is shown in figure 7.21. TR_1 is a phase-splitter stage fed with positive-going line sync. pulses at its base. Forward bias for the stage is provided

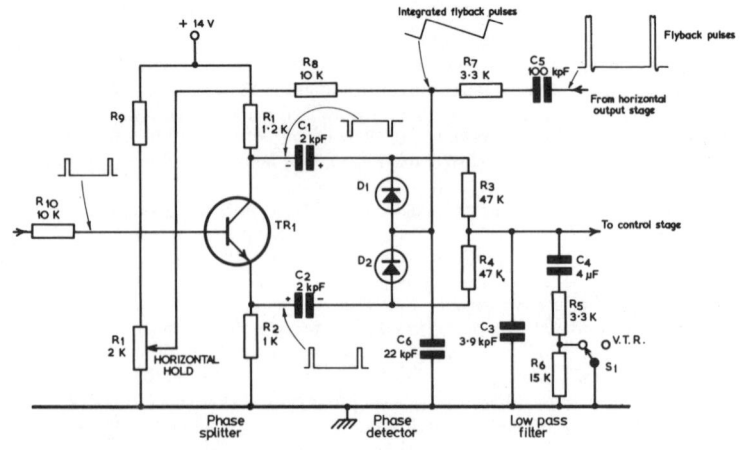

FIG. 7.21 DIODE PHASE DETECTOR

in the d.c. coupling from the previous stage. Anti-phase sync. pulses are produced at the collector and emitter and these are fed to the phase detector diodes D_1 and D_2 via C_1 and C_2. The pulses cause the diodes to conduct and to charge C_1 and C_2 equally with the polarity as shown. Between pulses the capacitors slowly discharge. With equal value resistors R_3 and R_4 the resultant voltage at the junction of these resistors will be zero (R_1 slider set to zero).

Horizontal flyback pulses from the horizontal scan output stage are used as the reference signal. The d.c. component of these pulses is blocked by C_5 and the pulses are then integrated by R_7, C_6. Thus, at the junction of the two diodes a sawtooth voltage waveform is produced having a positive-going flyback section sitting about zero voltage.

SYNC. PULSE AND SAWTOOTH IN PHASE

When the sync. pulse period occurs as the sawtooth waveform is passing through zero as in figure 7.22, the sync. pulse and sawtooth are in phase as regards the circuit operation which is desired. To determine the effect of the sync. pulse and sawtooth

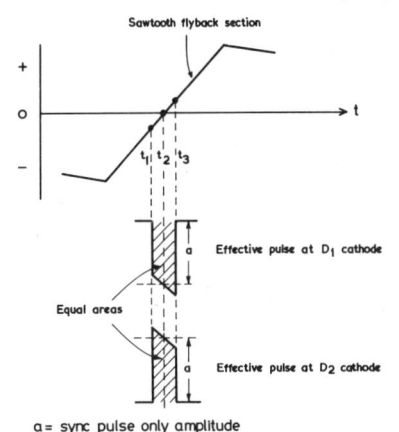

FIG. 7.22 SYNC. PULSE AND SAWTOOTH IN PHASE (ZERO OUTPUT FROM PHASE DETECTOR)

voltages applied to each diode one may consider the effective voltage applied to either the cathode or anode of each diode. For example, when the sync. pulse is commencing at D_1 cathode, the sawtooth is at a negative voltage on D_1 anode (instant t_1). This effectively reduces the conduction through D_1 which is equivalent to reducing the pulse amplitude at D_1 cathode. As far as D_2 is concerned at this instant, the negative voltage on its cathode due to the sawtooth increases conduction which is equivalent to increasing the pulse amplitude at D_2 anode. A little later at instant t_2 the sawtooth is zero and the effective voltage applied to each diode is the same and is that due to the sync. only. At instant t_3, when the sawtooth is positive, the effect at D_1 anode is to increase conduction whilst at D_2 cathode it decreases conduction. Accordingly, the effective pulse voltage at D_1 cathode is increased and that at D_2 anode is decreased. However, both pulses have equal areas thus C_1 and C_2 will receive equal charges and the resultant voltage at the junction of R_3, R_4 will be zero.

SYNC. PULSE AND SAWTOOTH OUT OF PHASE

If the sync. pulse period occurs when the sawtooth is going positive as in diagram (i) of figure 7.23 a larger effective pulse is applied to D_1 than to D_2. Thus, D_1 conducts harder than D_2, and C_1 receives the greater charge. This causes a net negative d.c. voltage to appear at the junction of R_3, R_4. If, on the other hand, the sync. pulse period corresponds to a time when the sawtooth is going negative as in diagram (ii), D_2 conducts harder than D_1 and C_2 receives the greater charge. As a result a net positive d.c. voltage is obtained at the junction of R_3, R_4.

Thus in both cases where there is an error in phase, an error voltage is obtained at the output which has an amplitude proportional to the initial phase error and a polarity depending on whether the sync. is early or late compared with the zero crossing of the sawtooth.

Although we have been talking in terms of phase errors, a small frequency error between the sync. and reference sawtooth results in a similar effect and this is shown in

176 INDUSTRIAL AND COMMERCIAL CCTV

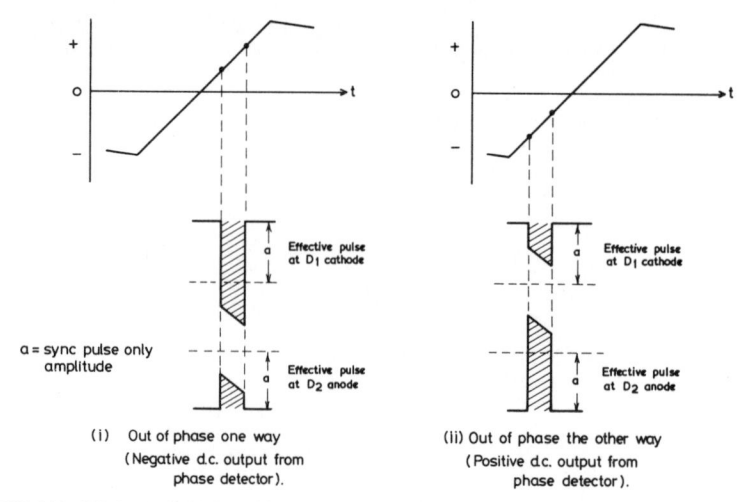

FIG. 7.23   SYNC. PULSE AND SAWTOOTH OUT OF PHASE (OUTPUT VOLTAGE OBTAINED FROM PHASE DETECTOR)

(i) Out of phase one way (Negative d.c. output from phase detector).

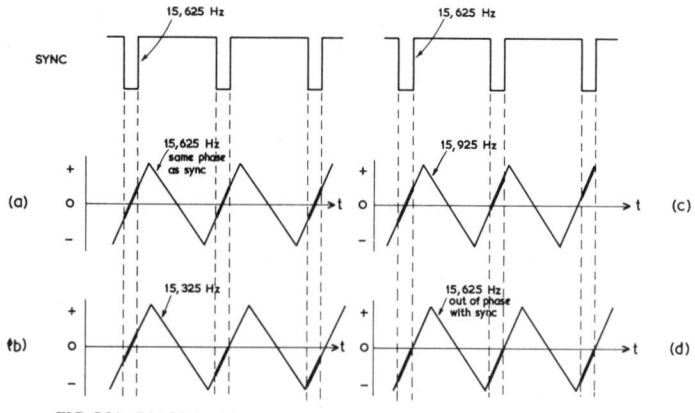

(ii) Out of phase the other way (Positive d.c. output from phase detector).

a = sync pulse only amplitude

FIG. 7.24   DIAGRAMS SHOWING FREQUENCY AND PHASE SHIFT ERRORS

figure 7.24. In diagram (a) the sawtooth has the same frequency and phase as the sync. which is assumed constant at 15,625 Hz. In diagram (b) the sawtooth frequency has been reduced by 300 Hz. Note now that the sync. pulse period corresponds to the negative part of the flyback period on the second and third flyback sections. The effect of increasing the frequency of the sawtooth by 300 Hz is shown in (c). Here the period of the sync. pulse corresponds to the positive part of the flyback period on the second and third flyback sections. In diagram (d) the sawtooth has the same frequency as the sync. but is slightly out of phase, *i.e.* the sync. period lines up with a negative part of the sawtooth flyback section.

HORIZONTAL HOLD

In the description of the circuit operation it has been assumed that the slider of the horizontal hold control had been set to zero so that the 'error' voltage was able to swing positive or negative with respect to zero. To allow the horizontal oscillator to be set to 15,625 Hz, R_1 is included. This control feeds a positive potential to the junction of the diodes and hence to the junction of R_3, R_4. By varying R_1 the positive d.c. fed to the control stage alters the frequency of the horizontal oscillator so that it may be set to the nominal horizontal frequency. This permits symmetrical operation of the circuit about 15,625 Hz. Thus, with sync. applied to the circuit the error voltage will now swing positive or negative with respect to the horizontal hold potential.

It will be noted that during adjustment of R_1 within the lock of the flywheel circuit, the visible effect on the monitor screen is to cause the picture to shift horizontally to the left and right. *i.e.* it operates like a picture shift control. It should not be used as such, as one is liable to set R_1 towards the end of the locking range and the circuit may be unable to cope with subsequent drift.

LOW PASS FILTER

R_3, R_4 and C_3 form the low pass filter. It is this time-constant which gives the arrangement its 'flywheel effect'. The longer the time-constant the more effective the filter is in smoothing out interference pulses. If it is too long however, the capture range is reduced which is important at switch-on or when changing cameras and can introduce severe problems with Helical VTRs (tension errors). Thus, a compromise value must be used. The series network C_4 and R_5 form a damping circuit to prevent ringing of the control voltage which may occur during the vertical blanking period when the normal sequence of line pulses is broken. The values of C_4, R_5 are usually quite critical. S_1 changes the damping network when operating the monitor in conjunction with a v.t.r.

CONTROL STAGE AND HORIZONTAL OSCILLATOR

A sine wave oscillator is generally used as the horizontal timebase generator since this type of oscillator has better frequency stability than a relaxation oscillator. An example is given in figure 7.25. In this monitor a Hartley oscillator is employed consisting of TR_2 and the tapped inductor L_1 together with its associated tuning capacitance. C_3 forms the main part of the tuning capacitance. Feedback to the base is via C_4 and D_1. By overdriving the transistor an approximate square wave is obtained at TR_2 collector which after suitable shaping is eventually used to switch the horizontal output transistor.

The frequency and phase of the oscillator is controlled by a reactance transistor stage TR_1. Part of the oscillatory voltage v_o is fed via an RC network C_3 and R_2 to the emitter of TR_1. The transistor and RC network present a capacitive reactance to the tuning inductor L_1. However, the transistor *itself* exhibits an inductive reactance in parallel with C_3. By varying the *gm* of the transistor with the error voltage from the phase detector, the effective inductance of TR_1 may be altered. Thus, it may be considered that the effective inductive reactance of TR_1 partially offsets the capacitive reactance of C_3, hence varying the effective capacitance 'seen' by L_1. The theory of the reactance stage is given in Appendix B. The magnitude of the effective capacitance seen by L_1 is determined by the d.c. error voltage and the setting of the line hold control. R_6 is the line hold control which feeds a variable d.c. to the phase detector together with a reference pulse derived from a winding on the horizontal output transformer.

An alternative form of control that is sometimes used is given in figure 7.26. Once again a Hartley oscillator is used with TR_2 being the oscillator transistor and L_1, C_2 forming the main tank circuit components. Feedback to the tank circuit is from TR_2 emitter which sustains the oscillation. Starting bias for the oscillator is provided from the potential divider R_6, R_5 via D_2. This stage drives the horizontal output transistor by way of a driver transformer T_1.

The frequency of the horizontal oscillator is kept in step with the line synchronizing pulses by the varactor diode D_1. This diode is supplied with reverse voltage from TR_1 collector via R_4. TR_1 amplifies and inverts the 'error' voltage from the phase detector, thus a unidirectional correcting voltage is fed to D_1. Therefore, as a result of a change

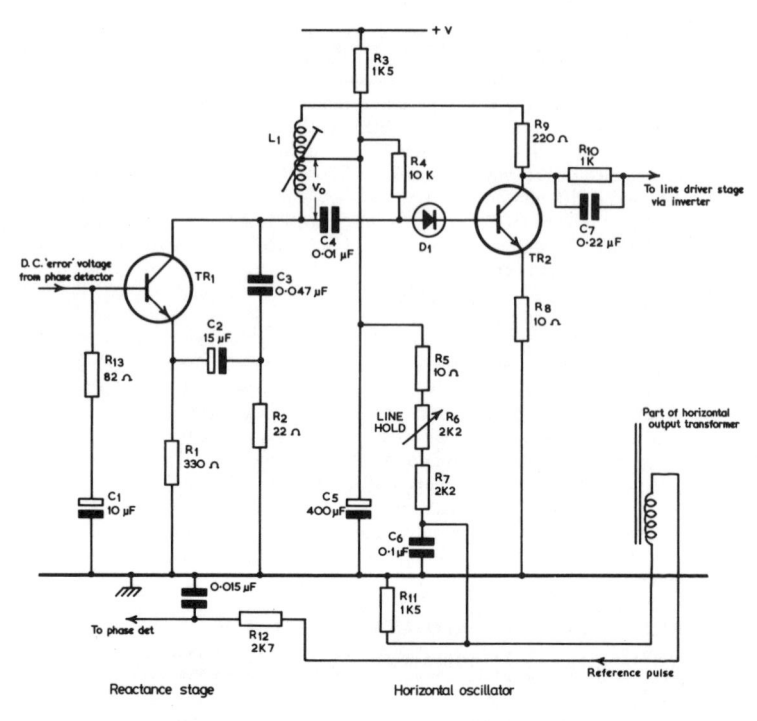

FIG. 7.25 USE OF REACTANCE TRANSISTOR STAGE AND SINE-WAVE HORIZONTAL OSCILLATOR

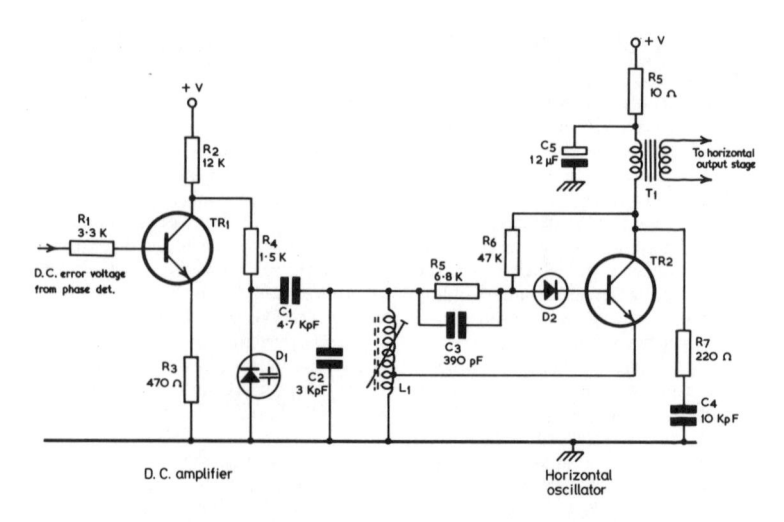

FIG. 7.26 USE OF D.C. AMPLIFIER AND VARACTOR DIODE

in reverse bias, the capacitance of D_1 is altered. Since D_1 is effectively in shunt with C_2 via C_1, the frequency of the tank circuit will be adjusted.

HORIZONTAL OUTPUT STAGE

The horizontal scan output stage in the monitor is basically similar to that of the camera. As is common with transistor output stages a shunt fed efficiency diode circuit is normally employed and the theory of this arrangement was described on page 107, Chapter 5. Since the monitor c.r.t. requires a much higher final anode voltage than the camera tube, and a greater deflection angle must be used in view of the c.r.t. dimensions, the scanning current will be of larger amplitude than in the camera.

A circuit which is respresentative of modern design is given in figure 7.27. TR_1 base is supplied with a square wave at line rate from the horizontal oscillator via an inverter stage. When the input goes positive, TR_1 conducts and current flows in the primary of the driver transformer T_1. The winding direction of T_1 is such that TR_2 base is driven negative thus keeping TR_2 in the OFF state. When the input to TR_1 goes negative TR_1 switches OFF and TR_2 base goes positive. As a result, TR_2 switches ON and the energy stored in T_1 supplies the base current for TR_2 which is driven into saturation for approximately 27 µs (almost half the line period). By arranging that TR_2 is ON when TR_1 is OFF, a fast switching action for TR_2 is ensured thus keeping the dissipation in TR_2 to a minimum. D_1 prevents the collector of TR_1 from swinging above the supply line when TR_1 switches OFF.

With TR_2 conducting, a square wave is applied across the scan coils and current flows, increasing exponentially by the LR ratio. At the end of the scan period, TR_2 is switched OFF and the energy in the scan coils circulates into the flyback tuning components of C_4, C_3 and T_3 primary to produce a sinusoidal oscillation. At the end of the first half-cycle, when TR_2 collector attempts to swing negative, the efficiency diode D_2 conducts, clamping the voltage to chassis. The stored energy now dissipates as current flows in the scan coils. The timing is such that as D_2 comes out of conduction, TR_2 is ON once more and current continues to rise in the scan coils. A saturable reactor L_1 (damped by R_3) is connected in series with the scan coils to counteract the LR effect. C_5, C_6 compensate for curvature of the c.r.t. face (S-correction). Any overvoltage transients are suppressed by D_3 and C_8. The half sine wave voltage developed at TR_2 collector during flyback is capacitively coupled to T_3 primary by C_7. Third harmonic tuning is employed in T_3 to reduce the peak voltage at TR_2 collector to a safe margin within the transistor ratings.

The high voltages required by the c.r.t. electrodes and their drive circuits are generated in the e.h.t. transformer during the flyback period. The winding w_1 steps up the flyback voltage and this is rectified by D_4 to provide a positive e.h.t. supply of 18 kV maximum. Smoothing of this supply is carried out by the capacitance between the final anode of the c.r.t. and its external aquadag coating. As this particular circuit was designed for use with different size c.r.t.s. the value of C_3 may be adjusted to give the required e.h.t. voltage (increasing the size of C_3 reduces the e.h.t.). Windings w_2 and w_3, together with the associated rectifiers D_5 and D_6, provide the additional high voltages required by the accelerating and focusing electrodes of the c.r.t.

E.H.T. REGULATION

E.H.T. systems using the flyback principle with a large e.h.t. winding have a high internal impedance and consequently suffer from poor regulation. Various stabilizing circuits may be employed to maintain a steady e.h.t. supply in spite of varying beam current demands. One example is shown in figure 7.28.

The regulator circuit comprises TR_3, TR_4 and associated components. Regulation is achieved by controlling the amplitude of flyback voltage via the regulator circuit which is driven by the c.r.t. beam current. The amplitude of the flyback pulse voltage across T_3 primary is controlled by two shunt capacitance paths. One path is via C_4 (figure 7.27) and the other via C_{13} (and D_7). The amplitude of the flyback pulse increases with a decrease in shunt capacitance. The pulse voltage across C_4 is fed to a

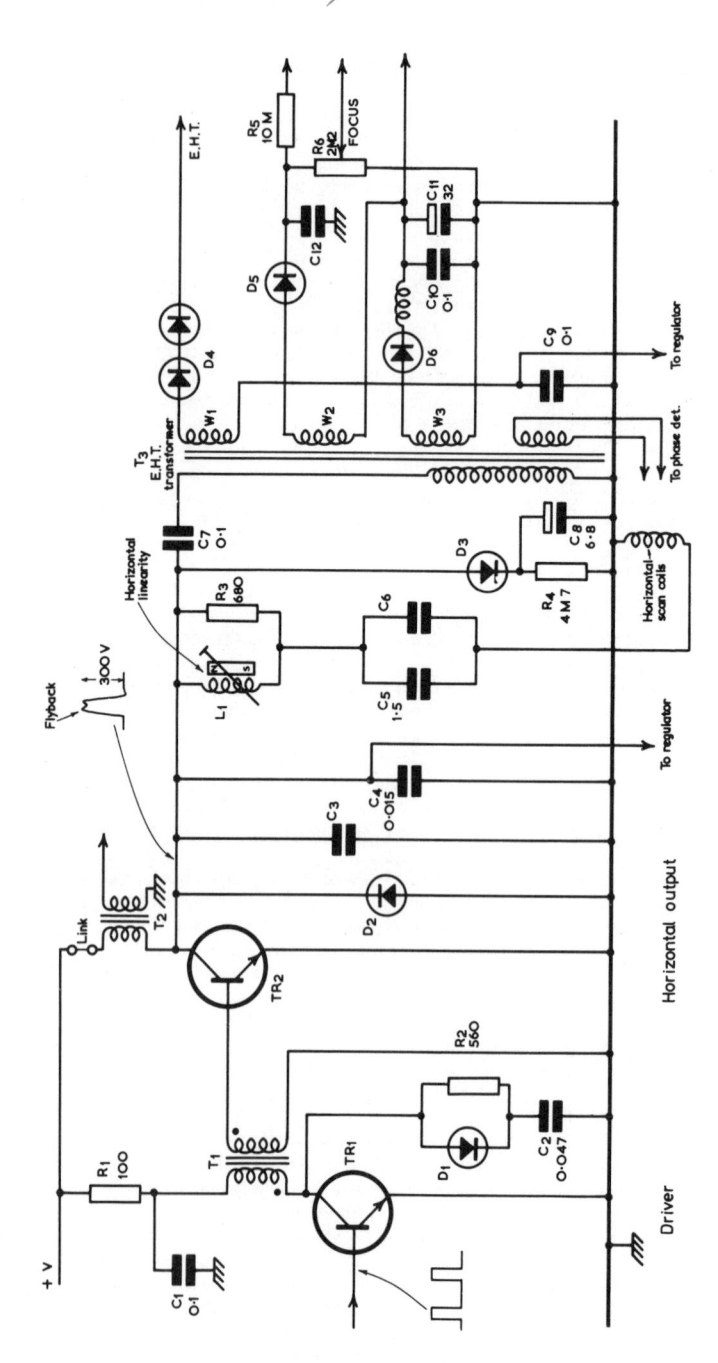

FIG. 7.27 HORIZONTAL SCAN OUTPUT STAGE WITH TRANSFORMER FOR GENERATING THE HIGH VOLTAGES REQUIRED FOR THE C.R.T.

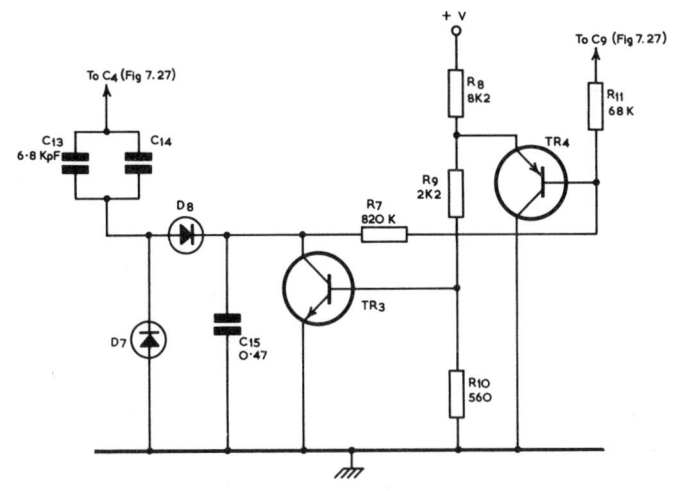

FIG. 7.28 E.H.T. REGULATOR USED WITH CIRCUIT OF FIG. 7.27

detector circuit comprising D_8, C_{15} and D_7 via C_{13}. D_8 conducts on the pulse voltage causing C_{15} to charge. With C_{15} fully charged, D_7 will conduct very little during flyback and in consequence C_{13} will have little effect on flyback tuning. With C_{15} discharged, D_7 will conduct for the whole of the flywheel period via D_8 and R_7 and C_{13} will be fully in parallel with T_3 primary.

With zero beam current TR_3 is biased from the potential divider R_8, R_9 and R_{10} to conduct heavily, keeping C_{15} almost fully discharged. As the beam current is increased, the d.c. voltage built up across C_9 (figure 7.27) goes more negative causing TR_4 to conduct harder. In consequence, due to the flow of TR_4 emitter current in R_8, the base voltage to TR_3 is lowered and TR_3 conducts less heavily, C_{15} can therefore charge to a higher voltage via D_8 and the conduction in D_7 is reduced. As a result, C_{13} has less effect on flyback tuning. This allows the flyback voltage to rise which drives the e.h.t. transformer harder to provide the extra power required by the c.r.t.

With this arrangement the e.h.t. is maintained at a relative constant voltage against changes in c.r.t. beam current up to about 350 μA d.c.

VERTICAL OSCILLATOR AND OUTPUT STAGE

Similar circuits to those found in the camera may be employed. The oscillator must supply the output stage with a drive waveform at 50 Hz which is locked to the incoming vertical sync. A linear sawtooth current must be delivered by the output stage to the field scan coils of adequate amplitude for vertical deflection of the c.r.t. beam. In the design of the deflection circuits particular attention is given to maintaining stable scan amplitude and linearity against supply voltage changes and temperature variations.

An example of a vertical oscillator using a thyristor with two gates is given in figure 7.29. The vertical oscillator comprises TH_1, C_2, R_3 and R_4. When the anode of TH_1 is positive to its cathode and both gates are disconnected the device is OFF and may be regarded as an open-circuit. The device may be brought ON by biasing either gate (g_2 positive to the cathode or g_1 negative to the anode). It will remain ON unless the current falls below its 'holding' value.

A fixed voltage is applied to g_2 from the potential divider R_5 and R_6, and negative-going vertical sync. is applied to g_1 via C_1. When the supply is first switched ON TH_1 is in the OFF state as the voltage at the junction of R_3, C_2 is at the supply line potential. C_2 now charges via R_3 and R_4 towards 11 V. As C_2 charges, the voltage across R_3, R_4 falls until g_2 becomes sufficiently positive with respect to the cathode and the device

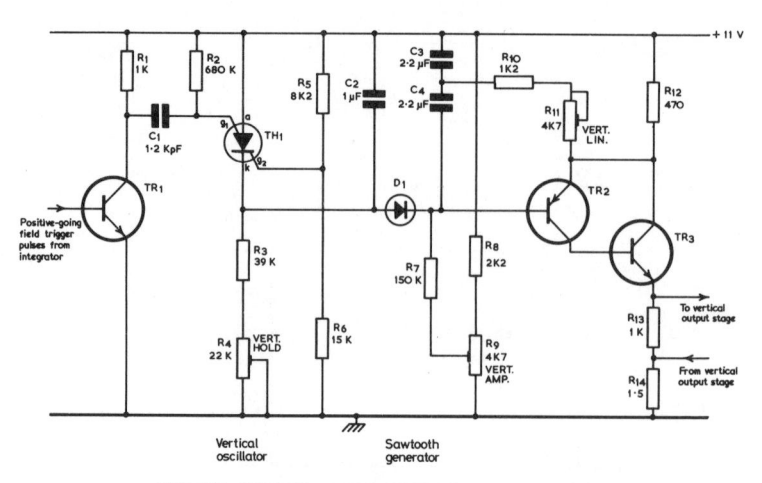

FIG. 7.29 VERTICAL OSCILLATOR USING A THYRISTOR

suddenly switches ON. C_2 now discharges via TH_1. Eventually, when the discharge current falls below the 'holding' value, TH_1 switches OFF. C_2 is then able to charge once again via R_3, R_4 and when the voltage on the cathode falls sufficiently below g_2 the device switches ON again causing C_2 to discharge. This sequence of events takes place continuously in the absence of any sync. input, *i.e.* the oscillator is free running. The repetition rate is determined by the time-constant C_2, R_3, and R_4; thus by varying R_4 the timebase frequency may be adjusted.

The discharge cycle of C_2 may be initiated by a negative-going field sync. pulse applied to g_1 so that the timebase is synchronized to the incoming sync. This trigger pulse is derived from the field pulse integrator after amplification and inversion in TR_1. The oscillation will be properly triggered only when the free-running frequency is lower than that of the incoming vertical sync. pulses.

The actual sawtooth fed to the vertical output stage is generated by a separate CR network C_4, R_7. At switch-on, C_4 charges up via R_7 but more slowly than the charging of C_2. As a result D_1 is in a reverse bias state. When TH_1 comes ON D_1 is turned ON and C_4 discharges via D_1 and TH_1. The charging of C_4 via R_7 constitutes the field scan and the discharge of C_4 via D_1 and TH_1 the flyback. The magnitude of the voltage at the junction of C_4, R_7 at the instant D_1 becomes conductive is determined by the setting of R_9 which thus serves as the vertical amplitude or height control.

At TR_2 base a sawtooth waveform is produced by the charging and discharging of C_4. TR_2 and TR_3 are connected in a special arrangement so as to obtain a high input impedance which is necessary to avoid loading the sawtooth forming network. Feedback via R_{11} and R_{10} to across C_3 is to linearise the waveform with R_{11} providing control over the degree of feedback. The sawtooth output from across R_{13} is then fed to the vertical scan output stage.

A vertical timebase working on similar principles to the thyristor circuit but using an astable (free-running) multivibrator is shown in figure 7.30. TR_1 and TR_2 form the multivibrator circuit with positive-going vertical sync. applied to TR_1 emitter. TR_2 collector is d.c. coupled to TR_1 base via R_8 and TR_1 is a.c. coupled to TR_2 base via C_1. During the vertical scan TR_1 is conducting and TR_2 is cut off due to the charge on its base. C_1 then discharges via R_9 and R_{10} until TR_2 base voltage rises sufficiently for TR_2 to come ON. The precise instant at which TR_2 turns ON, which is the start of the vertical flyback, is determined by the vertical sync. pulses applied to TR_1 emitter. These pulses will be positive-going at the junction of R_2, R_3 and also at the base of TR_2. As TR_2 comes ON TR_1 switches OFF. C_1 now quickly charges via R_2 and the base-emitter

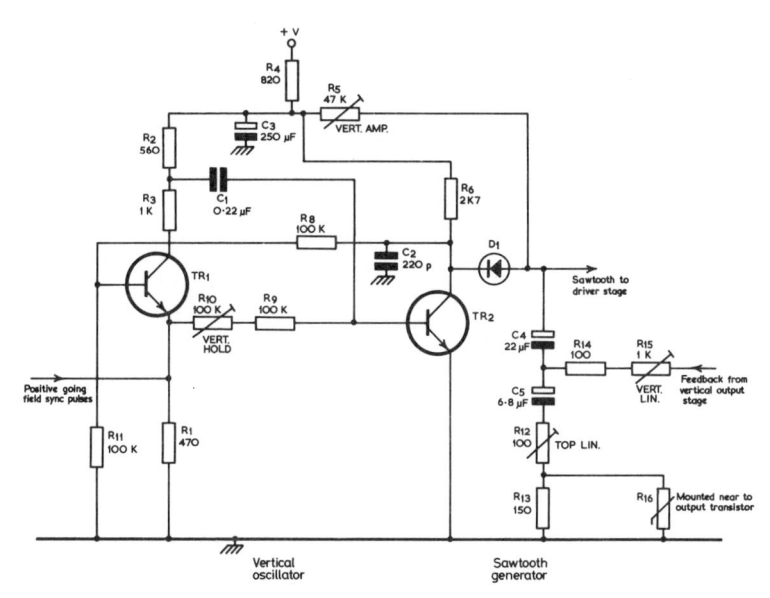

FIG. 7.30 VERTICAL OSCILLATOR WITH A STABLE MULTIVIBRATOR

junction of TR_2. When the base current of TR_2 falls below a certain value TR_2 switches OFF and the regenerative action switches TR_1 ON. This action takes place continuously with a frequency governed by the time-constant C_1, R_9 and R_{10}. The latter resistor thus serves as the vertical hold control.

During the interval that TR_2 is OFF, D_1 is also OFF allowing C_4 and C_5 to charge via R_5. When TR_2 turns ON, D_1 is biased ON and C_4 and C_5 discharge via D_1 and TR_2. R_5 determines the magnitude of the voltage across C_4, C_5 during the time that TR_2 is OFF and thus serves as the height control. Across C_4, C_5 a sawtooth voltage is developed which is fed to the output driver stage. Variable feedback from one of the output transistors is applied via R_{15}, R_{14} to across C_5. Overall linearity is adjusted by R_{15}. Top linearity adjustment is provided by R_{12} which mainly affects the linearity at the start of the field scan period. This is temperature compensated by a thermistor R_{16} which is in thermal contact with one of the vertical scan output transistors.

Many different circuit arrangements may be used for the vertical scan output stage and it is difficult to find examples which are common to monitors of different manufacture. However, a number of designs use a circuit which is similar to the complementary output stages found in modern audio amplifiers. One such example is given in figure 7.31.

TR_2 and TR_3 form the output stage using complementary transistors, connected in series as regards d.c. across the line supply voltage. Bias for the bases of the output transistors is obtained from across R_3, TR_4. The preset resistor R_3 sets the current in TR_4 and hence its collector-emitter voltage drop which thus determines the difference in base voltage between TR_2 and TR_3. The driver stage TR_1 is forward biased from the commoned-emitters of TR_2 and TR_3 via R_8 and R_9. These two resistors provide d.c. feedback to TR_1 to stabilize the mid-point voltage at the commoned-emitters of the output pair.

The sawtooth voltage applied to TR_1 base is amplified and developed across the collector load R_4. During the first half of the forward scan TR_3 is conducting and TR_2 is cut off. TR_3 thus supplies current to vertical scan coils. During the second part of the forward scan the conditions are reversed as TR_2 conducts and TR_3 cuts off (class–B

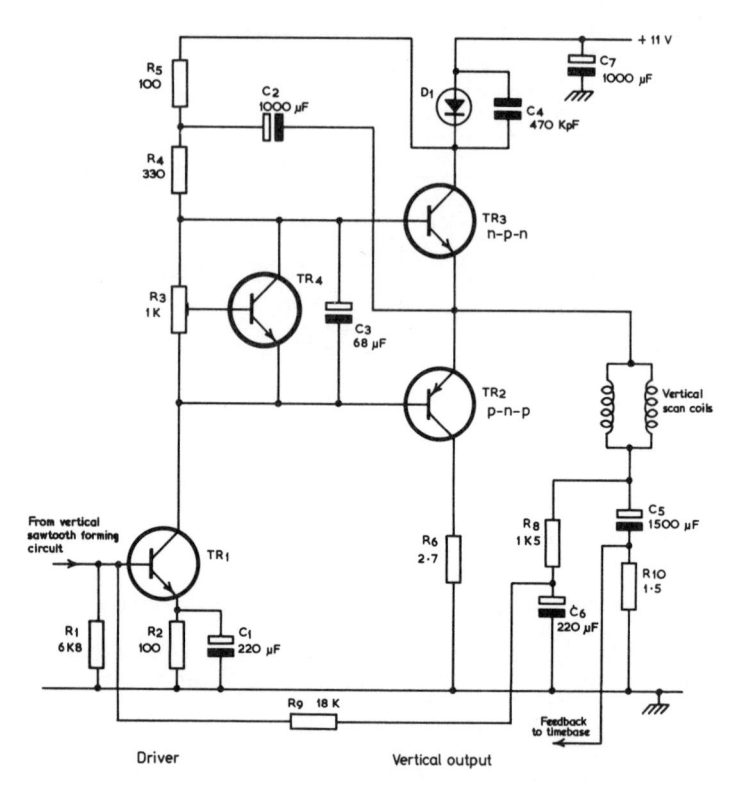

FIG. 7.31 VERTICAL SCAN OUTPUT STAGE USING COMPLEMENTARY PAIR

push-pull operation). TR_2 now supplies current to the deflector coils. R_3 is adjusted so that the changeover between TR_2 and TR_3 takes place at just the right moment. The bootstrap capacitor C_2 ensures that the drive is effectively applied between the base and emitter of TR_2 and TR_3.

During flyback the stored energy in the deflector coils takes the voltage at TR_3 emitter positive, and also TR_3 collector via C_2 and R_5. D_1 is then non-conductive. When the voltage attempts to reverse, D_1 conducts, clamping the voltage to the line supply. During the flyback the current direction in the deflector coils is reversed.

REDUCED SCAN

Some monitors incorporate a reduced-scan switch which when operated reduces the picture width and height to produce about half-an-inch border surrounding the picture. This facility is particularly useful when setting up the scan geometry of the camera. Figure 7.32 shows the necessary circuit arrangements used in one monitor.

S_1A is used for the switching in the horizontal output circuit and this is linked to S_1B in the vertical output stage. In the reduced scan position of S_1A an extra inductor L_1 is placed in series with the horizontal deflector coils thereby reducing the scan amplitude. L_1 is adjusted to give the desired degree of underscan. L_2 serves as the normal picture-width control. In the vertical circuit, switching S_1B to the reduced scan position increases the amount of feedback applied to earlier stages thus reducing the picture height. R_2 sets the amount of underscan and R_1 serves as the normal picture-height control.

(a) Horizontal (b) Vertical

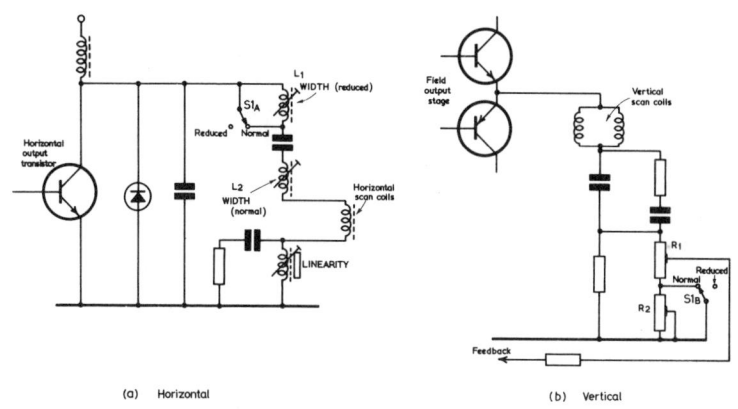

FIG. 7.32 REDUCED SCAN FACILITY

FLYBACK BLANKING

It is normal practice in the monitor to suppress the horizontal and vertical flyback which, if left unsuppressed, cause annoying bright-up lines on the monitor screen. This is usually carried out by feeding large negative-going pulses corresponding to the horizontal and vertical flyback periods to the grid of the c.r.t. A typical circuit is shown in figure 7.33.

Suitable flyback pulses are normally available at the vertical scan coils and the horizontal output transformer. The vertical pulses are fed to the base of TR_1 via C_2 and R_3. Horizontal pulses are supplied to the base of TR_1 via C_1, R_1 and D_1. Thus, across R_4 mixed suppression pulses are present. These positive-going pulses cause TR_1 to conduct thereby producing negative-going pulses at the collector. The negative pulses are then fed to the c.r.t. grid via a d.c. restorer circuit comprising C_3, D_2 and R_7. This restores the pulses negatively to the brightness potential fed to the grid from the junction of R_{10}, R_{11}. Each time a flyback pulse is applied to the c.r.t. grid, the grid potential is taken rapidly negative thereby cutting off the beam current and suppressing the flyback. The use of a d.c. restorer ensures that when the brightness is changed the suppression remains effective. R_9 and R_{11} are used for setting the maximum and minimum brightness voltage levels.

In some circuits instead of feeding negative pulses to the c.r.t. grid, positive suppression pulses are supplied to the tube cathode. These pulses momentarily lift the cathode potential above its steady voltage thereby cutting off the beam current. The suppression pulses are typically of 50–60 V in amplitude.

ELIMINATING SWITCH-OFF BURNS

When a monitor is switched off, the deflection fields collapse but as the c.r.t. heater takes some time to cool there is sufficient emission to produce a small beam until the e.h.t. smoothing capacitor has discharged. This will result in a small spot at the screen centre and over a period of time a burn mark will appear unless measures are taken to avoid this trouble. There are two methods open: (a) bias the electron gun at switch-off so that no current can reach the c.r.t. screen; or (b) cause the c.r.t. to conduct heavily at switch-off thereby discharging the e.h.t. capacitor before the deflecting fields have fully collapsed.

One example of method (a) was given on page 165 and another example is shown in figure 7.34.

The emitter-follower stage TR_1 feeds a brightness voltage and flyback suppression pulses to the c.r.t. cathode via R_2. During normal operation TR_2 is OFF and C_1 is charged to the +40 V rail via D_1. When the monitor is switched off the supply rail voltages die away but TR_2 collector-base junction conducts hard. This takes the tube

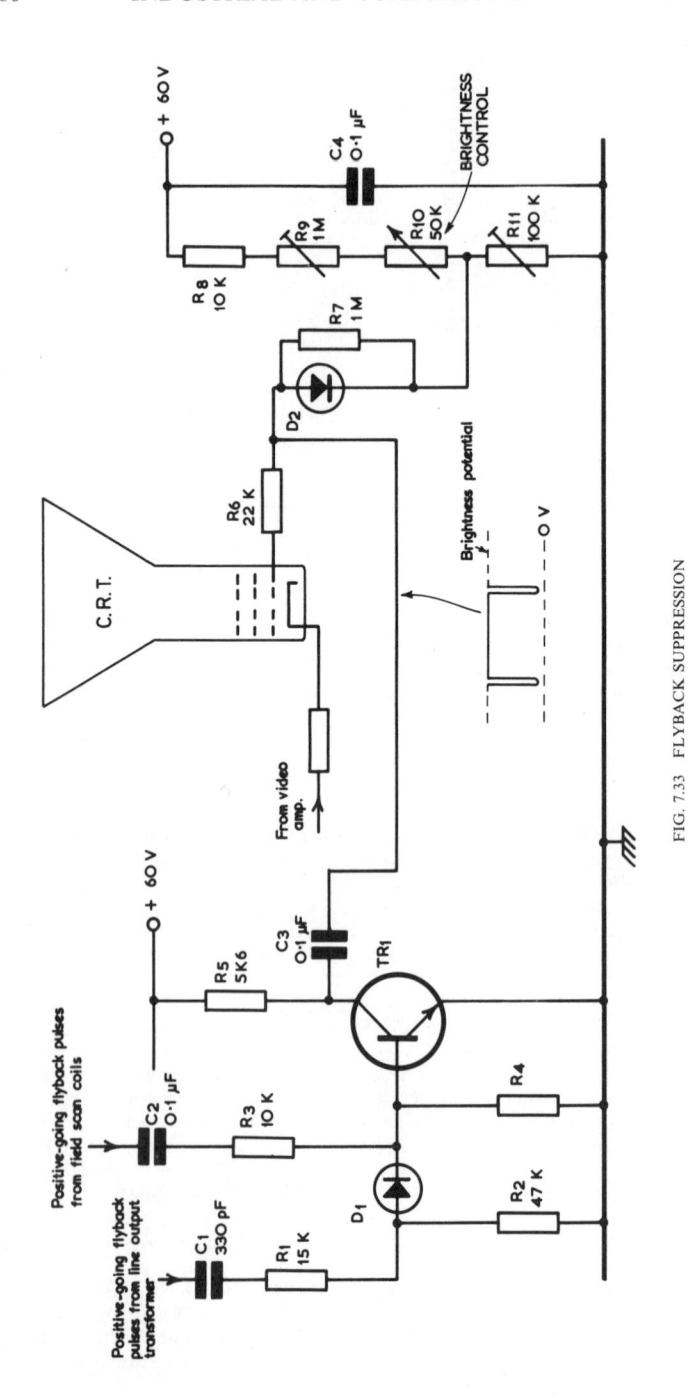

FIG. 7.33 FLYBACK SUPPRESSION

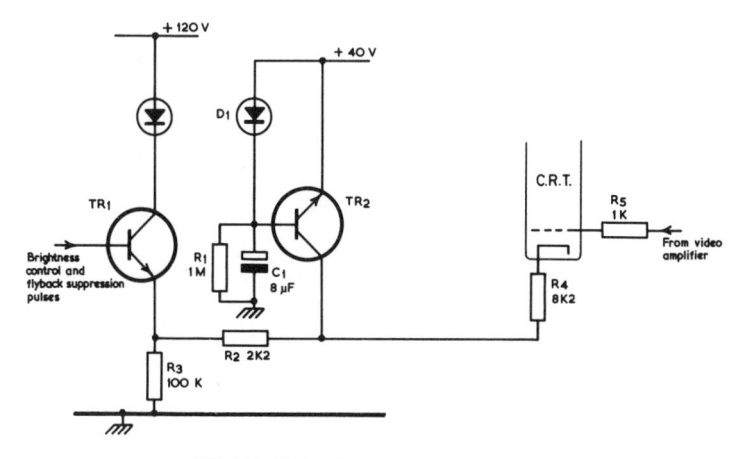

FIG. 7.34 SPOT BURN SUPPRESSION CIRCUIT

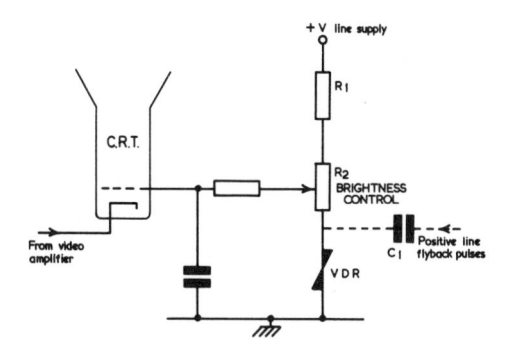

FIG. 7.35 USE OF VDR FOR SPOT SUPPRESSION

cathode potential to that across C_1 lifting it momentarily above the grid and suppressing the beam current. C_1 provides the current in TR_2 during this brief period when D_1 isolates the capacitor from the 40 V rail.

An example of method (b) is given in figure 7.35 which uses a voltage dependent resistor. When the monitor is switched off, the line supply voltage drops as the smoothing capacitors discharge. The cathode voltage of the c.r.t. also drops since it falls with the line supply rail. However, the fall in voltage at the c.r.t. grid is less due to the use of the v.d.r. since with a smaller current in the v.d.r., proportionately a greater drop occurs across the v.d.r. than across R_1 and R_2. Thus, the grid of the c.r.t. is made positive with respect to the cathode causing the tube to conduct hard and rapidly discharge the e.h.t. smoothing capacitor before the deflecting fields have fully collapsed.

Sometimes the v.d.r. is used as a rectifier fed with line flyback pulses *via* a capacitor (C_1), shown dotted. This produces a negative potential across the v.d.r. which results in a lower brightness voltage than the previous arrangement (which may be compensated for by adjusting the brightness control). At switch-off the line pulses rapidly die away and the negative potential across the v.d.r. fades away. This causes the grid to assume a higher potential for a brief period which rapidly discharges the e.h.t. smoothing capacitor.

VIDEO SIGNAL CLAMP

In place of a simple d.c. restorer circuit in the video channel a clamp circuit may be employed. A d.c. clamp which is operated by pulses occurring at regular intervals is capable of dealing more effectively with hum or extraneous interference in the video signal than a d.c. restorer. A typical circuit is given in figure 7.36.

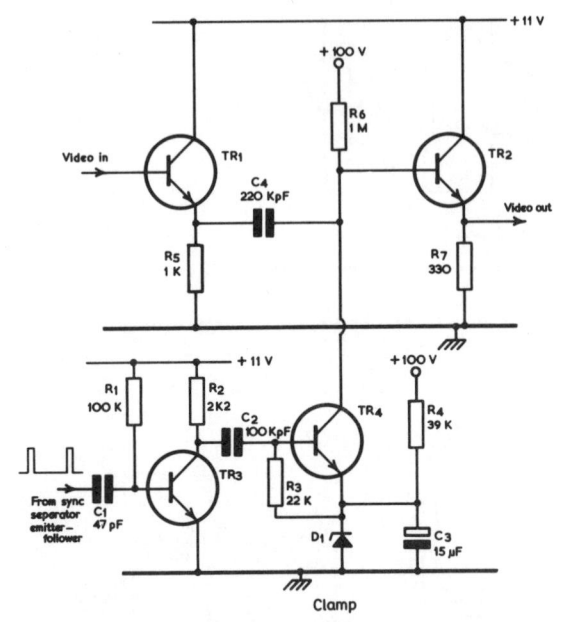

FIG. 7.36 D.C. CLAMP

TR_1 and TR_2 form part of the video channel with the composite video signal coupled from TR_1 emitter to TR_2 base via C_4. Line synchronizing pulses derived from the sync. separator stage are applied to a differentiating circuit comprising C_1, R_1. TR_3 amplifies the negative-going differentiatied pulse which causes a positive-going pulse at the collector with a duration confined to the back porch period. This pulse is fed to the base of TR_4 via C_2 turning the transistor hard ON. The 'bottoming' of TR_4 causes the junction of C_4, R_6 to be clamped to the emitter of TR_4. The zener diode D_1 sets the emitter voltage of TR_4 to +8·7 V. Thus, the junction of C_4, R_6 is clamped to +8·7 V during the back porch of the composite video signal applied through C_4. Between clamping pulses TR_4 is OFF but the voltage at TR_2 base does not change appreciably as a long time-constant C_4, R_6 is employed. Thus the black level at the base of TR_2 corresponds to a constant +8·7 V (approximately) in this particular amplifier. The d.c. coupling used in the following stages ensures that a constant black level is maintained on the modulating electrode (the cathode) of the c.r.t. within the d.c. stability of the coupling amplifiers.

COTRON 9" VIDEO MONITOR (PM SERIES) FEATURING HINGED PRINTED CIRCUIT BOARDS FOR ACCESS DURING MAINTENANCE OR ADJUSTMENT

(Photograph courtesy of Cotron Electronics Ltd., Coventry)

CHAPTER 8
BASIC ADJUSTMENTS TO CAMERA AND MONITOR

ANY distortion in the size or shape of the image displayed on the monitor can be due to maladjustment or incorrect operation in either the CAMERA or MONITOR. Therefore, camera and monitor are required to be set up separately. The monitor will be dealt with first as a correctly aligned monitor will be needed when making adjustments to the camera.

The following information, confined to general procedures, is not intended to supplant manufacturer's instructions which should always be carefully adhered to when carrying out adjustments.

MONITOR PICTURE GEOMETRY

The principal adjustments to be made to a monitor at installation or after a repair are concerned with the geometry of the picture, *i.e.* obtaining the correct height, width, linearity, picture centering, etc. Some form of linearity test signal is needed to ensure good results. A common type of test signal used for geometry adjustments is a cross-hatch or grating display, see figure 8.1. This is obtained from a cross-hatch generator and consists of vertical and horizontal white lines, the number of which can be controlled in either direction.

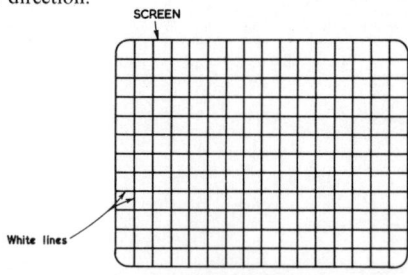

FIG. 8.1 CROSS-HATCH PATTERN

The generator supplies a pulse-type video signal for producing the displayed bars and composite sync. pulses to lock the vertical and horizontal timebases. An alternative signal source is a test card from a correctly aligned camera. While this is an acceptable method for non-critical work, errors within the specification of the camera will be present in the displayed test card so it will be assumed that a cross-hatch pattern is being displayed. When adjustments are made some interaction may occur between controls, *i.e.* between

    Width and Brightness
    Vertical height and linearity
    Horizontal hold and Picture centering

These interactions must be remembered when carrying out the following procedure:

(1) Apply a signal from the cross-hatch generator to the 'video-in' socket and set the sync. selector to 'internal'. Switch on the monitor and allow to warm up.

(2) Obtain a display using the brilliance and contrast controls. If the picture is not locked adjust the vertical and horizontal hold controls.

(3) Readjust the brilliance and contrast controls to obtain the sharpest cross-hatch display set against a dark background.

(4) At this stage any 'set h.t.' or 'set e.h.t.' procedures outlined in the manufacturer's service manual may be carried out. This might not always apply as in

BASIC ADJUSTMENTS TO CAMERA AND MONITOR 191

some monitors these voltages are fixed. It may also be necessary to carry out 'set c.r.t. a_1 voltage' or 'set max/min brightness range' adjustments at this point if these preset controls are fitted. Again, reference should be made to manufacturer's instructions. It is usually necessary to make these adjustments before setting the picture geometry because, if they are made afterwards, the geometry adjustments will most likely have to be repeated.

(5) The 'hold' controls may now be set more accurately. The vertical hold control is rotated in either direction until the picture goes out of lock. As the control is moved in one direction a point will be reached when the picture commences to slowly 'roll-down' the screen—note this setting. The control should now be rotated in the other direction until the picture starts to 'jump'—note the setting. The correct setting is midway between these two 'out of lock' settings to allow for frequency drift in the vertical oscillator.

Line flywheel sync. circuits usually incorporate two adjustments by way of a 'horizontal hold' control and a 'horizontal frequency' tuning slug. The former (a fine frequency control) is available as a user's control whilst the latter (a coarse frequency adjuster) is an internal preset. A common method of adjustment is as follows. Set the horizontal hold control to its mid position and set the sync. selector to 'external'. The picture will now be out of lock as the sync. separator has been removed. Now adjust the 'horizontal frequency' tuning slug with a non-metallic tool until an unbroken picture just 'floats through' or steadies in the horizontal direction, see figure 8.2. Reset the sync. selector to 'internal' and a locked picture will be restored with the adjustments optimised for dealing with slight drift.

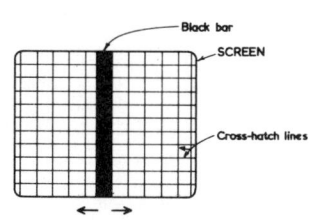

FIG. 8.2 PICTURE 'FLOATING THROUGH' DURING ADJUSTMENT TO 'HORIZONTAL FREQUENCY' TUNING SLUG

(6) Adjust the width control to slightly underscan, *i.e.* to produce vertical black bars at either side of the screen. Similarly adjust the height control to produce horizontal black bars at the top and bottom of the screen.

(7) Unlock the clamp securing the scan coils to the tube neck. Rotate the coils until the horizontal edges of the pictures are parallel with the top and bottom of the mask. Before locking the scan coil clamp ensure that the coils are pushed as far up the tube flare as they will go to avoid 'corner cutting'. Take special care when handling the scan coils as high voltages are present, and do not apply extreme pressure when turning the coils.

(8) If the vertical and horizontal linearity appears reasonable at this stage, the picture shift magnets may be adjusted to centralise the picture in the screen (or mask) of the c.r.t. If not, approximate adjustment (10) and then centralise the picture.

(9) Adjust the width and height controls until the picture just fills the screen. The raster correction magnets may now be set. These are mounted on the scan coil assembly and should be positioned until the edges of the raster are free from pin-cushion or barrel distortion.

(10) At this point the horizontal and vertical linearity may be set. First reduce the height and width to produce a $\frac{1}{4}''$ black border around the edges of the screen. Now adjust the horizontal and vertical linearity to give equal spacing of the vertical and horizontal lines of the cross-hatch display. There are usually two vertical controls, one for 'top linearity' adjustment and the other for 'overall linearity'. These controls are

interpendent to some extent so the adjustments may have to be repeated a few times, each time getting closer to the optimum setting.

It is rather difficult to detect small linearity errors from square to square without some sort of reference with which the eye can compare. The linearity of the monitor is often set with the aid of a transparent plastic mask. This has lines engraved on it to form rectangles and is fitted over the face of the screen as illustrated in figure 8.3. The mask must be fitted as close as possible to the phosphor layer to reduce parallax errors.

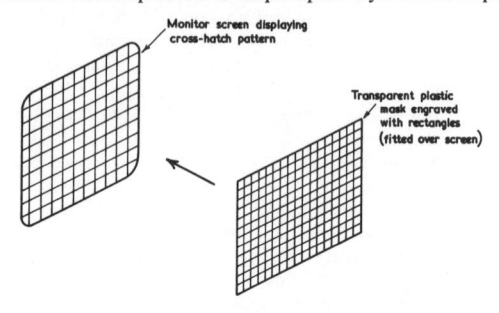

FIG. 8.3   USE OF PLASTIC MASK FOR CHECKING VERTICAL AND HORIZONTAL LINEARITY

The cross-hatch generator is adjusted to display the same number of rectangles as the mask. The two sets of rectangles are then compared and the linearity controls adjusted for best coincidence.

Manufacturers usually specify the maximum linearity errors to be expected in the specification of the monitor (2% for a good monitor) and these may be checked with the aid of the mask. Figure 8.4 shows a small area of the mask with the horizontal lines of the cross-hatch display immediately behind. The maximum vertical linearity error in

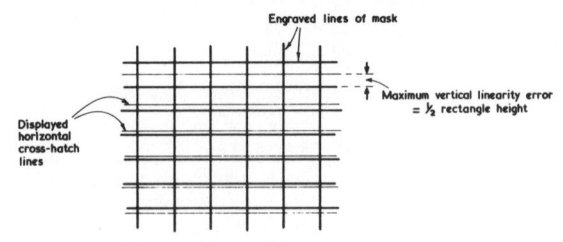

FIG. 8.4   ASSESSING A LINEARITY ERROR USING THE PLASTIC MASK

this case is equal to half the height of a rectangle. Thus with 50 rectangles (which is typical) the maximum error would be 1%. When assessing the error at a particular point on the screen, align the eye with a line of the plastic grating as the grating and display are not in the same vertical plane. Manufacturers sometimes recommend the use of a projected grating to avoid parallax errors. This may be obtained from a projected slide on which is an accurately marked grating. The grating is projected on the face of the phosphor layer through the screen glass. Since the projected grating and the display grating lie in the same plane, parallax errors are eliminated. Try to spread any linearity errors over the entire screen rather than allow them to be concentrated in a particular area. It is better to have small errors distributed evenly over the screen than a large error in one place.

(11) The width and height may now be reset to just fill the screen or to give a slight overscan (about $\frac{1}{4}''$) so that the edges of the picture cannot be seen under extreme brightness or contrast settings. Some small adjustment to linearity may now be required and steps (10) and (11) should be repeated as often as necessary.

(12) The focus control may now be adjusted to produce the best compromise in sharpness of the displayed image at the centre and corners of the screen.

SETTING THE CAMERA GEOMETRY

To set up a camera successfully some form of test rig is needed and figure 8.5 shows one arrangement. The basic requirement is a flat test bed which may be formed from

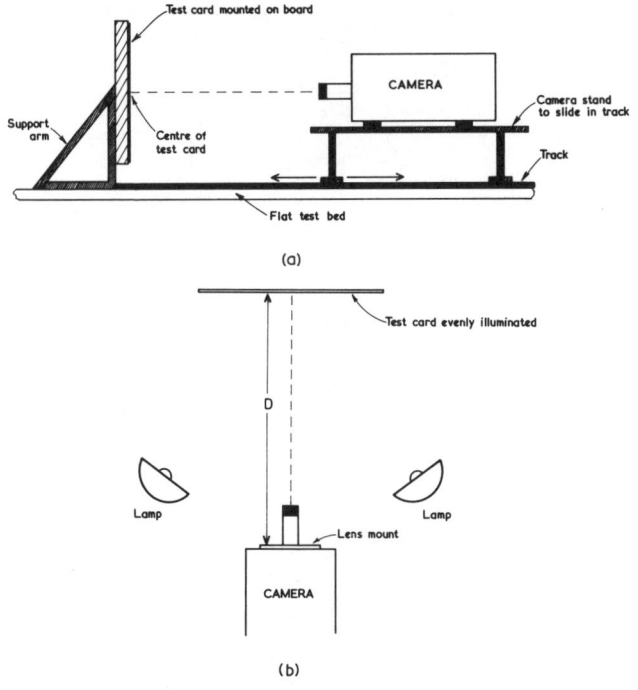

FIG. 8.5 CAMERA TEST BED
(a) Side View;   (b) Plan View

the surface of a solid bench with either a hard wood or formica finish. On this is mounted a track constructed of angle strip in which the camera stand may be accommodated. The camera stand may also be constructed of angle strip and dimensioned so that it will slide smoothly along the test bed track. The camera is clamped to the stand and some means must be provided for clamping the stand to the track once the distance from the test chart has been established, *i.e. via* a slot which runs the length of the track.

The test card which is usually of cardboard construction is mounted on a flat piece of solid timber which is clamped to the test bed by means of a rigid support arm. The test card must be mounted so that it lies parallel with the camera lens mount and with the centre of the chart at the same height above the test bed as the camera lens. It is useful to be able to slide the camera right up to the test card so that the lens is seen to coincide with the centre of the chart. If the test rig is to be used with different types of cameras the test card height will have to be adjusted and this should be borne in mind during construction. In any event, height adjustment is normally desirable to take up small variations in camera dimensions. The actual dimensions of the test rig will depend upon the test card size, camera dimensions and lens used. With a test card $12'' \times 9''$ and a 'normal' lens (25 mm) the distance D from the lens mount to the test card

ENGLISH ELECTRIC VALVE CO. LTD. CAMERA TUBE TEST CHART

BASIC ADJUSTMENTS TO CAMERA AND MONITOR

will be 2 feet. If a lens of longer focal length is used the distance D will have to be increased. Racking the camera over a distance up to about 3 feet will be suitable when using a 'normal' lens. It is important that the camera and test chart are square and level with each other (a spirit-level may be used for accurate alignment).

The test chart should be evenly illuminated by lamps placed facing the chart but away from the field of view of the camera lens. A pair of 150 W tungsten bulbs mounted in reflector lamps is suitable but ensure that direct light from the lamps does not fall on the camera lens.

An alternative to the optical bench for setting the camera geometry is the diascope. A suitable instrument for CCTV use is one made by Isopelem in France and imported by Rank-Taylor-Hobson and marketed under the name 'Vidicon Quick Tester'. This diascope is about the size of a small zoom lens and screws into the lens 'C' mount in place of the camera lens. Inside the instrument, which contains its own optical system, is a test chart transparency illuminated by a small bulb. The diascope throws an accurately sized image on to the faceplate of the camera tube so that it exactly fills the active area of the target layer. The instrument is supplied with a separate power unit which contains a control to permit the intensity of the light falling on the camera tube to be adjusted. Interchangeable test chart transparencies are available to suit different sized camera tubes.

TEST CHARTS

General line-up charts are available in a variety of patterns and sizes but usually they contain similar information for checking the operation of the camera and for overall system appraisal. A chart which was designed for CCTV systems using vidicon cameras is shown opposite. This test chart uses a 4:3 aspect ratio and is recommended for use with a front illumination intensity of between 100 and 200 foot-candles. The chart contains the following features:

(1) The general background of the chart is a neutral grey equivalent to the average brightness of a typical scene.

(2) The centre-line of the castellated surround defines the active picture area. Setting to the inner or outer edges of the castellations indicates a small amount of under or over scanning of the picture area.

(3) The coarse black grating pattern of horizontal and vertical black lines allows a qualitative assessment to be made of the linearity of the camera scanning generators.

(4) The large central circle and the four corner circles are for checking picture geometry in addition to the grating mentioned in (3). The eye is very sensitive to small distortions in a circle and this feature is an aid in setting the scan amplitudes and linearity.

(5) The fine detail resolving power of the camera or CCTV system can be noted from the resolution blocks and wedges in the test chart. The VERTICAL bars within the circle are arranged in blocks to give the HORIZONTAL resolution of the system in terms of the number of lines per active picture height. The vertical resolution wedges enable limiting picture centre resolution to be observed. The corner resolution wedges enable both vertical and horizontal resolutions to be assessed. When adjusting the optical and electrical focus of the camera for optimum setting, the centre and corner wedges may be used.

The following relationship exists between bandwidth and limiting resolution:

BANDWIDTH	LIMITING RESOLUTION
3 MHz	234 lines per active picture height
4 MHz	312 lines per active picture height
5 MHz	390 lines per active picture height
6 MHz	468 lines per active picture height
8 MHz	624 lines per active picture height
10 MHz	780 lines per active picture height

(6) A five-step grey-scale is included for setting-up of the camera tonal range and enables final adjustment to be made to the brightness and contrast levels.

(7) Blocks of black on white and white on black permit observation of video low and middle frequency response. Streaking of either black or white after black indicates that the camera equipment has a low frequency phase shift.

ADJUSTING THE CAMERA SCANS

There are a number of different methods used for setting the vertical and horizontal scan amplitudes of the camera. The use of an optical bench with the test chart placed at the correct distance from the camera or the employment of a diascope ensures that an image of correct size and position is formed on the target layer of the camera tube. However, the portion of the image area that is displayed on the monitor screen depends upon the amplitudes of the camera scans and the centering of the camera tube electron beam. It is important that the camera scans are set correctly for a particular camera and here reference should be made to the manufacturer's instructions.

A correctly set-up monitor will be required to see the effect of adjustments to the camera geometry controls. It is helpful if the height and width of the monitor picture are reduced so that the edges of the raster are just visible whilst maintaining the normal 4:3 aspect ratio. If a reduced-scan facility is available on the monitor this should be used instead. The width and height of the camera are adjusted so that the castellated edges of the test chart being viewed by the camera just fill the raster on the monitor. It will be found that the camera scan amplitude controls appear to work opposite to those of the monitor. For example, when the camera vertical scan amplitude is reduced the effect on the monitor is similar to increasing the picture height of the monitor.

Some camera tubes are fitted with a mask as an aid to setting the camera scans. The mask is made from opaque plastic and has a 4:3 aperture cut out with dimensions to suit the size of camera tube in use, figure 8.6. The mask is permanently clamped into position over the face of the tube faceplate and thus accurately defines the size of the

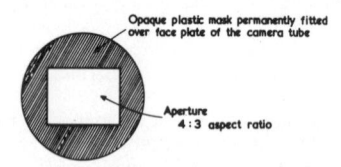

FIG. 8.6 SCAN DEFINING MASK FITTED TO SOME CAMERA TUBES

image formed on the target layer. If the camera is over-scanned as in diagram (a) of figure 8.7 the effect on the monitor is to produce a black border around the reproduced image, diagram (b). The black border may be used to centralise the camera scanning beam by adjusting the shift-magnets until the reproduced image on the monitor lies centrally within the black border. Also, the black border produced by the mask may be used for setting the camera scans. When the camera is under-scanning as in diagram (a)

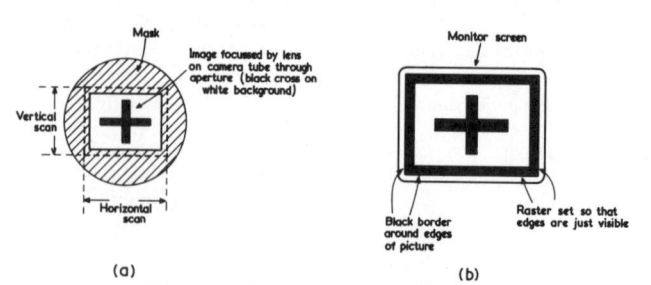

FIG. 8.7 OVER-SCANNING OF CAMERA TUBE

BASIC ADJUSTMENTS TO CAMERA AND MONITOR 197

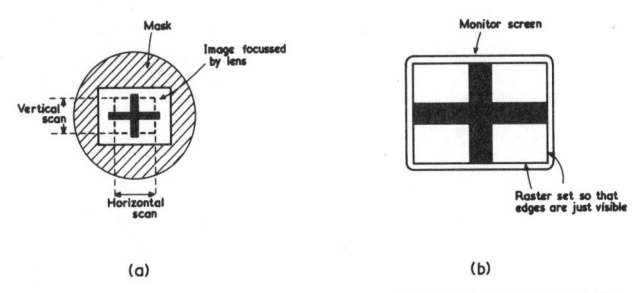

(a)   (b)

FIG. 8.8 UNDER-SCANNING OF CAMERA TUBE (COMPARE WITH FIG. 8.7)

of figure 8.8 the effect on the monitor is for the black border to disappear and for a smaller portion of the camera image to be displayed, diagram (b).

To set the scans correctly, initially adjust the camera vertical and horizontal scan amplitude controls to produce the over-scan condition of figure 8.7, *i.e.* with the black border surrounding the monitor picture. Now reduce the amplitude controls until the black border just disappears over the edges of the raster at the top, bottom and sides. The scans are now set correctly, see figure 8.9.

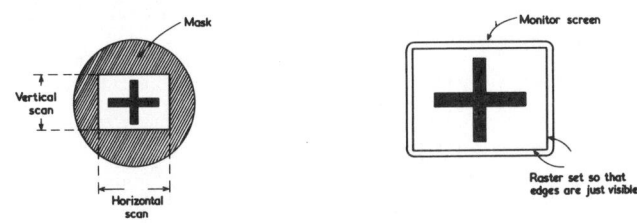

(a)   (b) Black border just disappears

FIG. 8.9 CAMERA TUBE SCANNING SET CORRECTLY (COMPARE WITH FIGS. 8.7 & 8.8)

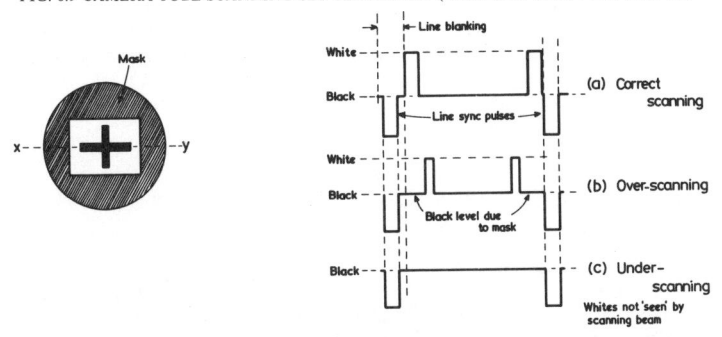

FIG. 8.10 COMPARISON OF VIDEO OUTPUT WAVEFORMS DURING ONE LINE CORRESPONDING TO A LINE SCAN X-Y

The waveforms of figure 8.10 show the effect on the video waveform over one line period during the three conditions illustrated in figures 8.7, 8.8. and 8.9. The reason for the appearance of the black border on the monitor in the over-scan condition may be seen from waveform (b). Each time the camera scanning beam goes beyond the edges of the mask aperture, the video output falls to black level.

Another method of accurately fixing the camera scans is with the aid of an engraved disc, figure 8.11. The disc, which fits over the faceplate of the camera tube, has two

198   INDUSTRIAL AND COMMERCIAL CCTV

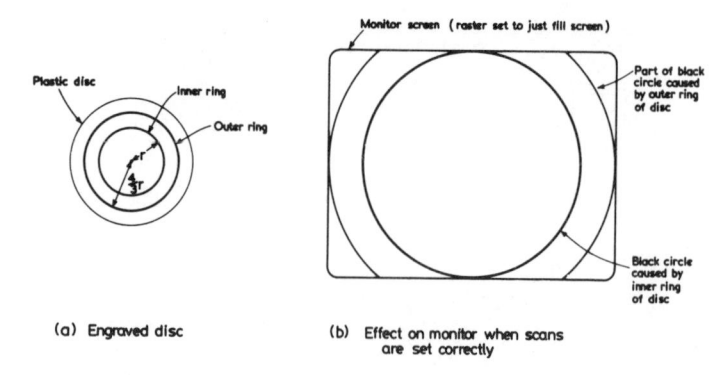

(a) Engraved disc

(b) Effect on monitor when scans are set correctly

FIG. 8.11   USE OF ENGRAVED DISC FOR SETTING SCANS

engraved circles filled with a black pigment. The inner circle sets the vertical scan and the outer circle sets the horizontal scan limits. When the camera tube is illuminated the engraved circles show up on the monitor as in diagram (b). If the camera scans are set correctly, the inner circle just fills the height of the monitor screen and the outer circle just touches the sides of the raster. When using a mask or disc to set the camera scans, a test chart is not necessary as the scans can be accurately set on a normal scene. However, a test chart would normally be employed as this contains additional features which assist in making other adjustments to the camera.

TYPICAL SETTING-UP PROCEDURE

In the following it is assumed that a correctly set-up monitor is available, that an optical bench is being used and that the camera tube is fitted with an image defining mask.

Camera and Monitor off and Lens Cap fitted

(1) Position the test card so that it is viewed squarely by the camera. Set the distance between the camera lens mount and the test chart to that specified in the instruction manual. If this information is not available, the distance may be calculated using the expression given on page 31, Chapter 2.

(2) Evenly illuminate the test chart to the intensity recommended in the service manual. A light-meter may be used to check the intensity.

(3) Set the optical focus on the lens barrel to correspond with the distance between test chart and camera.

(4) Set the 'target sensitivity' and 'beam' controls to minimum and the electrical focus to mid position.

Switch on Camera and Monitor

(5) Adjust the brightness control of the monitor until the scanning lines are just visible. Set the contrast to about $\frac{3}{4}$ of the maximum setting and operate the reduced-scan switch. Adjust the monitor horizontal and vertical holds to give a locked raster.

(6) Remove the camera lens cap. Open up the lens 'iris' or 'aperture'. Increase the 'target sensitivity' to about $\frac{3}{4}$ of the maximum and increase the beam control setting until a picture appears on the monitor screen (it will probably appear blurred at this point).

(7) Adjust the 'optical focus' and the 'electrical focus' to provide the clearest possible picture at this stage. If the iris is opened too far focusing will be more critical with this short subject-to-camera distance.

(8) Readjust the 'target sensitivity' and 'beam' controls gradually until a picture with satisfactory contrast is obtained. The 'beam' control should be advanced to a

BASIC ADJUSTMENTS TO CAMERA AND MONITOR

point at which the highlights of the picture are just reproduced; if it is not the picture will appear 'flat' or 'clipped'. If the picture on the monitor is not of good contrast and the target sensitivity has been advanced to a point at which the background 'noise' is objectionable, it will be necessary to increase the lens aperture.

It is important that the camera is not operated for any length of time with insufficient beam current otherwise the image may become permanently 'burnt-in' on the camera target.

Readjust the electrical focus to obtain the clearest possible picture, which at this stage should be of good contrast. We are now ready to adjust the camera geometry and while preparing for the following steps, refit the lens cap.

(9) Remove lens cap. Increase the camera vertical and horizontal amplitude controls until a black border appears around the image of the test chart on the monitor screen.

(10) Rotate the deflector coil assembly and camera tube until the sides of the test card are parallel to the sides of the monitor raster (it will be necessary to release a locking screw to do this: see manufacturer's instructions).

(11) The camera tube is then rotated (again it will be necessary to unlock some form of clamping screw) until the black border on the monitor screen is squarely around the test card.

(12) Adjust the alignment magnets until the monitor picture rotates axially about the centre with adjustment of the electrical focus control (this aligns the electron beam in the focusing field).

(13) Set the camera vertical and horizontal scan linearity controls until the squares of the test chart grid pattern are of equal size.

(14) Decrease the camera vertical and horizontal scan amplitude controls until the black border (caused by the mask) just disappears over the edges of the raster. Recheck and adjust the linearity control as necessary.

The camera geometry is now set.

A full setting-up procedure for a camera will include such adjustments as 'video peaking', 'scanning generator speeds', 'video signal levels', 'focus preset', etc. Such adjustments cannot be considered here as methods differ between manufacturers, and according to the type of circuits used.

CHAPTER 9

LIGHTING

ONLY too often this aspect of a CCTV installation is overlooked resulting in poor pictures in spite of, perhaps, the use of quite expensive camera and monitor equipment. Inadequate lighting causes noisy pictures, loss of contrast and pictures that are brighter in the centre than at the edges. The wrong type or position of the light source causes shadows, flat look, loss of important scene detail or excessive contrast between the subject and background, etc. Studio lighting is an involved subject and is outside the scope of this volume. However, much useful information can be obtained from the way a professional lighting engineer sets out and uses the lights in a mini-studio, which will be considered later.

LIGHT SOURCES
(1) Incandescent
There are two main types of modern light sources available; the incandescent lamp and the tungsten halogen lamp. The incandescent lamp is the most common and is the one most people use in their homes. An incandescent lamp has a tungsten filament which emits visible light when the filament temperature is raised above about 500°C. Low wattage lamps up to about to about 40 W are normally of the vacuum type. Above this wattage they are usually gas-filled (argon and nitrogen) which reduces the evaporation rate of the tungsten filament enabling the bulb to be run at a higher temperature for an equivalent life. In normal domestic situations lamps of up to 200 W are used, but for television and other applications wattages in the range of 500 W to several kilowatts are available. The main advantages of the incandescent lamp are its relative cheapness, robustness and freedom from excessive heat. However, this type of lamp deteriorates with use due to the evaporation of the tungsten filament which blackens the inside of the bulb. Also, the lamp is relatively of low efficiency as a large portion of the electrical power that is supplied is dissipated as heat.

An incandescent lamp can be made in various forms to give directional beam properties. In one type, the lamp is constructed so that the filament forms the focal point of a parabolic glass surface which is blown to form an integral part of the bulb and coated with a highly reflective material such as aluminium. Where accurate beam characteristics are required a 'sealed beam' construction is used. In the manufacture of this type of lamp the filament is positioned relative to the reflecting surface with a high degree of accuracy. A pressed glass construction is used and the front of the lamp may have a glass cover or be fitted with a lens to produce a narrow, medium or wide beam of light.

The incandescent lamp may be fitted in floodlamps, profile spot-lights or fresnel spot-lights. A floodlight is a lantern with a diffuse beam of angle 100° or more. It usually employs a simple polished metal or silvered glass reflector and has no shielding of direct light from the lamp, diagram (a) of figure 9.1. A profile spot-light, diagram (b), projects an intense circular beam the diameter of which may be controlled by an iris. Light from the reflector is collected by a lens, the position of which determines whether the beam is brought to a 'hard' or 'soft' focus, A fresnel spot-light diagram (c) employs a fresnel lens (figure 9.2) which is used to produce a soft-edged beam. Its optical system comprises a spherical reflector, lamp and lens. Movement of the lamp and reflector towards or away from the lens varies the beam angle (usually over a range of about 15° to 60°). Thus the fresnel spot is really a combination spot-light/floodlight.

LIGHTING

(a) Floodlight
(wide angle diffuse beam)

(b) Profile spotlight
(concentrated circular beam)

(c) Fresnel spot-light
(soft-edged spot or flood)

FIG. 9.1 LUMINAIRES

(a) Parabolic lens

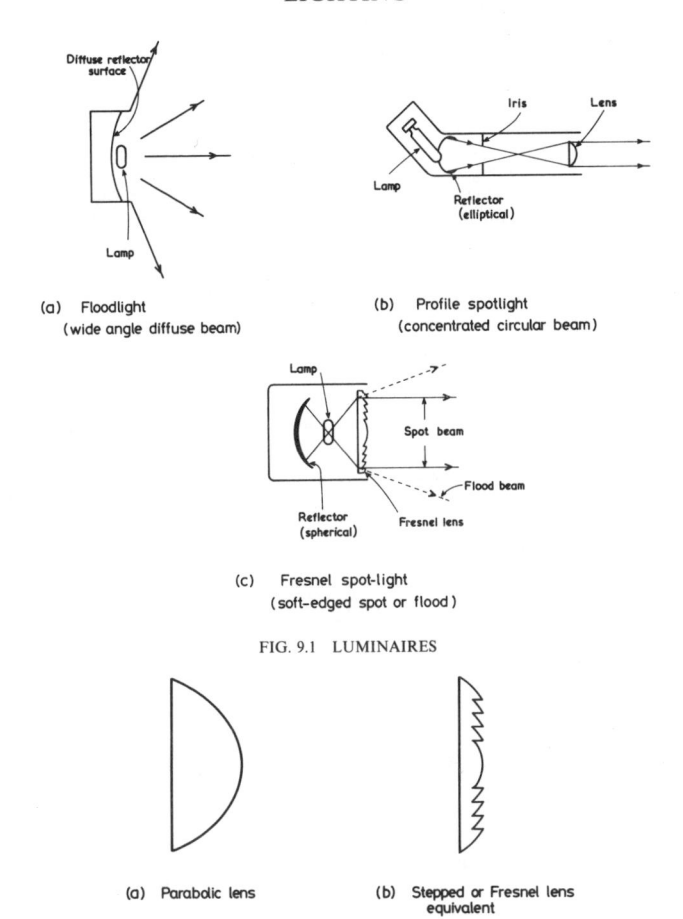

(b) Stepped or Fresnel lens equivalent

FIG. 9.2 THE FRESNEL LENS
(A parabolic lens which is used to reduce spherical aberration is heavy and bulky. Its size may be reduced by using a stepped or ridged construction)

(2) Tungsten Halogen (Quartz-Iodine)

As mentioned, the incandescent lamp loses filament material by evaporation, much of it being deposited on the bulb wall. If a halogen such as iodine is added to the filling gas the iodine vapour causes the evaporated tungsten to be deposited back on the filament where it becomes available for further light emission (and evaporation). This regenerative cycle requires a high operating temperature for the filament (about 1700°C) and the bulb wall (about 250°C). Because of the high temperature the lamp envelope is usually formed from fused silica (quartz). This type of lamp is known as the Tungsten-Halogen lamp but is often referred to as Quartz-Iodine or Quartz-Halogen. These lamps are smaller than the incandescent types, have a higher efficiency, are not blackened by age and produce a 'white' light that remains constant throughout their working life. They are, however, more expensive, create high temperatures and should not be handled by the naked hand (to prevent skin oils reacting with the quartz envelope) and they should never be touched when hot.

The light emitted from a tungsten-halogen lamp can be given directional beam properties in the same way as an incandescent lamp by the use of reflectors and lenses.

Moreover, because of its small size and robust construction, the tungsten-halogen lamp has allowed luminaires and optics to be miniaturised and made less expensive. For these reasons, coupled with the constancy of the white light emitted, this type of light source is the main one used by the national television broadcast companies.

LIGHTING THE SUBJECT

Consider as an example the lighting required for a miniature studio when the camera is viewing a person who is taking part in a discussion or lecture and looking straight into the camera lens, figure 9.3. Here the system of lighting employed can be broken down into four distinct parts: Key, Fill, Back and Background.

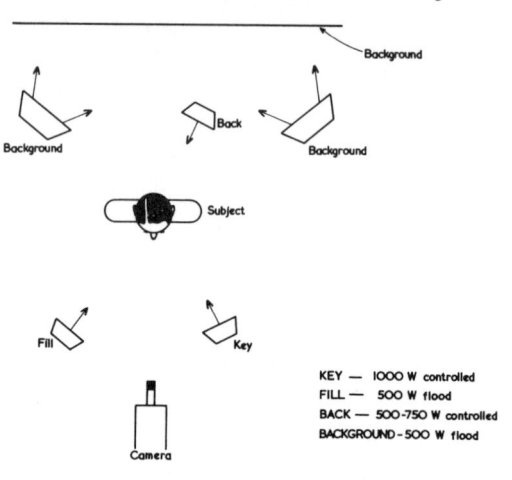

KEY — 1000 W controlled
FILL — 500 W flood
BACK — 500-750 W controlled
BACKGROUND - 500 W flood

FIG. 9.3 AN IDEALISED MINI-STUDIO SET-UP

The Key light is the essential light source for illuminating the set, providing small areas of light. If it were applied at the same angle as the camera, or at 90° to the subject, detail may be lost due to the lack of contrast on the face—the subject looking 'flat'. It is therefore applied at an angle to the subject to create an illusion of depth and to provide a presentable television picture. At an angle, however, the Key light causes shadows and if these are too harsh, some sort of 'fill' is needed. The Fill light softens the shadows and brings out detail that otherwise might be lost. The Back light is added to give a natural look and to create separation. It provides a halo around the head and shoulders of the subject and gives attractive highlighting to the hair. Finally, the Background lighting removes excessive contrast between the subject and the background, depending upon the nature and colour of the background surface. It should be mentioned that the front lighting of the subject should be applied at an angle to avoid dazzling the person. An angle of about 60° between the face, lamp and floor is desirable.

In a professional television studio, during the televising of the news for example, sometimes up to twenty lamps may be used to illuminate the set of the news reader. However, the five lamps used in the mini-studio provides the basic concept which is *splitting the light source* to achieve a picture with apparent depth, good detail, no harsh shadows and to modify the contrast range to a reasonable level.

LIGHTING IN INDUSTRY AND COMMERCE

In a television studio the contrast range may be limited to about 20:1 whereas outside the range is of the order of 100:1. Industrial installations often have to contend with such a range of contrast values. Inside a factory there may be adequate lighting for

LIGHTING

normal work purposes in some areas, but poor light in others. Thus an industrial CCTV set-up may start with some general form of lighting, *e.g.* fluorescent or incandescent but steps must be taken to improve the intensity where this is inadequate for television. A minimum illumination level of about 20–30 lumens per square foot (approximately 200–300 lux) should be aimed at.

By trial and error and careful observation, techniques similar to those used in the studio may be tried to modify the contrast range by splitting the same amount of light between lamps. Key and Fill lights may be all that is required, for by experimenting with lamp positions there may be sufficient direct or reflected light to illuminate the background. Often only a key light may be all that is provided and here again experimenting with the lamp position may provide sufficient reflected light to act as a 'fill'. Try for a picture with a moderate contrast range; remember that vidicon cameras viewing static contrasty scenes over long periods usually suffer from 'burn-in'. The high luminance of odd items in the scene, *i.e.* polished metal parts may be subdued by the use of paint.

In industrial situations where steam is present between the camera and subject, the light is best placed after the steam or directed down through the steam so that reflection off the steam or attenuation by the steam is kept to a minimum.

For compound or yard lighting where CCTV is used for security purposes it is the vertical surfaces that normally must be made visible. Odd pools of light, acceptable to the naked eye provide too much contrast for the camera under night-time conditions. Even overall floodlighting is required to lift out the dark areas and provide a more acceptable contrast range.

After establishing the position of the lamp it must be firmly fixed. Portable stands are not really suitable for industrial situations as they are easily knocked over. Suitable mounting brackets are available for fixing the lamp head to ceiling, wall, girder or some other rigid support in proximity to the subject. Stout extension poles may be used to place the light in the best position.

When an interview is being recorded on video tape or is televised for direct viewing, it may be laid down that no extra light source is to be used so that the person being interviewed does not have to face the glare of camera lamps. One has then to make do with the available lighting. Fluorescent lamps make a satisfactory light source provided they are sufficient to set up the minimum illumination level. Alternatively, the 'bounce principle' may be used: a flood-light is directed at the ceiling which, if light in colour, will provide a soft even 'fill' by reflection.

CHAPTER 10

THE SIGNAL CABLE

THE feeder cable in a CCTV system, diagram (a) of figure 10.1, is to convey the video signal energy from camera to monitor and should do so with negligible loss of signal power. We may represent the arrangement of diagram (a) by an electrical circuit in which the camera is replaced by a signal generator and the monitor is represented by a load resistor, as in diagram (b). Although in this type of diagram it is normal practice to represent the feeder by a pair of lines or an 'inner' conductor and screened 'outer' conductor we have to bear in mind the frequency of the signal to be conveyed along it. When the wavelength of the signal to be sent becomes comparable with the length of the line, there may be appreciable time difference (phase difference) between the voltage at one point and the voltage at another point along the line. This is because a pair of feeder wires possess distributed inductance and capacitance, which is usually ignored in the distribution of power frequency (50 Hz) voltages over normal distances met with in factories and homes. Diagram (c) shows an approximate equivalent circuit for a pair of lines or coaxial feeder, where C = distributed capacitance (farads) per unit length; L = distributed inductance (henrys) per unit length; R_1 = ohmic resistance per unit length; and G = leakance in Siemens per unit length. These quantities are evenly distributed along the length as an infinite number of infinitely small 'lumps'.

Suppose that a high frequency signal voltage is applied to the input of the line in diagram (c). The voltage will cause a current to flow in the first capacitor but the current will lag the input voltage because of the inductance in the wires. As the capacitor charges, a voltage will be built up across it but delayed on the input voltage. This voltage will in turn feed current into the next section causing the second capacitor

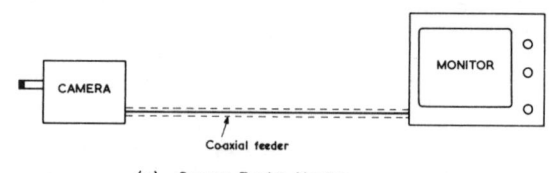

(a) Camera–Feeder–Monitor

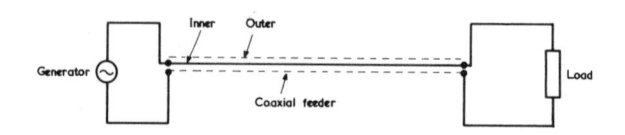

(b) Generator–Feeder–Load

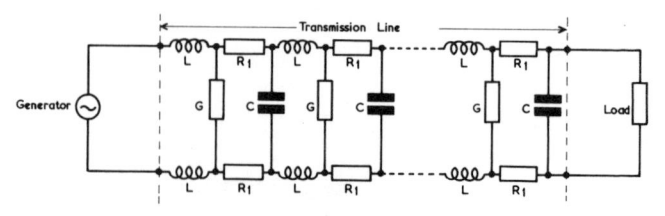

(c) Electrical representation of (b)

FIG. 10.1 TRANSMISSION LINE

THE SIGNAL CABLE

to charge up, and the voltage across it will be further delayed on the input voltage. In its turn the second capacitor voltage supplies current to the next section, and so on. In this way the voltage is transmitted along the cable until it reaches the load at the far end. The amount of phase lag depends upon the distributed capacitance and inductance per unit length and is related to them by

$$\text{Phase change coefficient } \beta = \omega\sqrt{LC} \text{ measured in radians per unit length}$$
$$\text{(when the line has no losses)}$$

The length of line at which either the current or voltage for the first time is in phase with the supply end is called the wavelength of the line, which is slightly shorter than the 'free-space' wavelength.

CHARACTERISTIC RESISTANCE (R_0)

(a) Infinite length of transmission line

(b) R.M.S. voltage and current along an infinite length of transmission line of $R_0 = 60\,\Omega$ (no losses-ideal case)

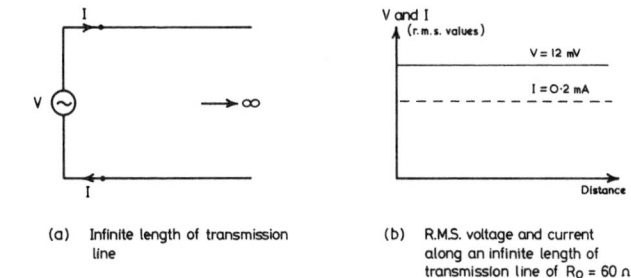

FIG. 10.2 CHARACTERISTIC RESISTANCE (R_0)

If we take an infinite length of feeder (with negligible losses) as in figure 10.2, diagram (a) and apply a signal voltage V, a current I will flow in the line (remember that the distributed components provide a current path). The ratio

$$\frac{V}{I} \quad \text{(r.m.s. values)}$$

is called the characteristic resistance of the feeder. Neglecting losses, the characteristic resistance may be found from

$$R_0 = \sqrt{\frac{L}{C}}$$

Note that frequency does not come into the expression, *i.e.* R_0 is the same for all frequencies. As an example suppose that a feeder has a distributed inductance of 1 mH per mile and a distributed capacitance of 0·25 µF per mile, then

$$R_0 = \sqrt{\frac{1 \times 10^{-3}}{0 \cdot 25 \times 10^{-6}}} \text{ ohms}$$

$$= 63 \text{ ohms (approx.)}$$

If we approximate R_0 to 60 ohms, then if a signal voltage of 12 mV is applied at the input of a feeder of this characteristic resistance, the current in the line will be

$$\frac{12 \times 10^{-3}}{60} = 0.2 \text{ mA},$$

diagram (b) of figure 10.2. Since R_0 is constant no matter where we 'look' into the line [diagram (a) of figure 10.3], any of the sections shown in the diagram may be removed and replaced by a resistor of value R_0 ohms as shown in diagram (b). Thus a finite line terminated in R_0 'looks' like an infinite length of line. This is the 'matched' or 'reflection-free' condition where all the power that is fed into the feeder is dissipated in the load R_0.

(a) R_0 is 'seen' wherever we 'look' into an infinite length of line

(b) Any of the sections of (a) may be removed and replaced by a resistor of value R_0 ohm

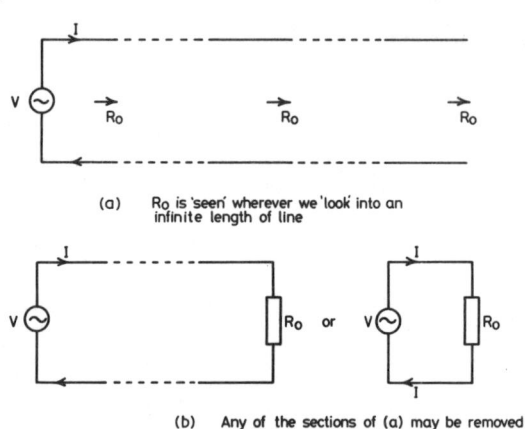

FIG. 10.3 FINITE LINE TERMINATED IN R_0 'LOOKS' LIKE AN INFINITE LINE

CHARACTERISTIC IMPEDANCE (Z_0)

In all practical transmission lines there are losses. These make the line reactive so it is more correct to speak of characteristic impedance Z_0. Typical characteristic impedance for a feeder cable used in CCTV is 75 ohms. Due to the losses [R_1 and G of figure 10.1(c)] in a practical feeder, V and I progressively get smaller as they travel down the line, see figure 10.4(a). The attenuation follows a logarithmic law and it is usually to express the attenuation in decibels per unit length (normally per 100 ft). Diagram (b) shows the attenuation for a low-loss cable plotted against feeder length. The losses increase with frequency, *e.g.* at 10 MHz the attenuation may be 0·6 dB per 100 ft whereas at 100 MHz it may be 2·4 dB per 100 ft. With signal distribution at video

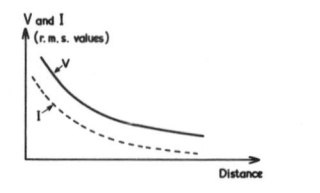

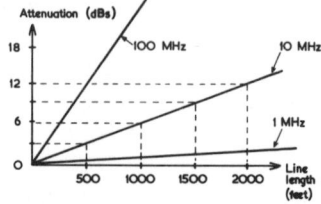

(a) Attenuation on a practical lossy line.

(b) Attenuation in dBs against distance for a low loss line at 1 MHz, 10 MHz and 100 MHz.

FIG. 10.4 CABLE ATTENUATION

THE SIGNAL CABLE

frequencies we are mainly interested in attenuation up to, say, 6 MHz for a camera of average quality; or perhaps 10 MHz for a camera of high quality. Because of the increasing attenuation at high frequencies, some equalisation should be used in the distribution of video baseband signals over long cables. It is less troublesome if the video signal is modulated on to an h.f. carrier of, say, 200 MHz as the relative losses across the band will be less. Cable losses at 200 MHz will, of course, be greater than in the video baseband, thus line amplifiers would be required to make up the losses in an h.f. distribution system.

Let us consider an example to show the effect of cable length on signal levels. Suppose that the camera output voltage when supplying a monitor which is correctly terminated in Z_0 is 1V. Let us say that the feeder length is 800 ft (not uncommon in industry) and that the attenuation of the cable is 0·3 dB per 100 ft at 6 MHz (assumed to be the highest video frequency).

$$\text{Total feeder attenuation} = 8 \times 0 \cdot 3 \text{ dB} = 2 \cdot 4 \text{ dB}$$

Now attenuation (dB) $= 20 \text{ Log } \dfrac{v_o}{v_i}$ (where v_o is the camera output voltage and v_i is the monitor input voltage)

$$\therefore 2 \cdot 4 = 20 \text{ Log } \frac{1}{v_i}$$

$$\text{or } \frac{2 \cdot 4}{20} = \text{Log } \frac{1}{v_i}$$

$$\text{or } \frac{1}{v_i} = \text{antilog } \frac{2 \cdot 4}{20}$$

$$\frac{1}{v_i} = \text{antilog } 0 \cdot 12$$

$$\frac{1}{v_i} = 1 \cdot 318$$

$$v_i = \frac{1}{1 \cdot 318}$$

$$\underline{v_i \simeq 0 \cdot 75 \text{ V}}$$

Note that for every 6 dB of attenuation the voltage is halved. Thus, with a camera voltage of 1 V and 6 dB attenuation the monitor voltage is 0·5 V, with 12 dB attenuation it is 0·25 V and for 18 dB it is 0·125 V, etc.

REFLECTIONS

If the termination is not equal to Z_0 the feeder is said to be mismatched and reflection occurs at the termination. An o/c or s/c termination, figure 10.5, will cause all of the arriving energy to be reflected back towards the sending end. As the wave travels back down the line it is retarded in phase and attenuated in precisely the same manner as when it is travelling in the forward direction. The forward and reflected waves travelling in the line give rise to a standing-wave pattern. At some points along the line the two waves cancel to produce current and voltage minima (nodes) whereas at other points addition occurs creating maxima (anti-nodes). The maxima and minima repeat themselves regularly at $\lambda/2$ intervals. In diagram (a) the voltage at the termination must be zero since it is a s/c whereas in diagram (b) the current is zero at the termination as it is o/c. Both situations represent a gross mismatch of the line.

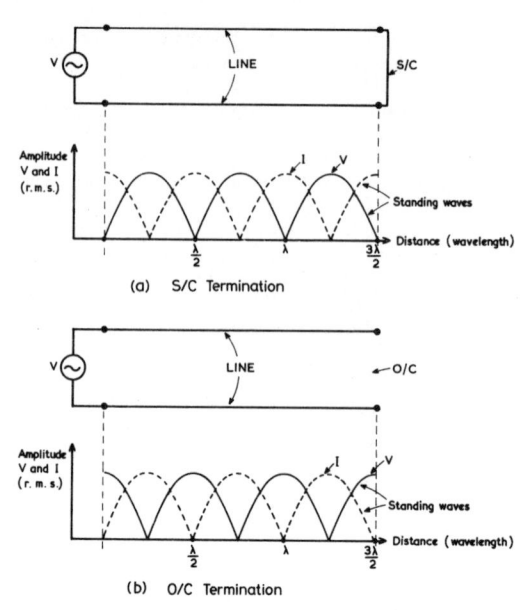

FIG. 10.5 STANDING WAVES ON MIS-MATCHED CABLE LINE

The reason that reflection occurs at the termination when it is mismatched to the line is that the wave does not 'know' what the termination is until it arrives there. A parallel situation occurs when a sound wave is directed at a wall. If the wall is made of an acoustically absorbent material the sound energy will be absorbed; this is the matched condition. If the wall is made of brick, reflection will occur and the sound wave will travel back to the source.

Of course, the s/c and o/c terminations are extreme examples of mismatch. Intermediate impedance values of termination will also cause a standing-wave pattern on departure from the nominal ideal loading of Z_0. Mismatch causes a loss of signal power and leads to radiation from the feeder, causing interference and patterning, particularly in high level r.f. cable systems. The degree of mismatch may be assessed by measuring the voltage (or current) peaks and troughs. The ratio of

$$\frac{V_{max}}{V_{max}} \quad \text{or} \quad \frac{I_{max}}{I_{max}}$$

is called the standing wave ratio (s.w.r.) of the feeder. For the matched line the s.w.r. becomes 1·00 S.W.R. values greater than unity indicate mismatch. The table shows the proportion of power lost for various s.w.r.s.

S.W.R.	PROPORTION OF POWER LOST
1·00	0
1·25	0·012
1·5	0·04
2·0	0·11
3·0	0·25
4·0	0·36
5·0	0·44
10·0	0·67

THE SIGNAL CABLE

GHOSTING

Under mismatch conditions in a camera-cable-monitor system, energy arriving at the monitor may be reflected back towards the camera. At the camera the video signal may again be reflected and travel back to the monitor. In travelling up and down the feeder line, some loss will occur but if the reflected signal is of sufficient amplitude and the feeder length appreciable a second or 'ghost' image may appear on the monitor screen. For example, consider a 100 metre length of coaxial cable employing a solid polythene dielectric which is mismatched at the monitor.

Now the velocity of a wave along the cable may be found from

$$v = \frac{300 \times 10^6}{\sqrt{\mu_r e_r}} \text{ m/s} \quad \text{(where } \mu_r \text{ and } e_r \text{ are the relative permeability and permittivity of the dielectric)}$$

$$\text{or } v \simeq \frac{300 \times 10^6}{\sqrt{e_r}}$$

For polythene $e_r = 2 \cdot 3$

Therefore

$$v = \frac{300 \times 10^6}{\sqrt{2 \cdot 3}} \text{ m/s}$$

$$v = 197 \cdot 8 \times 10^6 \text{ m/s}$$

Thus the time taken to travel 200 metres (100 metres each way)

$$= \frac{1}{v} \times 200 \text{s}$$

$$= \frac{200}{197 \cdot 8 \times 10^6} \text{s}$$

$$\simeq 1 \text{ }\mu\text{s}$$

On a monitor screen of width, say, 10" the ghost image will be displaced

$$\frac{1}{52} \times 10"$$

to the right of the main image (note active line period is 52 μs) which is approximately 0·2".

When the cable length is short (about 30 metres or less) it will be difficult to see the ghost image but a loss of definition will result.

COAXIAL CABLE

Diagram (a) of figure 10.6 shows the usual form of construction for a flexible coaxial cable of the type used in CCTV. The cable consists of an 'inner' solid copper conductor (or twisted copper wires) surrounded by a flexible insulating material such as polythene. The insulating medium may be of solid polythene or of cellular construction to reduce the losses at high frequencies. The 'outer' conductor consists of

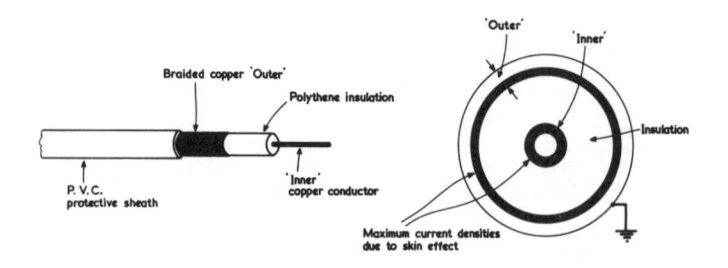

(a) Cable construction (b) Skin effect

FIG. 10.6 COAXIAL CABLE

a copper braid wound over the insulation. To protect the cable from moisture and damage it is covered with a tough p.v.c. sheath.

At high frequencies, the currents in a conductor tend to flow in its outer skin (skin effect). As the frequency is raised the thinner the skin becomes and the smaller the cross-sectional area. Since the resistivity of a conductor is inversely proportional to its cross-sectional area, the a.c. resistance of a cable increases with frequency (proportional to $\sqrt{f}$). With a coaxial cable the skin is formed on the outside of the 'inner' conductor but on the inside surface of the 'outer', see diagram (b). The penetration of current on the outer braiding is negligible (except under high s.w.r. conditions) thus the screen may be earthed. The earthing of the screen helps to suppress mutual interference between different feeder cables and the general pick up of unwanted signals.

As it is essential to maintain correct matching in a CCTV installation and continuous screening of the feeder cable, lengths of cable cannot be joined simply by ensuring a sound electrical connection only by way of a junction box or barrier strip as for ordinary electrical wiring. A properly designed 'line coupler' (figure 10.7) must be used which maintains the characteristic impedance of the cable at the joint and

Coaxial plug   Coaxial socket (surface mounting)   Coaxial socket (flush mounting)

Coaxial cable Line connector

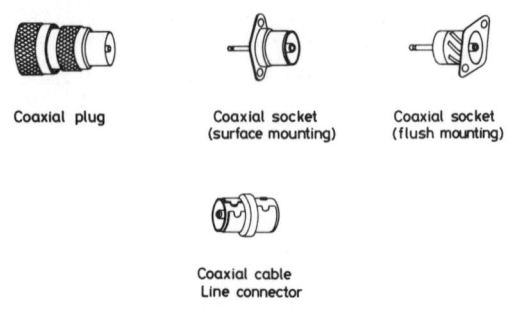

FIG. 10.7 COAXIAL CABLE PLUGS AND SOCKETS

provides a continuation of the screening. Because of the possibilities of signal loss and ghost images, feeders of different characteristic impedances should not be joined together. It is also important that all plugs and sockets should only be used with the cable Z_0 they were designed for.

When using a c.r.o. for fault-finding or adjustments on a camera it should be correctly terminated as shown in figure 10.8. When the c.r.o. is connected to the video output socket or other 75 Ω test-point the nominal 75 ohm impedance is maintained in the c.r.o. test lead and at the terminating resistor. This avoids reflections which may

THE SIGNAL CABLE

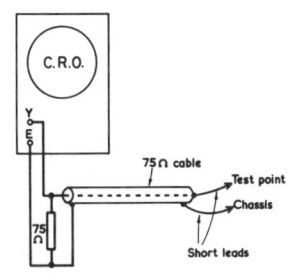

FIG. 10.8 TEST LEAD AND TERMINATION WHEN USING C.R.O. TO MONITOR VIDEO WAVEFORM AT VIDEO OUTPUT

produce incorrect signal levels and false results when carrying out adjustments to the signal. It is important to note that the 75 Ω termination on the c.r.o. only applies when measuring levels at the INPUT and OUTPUT of video equipment. A termination on the c.r.o. will make many measurements on the internal circuits of camera or monitor meaningless, especially high impedance ones. This is a common cause of measurement errors in practice.

VIDEO SWITCHING BOX

It is often desirable in a CCTV installation to be able to select the output from any of several cameras for display on the viewing monitor. This is a common requirement in process control where several cameras are strategically placed around the plant and feed signals to a control desk which for economical reasons and ease of operation houses only one monitor. Thus some form of video switching is needed and the basic form of one arrangement is shown in figure 10.9.

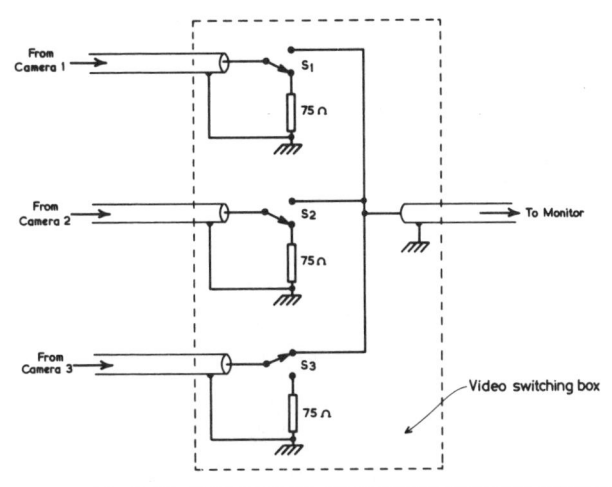

FIG. 10.9 USE OF A VIDEO SWITCHING BOX FOR FEEDING THE OUTPUT OF ANY OF SEVERAL CAMERAS TO THE DISPLAY MONITOR

Here the switches S_1–S_3 are mechanically interlocked so that operating one set of switch contacts disengages the others. In the position shown, the output of camera 3 is being fed to the display monitor. Cameras 1 and 2 are correctly terminated by 75 ohm load resistors which matches the camera feeder cables and thus avoids the setting-up of standing waves and mutual interference between the switching box output and the unused camera lines.

CHAPTER 11

FAULT-FINDING CHARTS

WHEN fault-finding on CCTV equipment a logical approach is required as with any other form of complex electronic equipment. In the charts that follow it will be assumed that the installation follows the general layout of figure 11.1. Here, the output from any of several cameras is switched to the input of the display monitor *via* a video switching box.

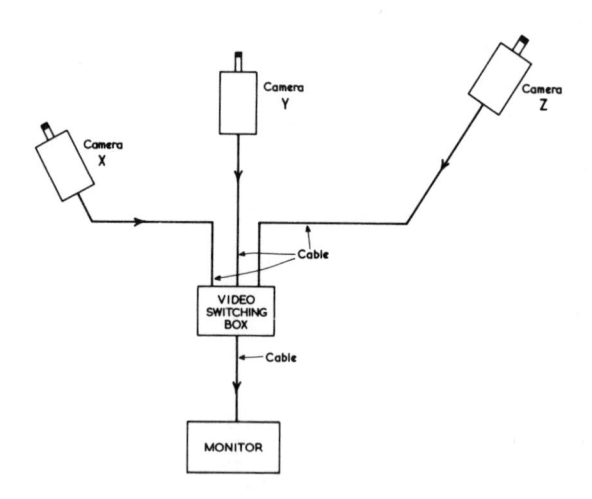

FIG. 11.1 CCTV LAYOUT TO BE USED IN FOLLOWING CHARTS

When a fault is present its effect is usually immediately apparent on the monitor screen, except for obscure fault conditions which only show up on particular types of picture information, *e.g.* test cards or with the aid of specialized test equipment. Thus the initial symptoms given in the chart are those observed on the monitor screen when attempting to display the ordinary camera signal. These symptoms form the starting point for the logical steps to be taken in narrowing down the fault to the camera, cable, switching-box or monitor.

The essential test equipment required consists of a good quality c.r.o. (Y sensitivity of 2 mV per cm and bandwidth of 0–10 MHz) with a terminated test probe, an electronic voltmeter/ohmmeter and a test signal source such as a grating generator or l.f. signal generator. The c.r.o. may be used for waveform checks in following the signal path in either the camera or monitor using the camera image as the signal source, or the c.r.o. may be used in conjunction with the l.f. generator in the camera/monitor for localising the fault to a particular stage or area. A high impedance electronic voltmeter is required for d.c. measurements on the various transistor stages etc. in attempting to narrow down the fault to a particular component.

# FAULT-FINDING CHARTS

214   INDUSTRIAL AND COMMERCIAL CCTV

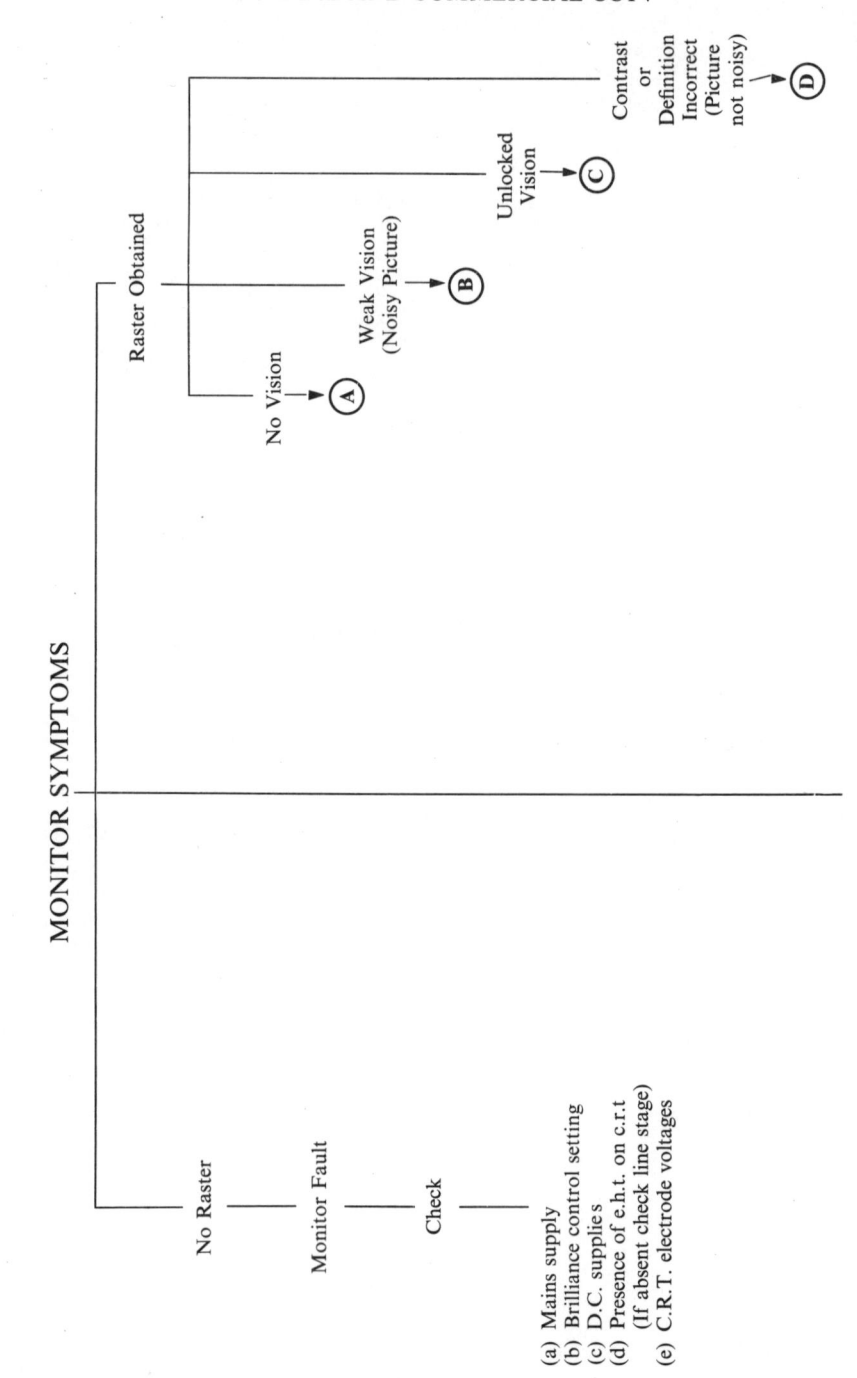

MONITOR SYMPTOMS

- Raster Obtained
  - No Vision → (A)
  - Weak Vision (Noisy Picture) → (B)
  - Unlocked Vision → (C)
  - Contrast or Definition Incorrect (Picture not noisy) → (D)

- No Raster — Monitor Fault — Check
  - (a) Mains supply
  - (b) Brilliance control setting
  - (c) D.C. supplies
  - (d) Presence of e.h.t. on c.r.t (If absent check line stage)
  - (e) C.R.T. electrode voltages

FAULT-FINDING CHARTS

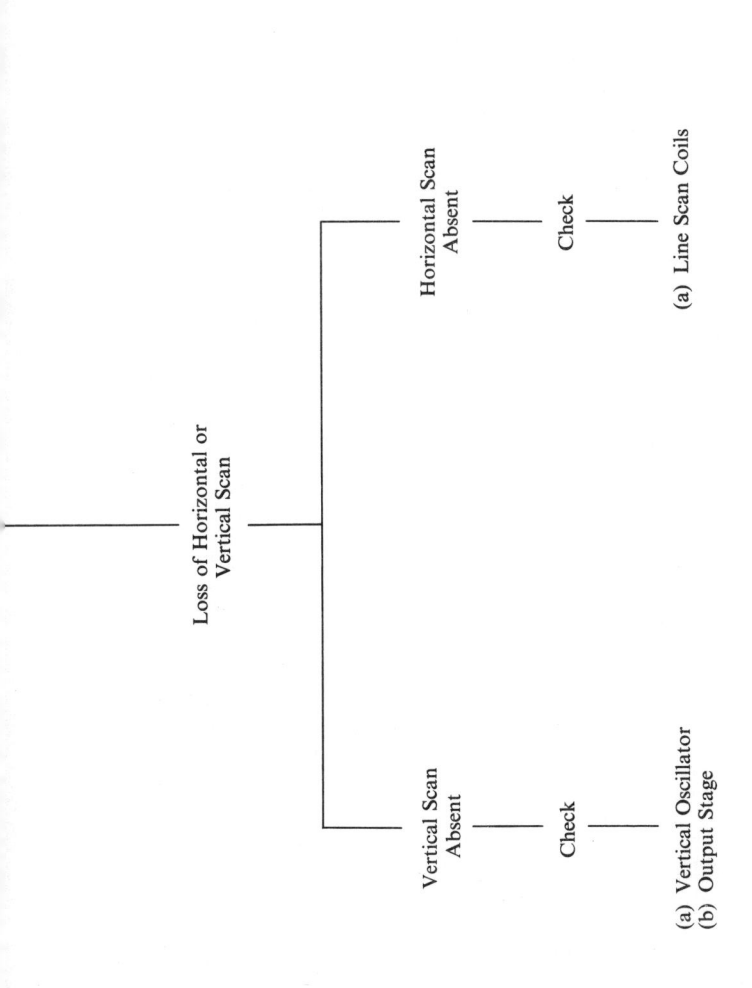

Loss of Horizontal or Vertical Scan
├── Horizontal Scan Absent — Check — (a) Line Scan Coils
└── Vertical Scan Absent — Check — (a) Vertical Oscillator
                                    (b) Output Stage

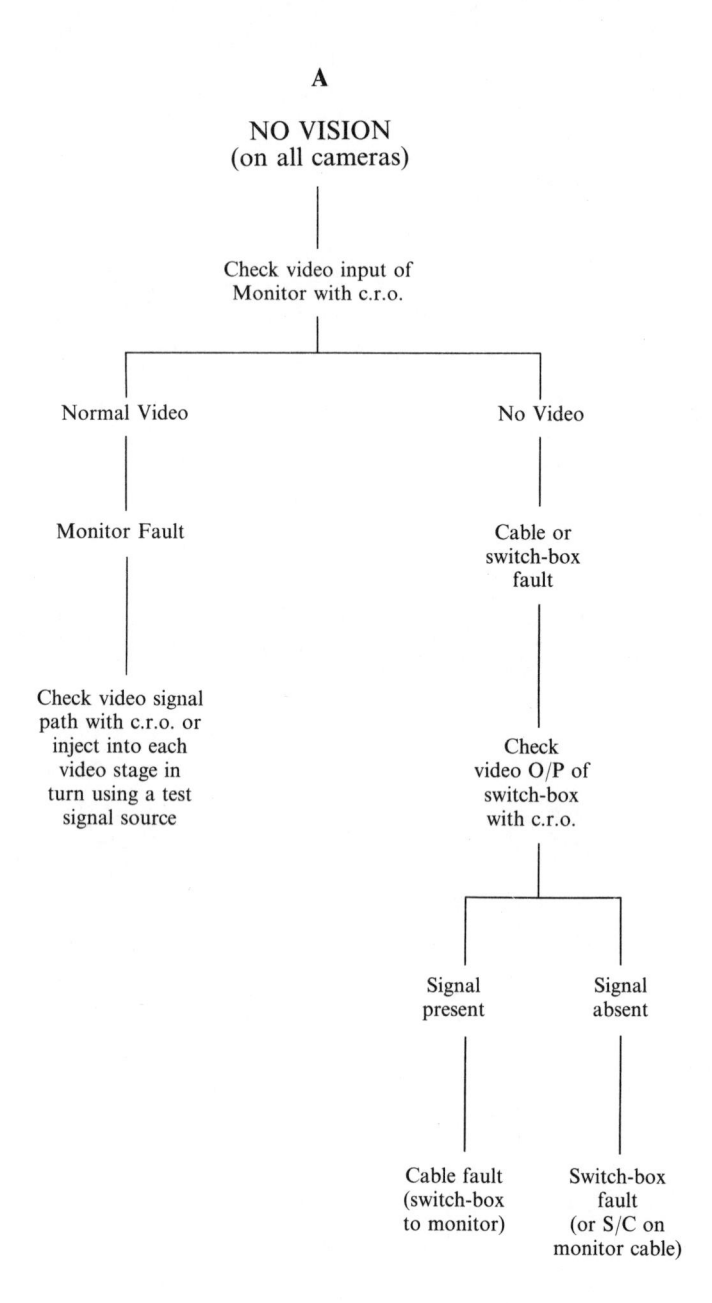

A

NO VISION
(on all cameras)

Check video input of Monitor with c.r.o.

- **Normal Video**
  - Monitor Fault
    - Check video signal path with c.r.o. or inject into each video stage in turn using a test signal source

- **No Video**
  - Cable or switch-box fault
    - Check video O/P of switch-box with c.r.o.
      - **Signal present**
        - Cable fault (switch-box to monitor)
      - **Signal absent**
        - Switch-box fault (or S/C on monitor cable)

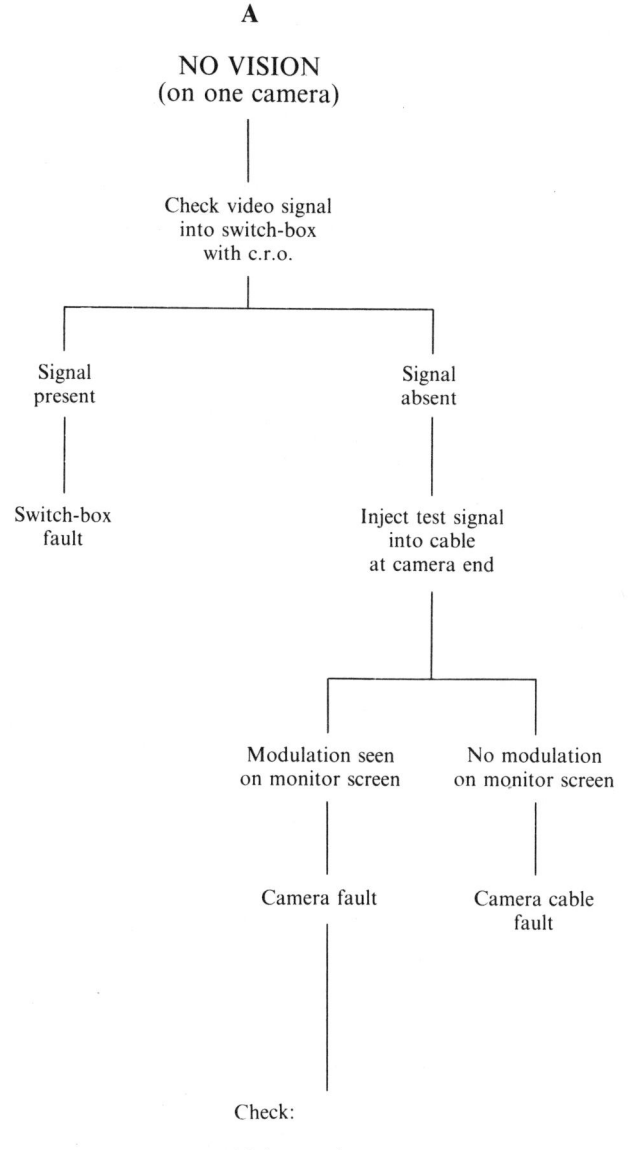

A

NO VISION
(on one camera)

Check video signal into switch-box with c.r.o.

- **Signal present**
  - Switch-box fault
- **Signal absent**
  - Inject test signal into cable at camera end
    - **Modulation seen on monitor screen**
      - Camera fault
        - Check:
          - (a) Mains supply
          - (b) Lens cap
          - (c) Iris setting
          - (d) Video path through camera
    - **No modulation on monitor screen**
      - Camera cable fault

INDUSTRIAL AND COMMERCIAL CCTV

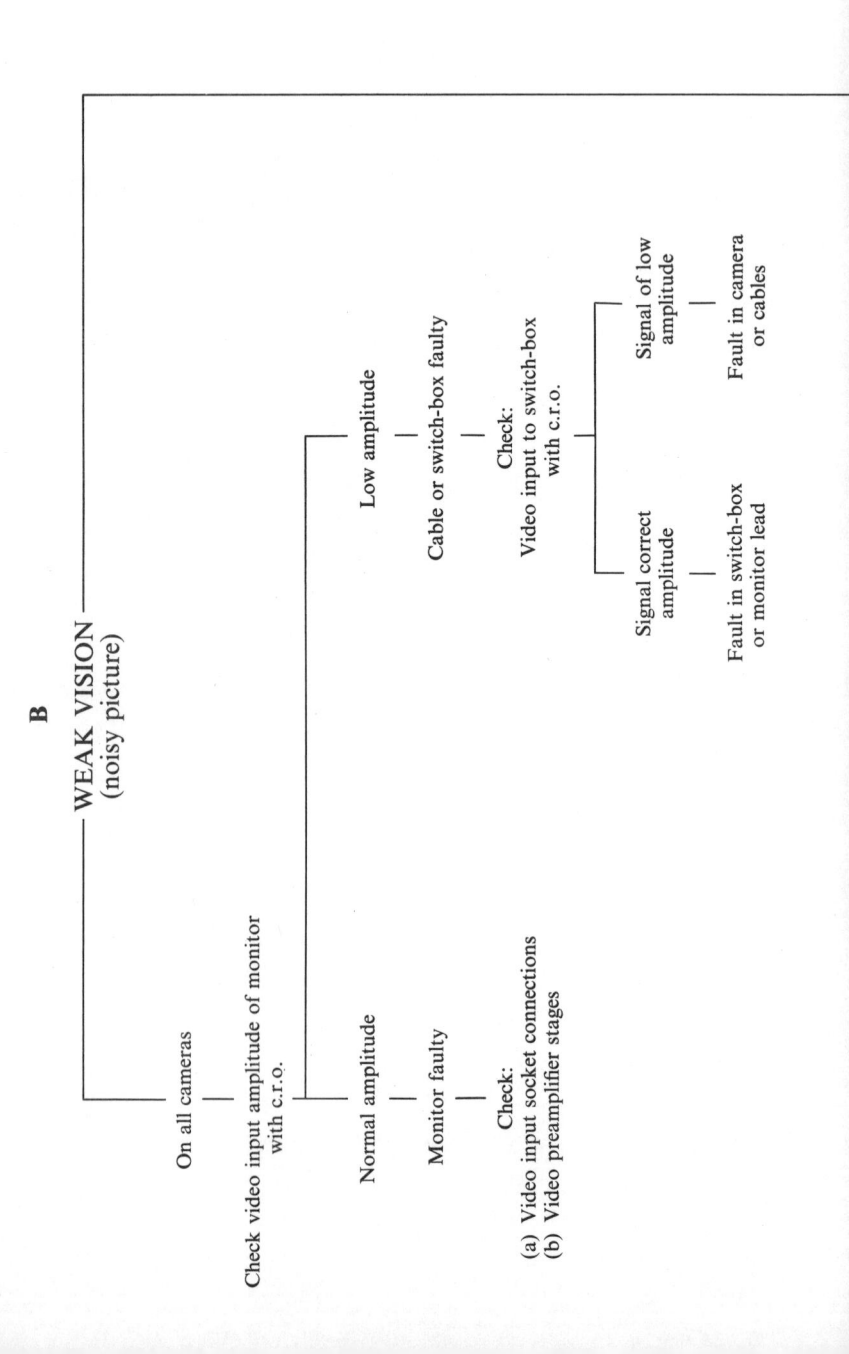

B — WEAK VISION (noisy picture)

- On all cameras
- Check video input amplitude of monitor with c.r.o.
  - Normal amplitude
    - Monitor faulty
    - Check:
      (a) Video input socket connections
      (b) Video preamplifier stages
  - Low amplitude
    - Cable or switch-box faulty
    - Check: Video input to switch-box with c.r.o.
      - Signal correct amplitude → Fault in switch-box or monitor lead
      - Signal of low amplitude → Fault in camera or cables

FAULT-FINDING CHARTS

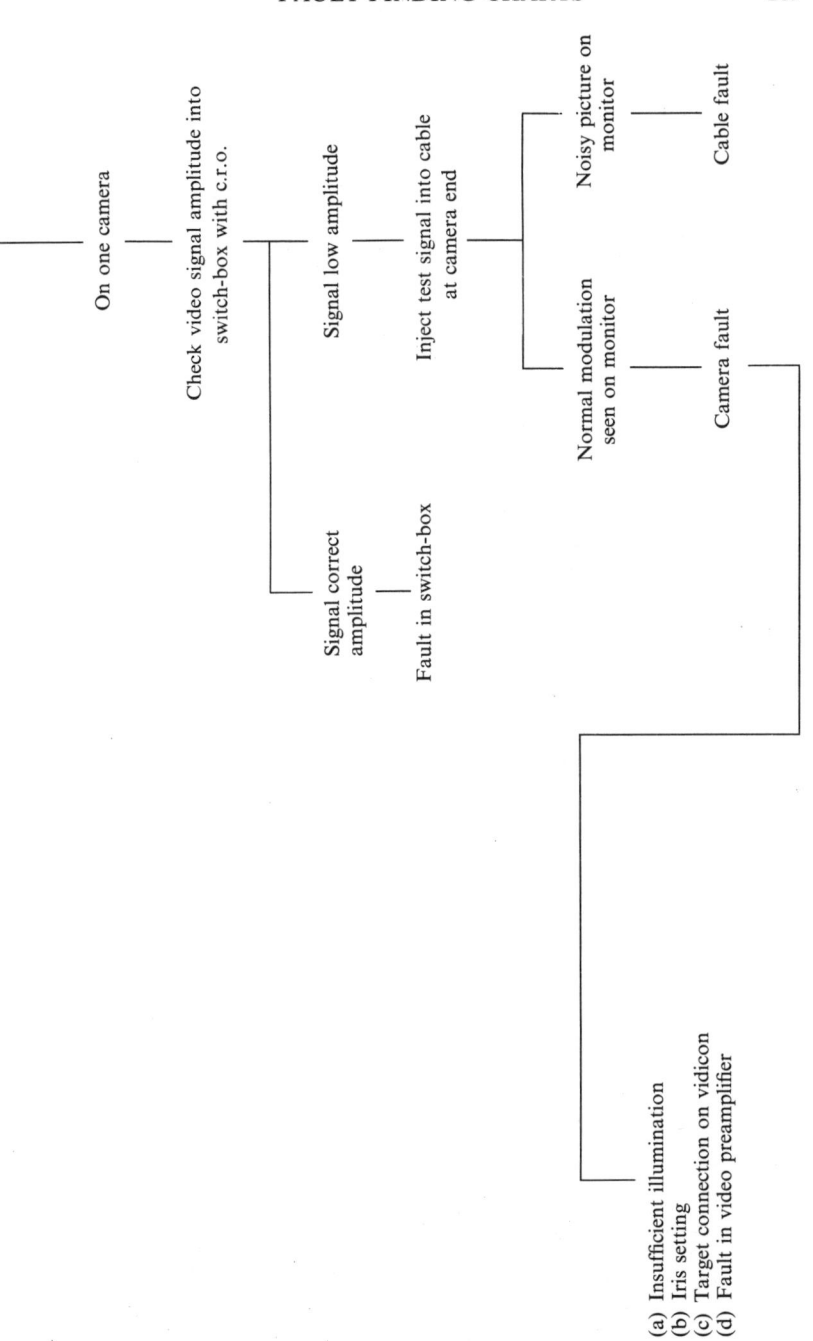

On one camera — Check video signal amplitude into switch-box with c.r.o.

- Signal correct amplitude — Fault in switch-box
- Signal low amplitude — Inject test signal into cable at camera end
  - Noisy picture on monitor — Cable fault
  - Normal modulation seen on monitor — Camera fault
    - (a) Insufficient illumination
    - (b) Iris setting
    - (c) Target connection on vidicon
    - (d) Fault in video preamplifier

C
UNLOCKED VISION

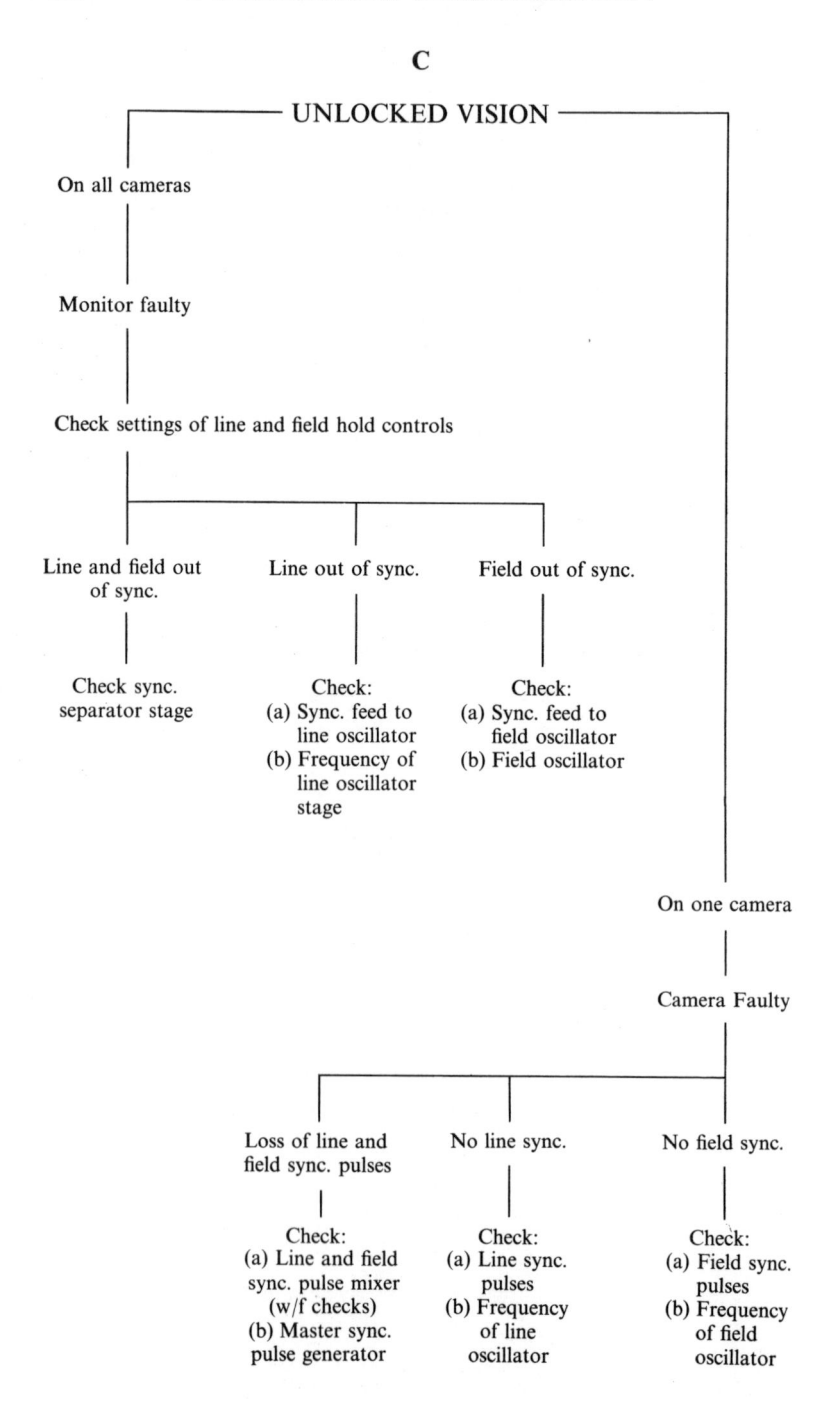

On all cameras

Monitor faulty

Check settings of line and field hold controls

- **Line and field out of sync.**
  Check sync. separator stage

- **Line out of sync.**
  Check:
  (a) Sync. feed to line oscillator
  (b) Frequency of line oscillator stage

- **Field out of sync.**
  Check:
  (a) Sync. feed to field oscillator
  (b) Field oscillator

On one camera

Camera Faulty

- **Loss of line and field sync. pulses**
  Check:
  (a) Line and field sync. pulse mixer (w/f checks)
  (b) Master sync. pulse generator

- **No line sync.**
  Check:
  (a) Line sync. pulses
  (b) Frequency of line oscillator

- **No field sync.**
  Check:
  (a) Field sync. pulses
  (b) Frequency of field oscillator

FAULT-FINDING CHARTS

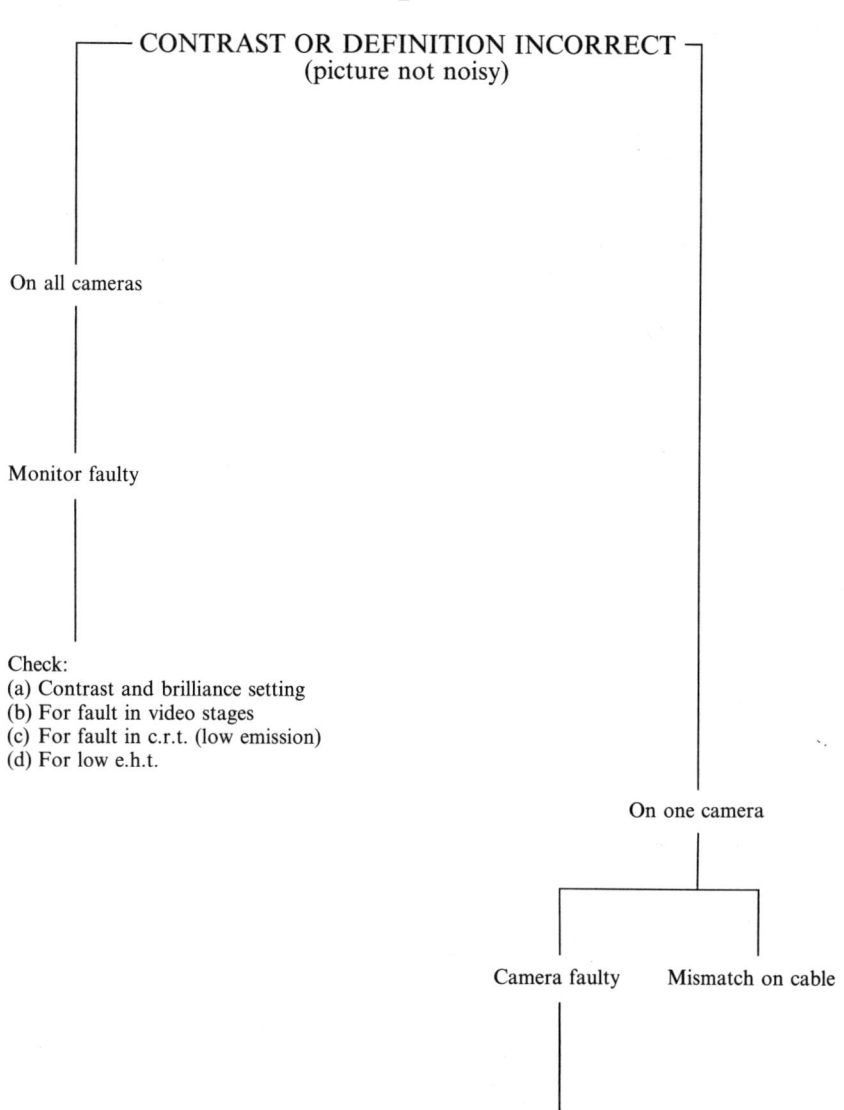

D

CONTRAST OR DEFINITION INCORRECT
(picture not noisy)

On all cameras

Monitor faulty

Check:
(a) Contrast and brilliance setting
(b) For fault in video stages
(c) For fault in c.r.t. (low emission)
(d) For low e.h.t.

On one camera

Camera faulty   **Mismatch on cable**

Check:
(a) Beam current setting
(b) Electrical focus
(c) For fault in video stages
(d) For fault in auto-target circuitry

APPENDICES

APPENDIX A

EFFECT OF SYNC. SEPARATION INTEGRATING NETWORK ON COMPOSITE SYNC. PULSE TRAIN

Before considering the output waveform of the integrating network during the period of the equalizing and vertical sync. pulses it is required to determine the voltage across the integrating capacitor prior to the receipt of the first equalizing pulse.

The complete sync. pulse train is fed into the integrator, thus prior to the receipt of equalizing and vertical sync. pulses, a series of line sync. pulses pass into the integrating

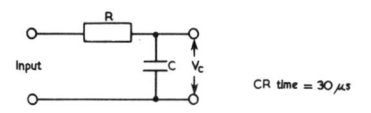

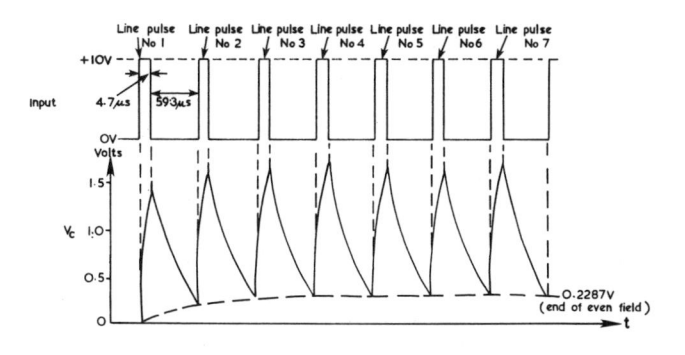

FIG. A.1 WAVEFORM ACROSS INTEGRATING CAPACITOR DURING LINE PULSE TRAIN

network. A few of these pulses are shown in figure A.1. The voltage across C at the end of each pulse may be found from

$$V_C = V_1 + V_D\left(1 - e^{\frac{-t_1}{CR}}\right) \qquad \text{(i)}$$

Where V_1 = voltage across capacitor prior to start of charge, V_D = the voltage difference between the pulse amplitude and V_1, t_1 = pulse duration, CR = time-constant of integrating network and e = base of natural logarithms (2·718).

The voltage across the capacitor at the end of the discharge period may be determined from

$$V_C = V_2\left(e^{\frac{-t_2}{CR}}\right) \qquad \text{(ii)}$$

Where V_2 = voltage across the capacitor prior to commencement of discharge and t_2 = interval between pulses.

Line pulse No. 1

As the terms in brackets in expressions (i) and (ii) are constant during the period of the line pulses, they may be determined initially and substituted as the calculation proceeds for each pulse. Thus:

$$\left(1 - e^{\frac{-t_1}{CR}}\right) = 1 - \frac{1}{e^{\frac{4\cdot7}{30}}} = 1 - \frac{1}{1\cdot17} = 1 - 0\cdot855 = \underline{0\cdot145}$$

$$\text{and } \left(e^{\frac{-t_2}{CR}}\right) = \frac{1}{e^{\frac{59\cdot3}{30}}} = \frac{1}{7\cdot219} = \underline{0\cdot139}$$

Charge

At switch on C is initially uncharged and $V_1 = 0$ V

$\therefore V_C = 0 + 10\,(0\cdot145)$

$V_C = \underline{1\cdot45 \text{ V}}$

Discharge

$V_C = 1\cdot45\,(0\cdot139)$
$V_C = \underline{0\cdot202 \text{ V}}$

Line pulse No. 2.

Charge $V_C = 0\cdot202 + 9\cdot798\,(0\cdot145)$
$V_C = \underline{1\cdot6227 \text{ V}}$

Discharge $V_C = 1\cdot6227\,(0\cdot139)$
$V_C = \underline{0\cdot225 \text{ V}}$

Line pulse No. 3.

Charge $V_C = 0\cdot225 + 9\cdot775\,(0\cdot145)$
$V_C = \underline{1\cdot6423 \text{ V}}$

Discharge $V_C = 1\cdot6423\,(0\cdot139)$
$V_C = \underline{0\cdot2282 \text{ V}}$

Line pulse No. 4.

Charge $V_C = 0\cdot2282 + 9\cdot7718\,(0\cdot145)$
$V_C = \underline{1\cdot6451 \text{ V}}$

Discharge $V_C = 1\cdot6451\,(0\cdot139)$
$V_C = \underline{0\cdot2286 \text{ V}}$

Line pulse No. 5.

Charge $V_C = 0\cdot2286 + 9\cdot7714\,(0\cdot145)$
$V_C = \underline{1\cdot6454 \text{ V}}$

Discharge $V_C = 1\cdot6454\,(0\cdot139)$
$V_C = \underline{0\cdot2287 \text{ V}}$

Line pulse No. 6.
    Charge $V_C = 0.2287 + 9.7713\,(0.145)$
              $V_C = \underline{1.64553 \text{ V}}$

    Discharge $V_C = 1.64553\,(0.139)$
              $V_C = \underline{0.228729 \text{ V}}$

Line pulse No. 7.
    Charge $V_C = 0.228729 + 9.771271\,(0.145)$
              $V_C = \underline{1.64556 \text{ V}}$

    Discharge $V_C = 1.64556\,(0.139)$
              $V_C = \underline{0.2287332 \text{ V}}$

Thus, after about five line pulses the voltage across the integrating capacitor settles down to approximately 0·2287V at the end of each discharge period. This voltage level may be used as the starting level at the end of an *even* field. At the end of an *odd* field there is a half-line difference between the last line pulse and the first equalizing pulse, thus the voltage across the integrating capacitor will be

$$V_C = 1.6456 \left(e^{\frac{-27.3}{30}} \right)$$ where 27·3 μs is the interval between line pulse and first equalizing pulse

$$V_C = 1.6456\,(0.4025)$$

$$V_C = 0.662 \text{ V}$$

There is thus a higher starting level at the end of an odd field.

Consider now the effect of the integrator during the equalizing and vertical pulses.

Fig. A.2 (overleaf) shows the voltage across integrating capacitor (even field).

228   INDUSTRIAL AND COMMERCIAL CCTV

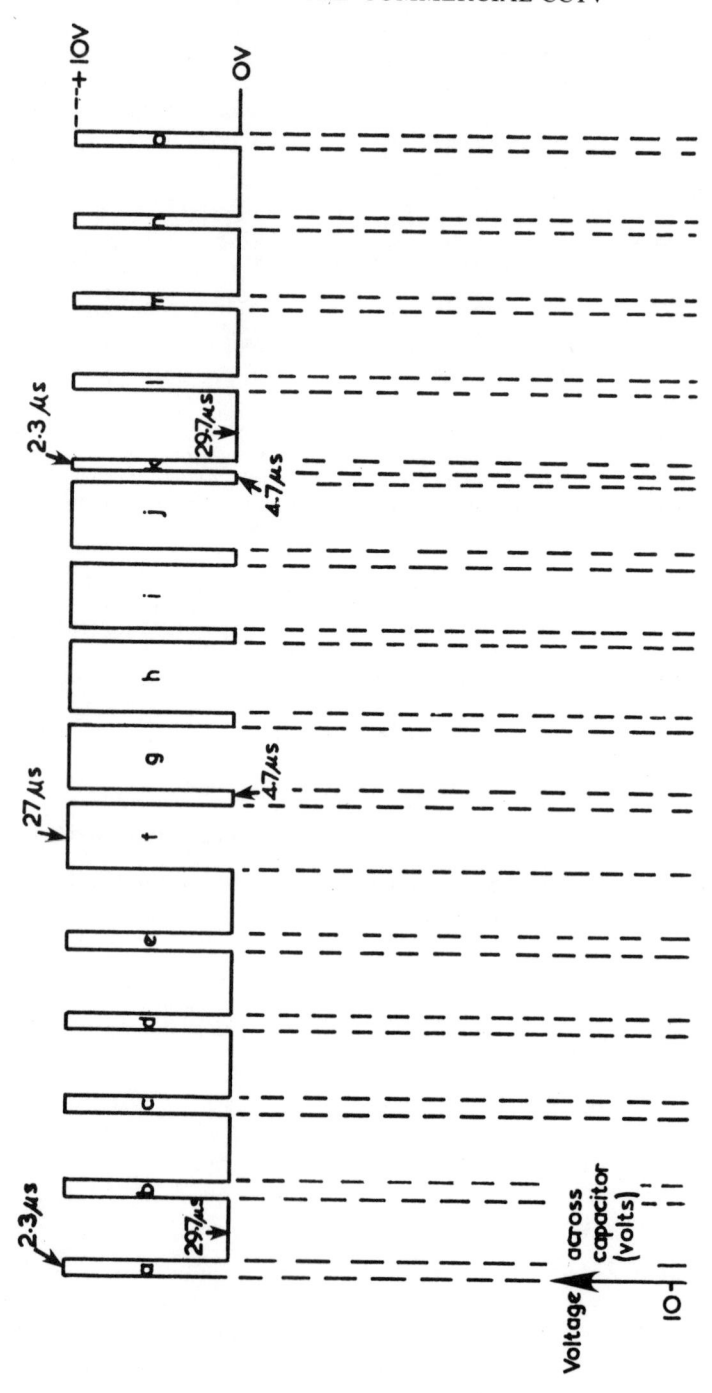

APPENDICES 229

FIG. A.2 VOLTAGE ACROSS INTEGRATING CAPACITOR (EVEN FIELD)

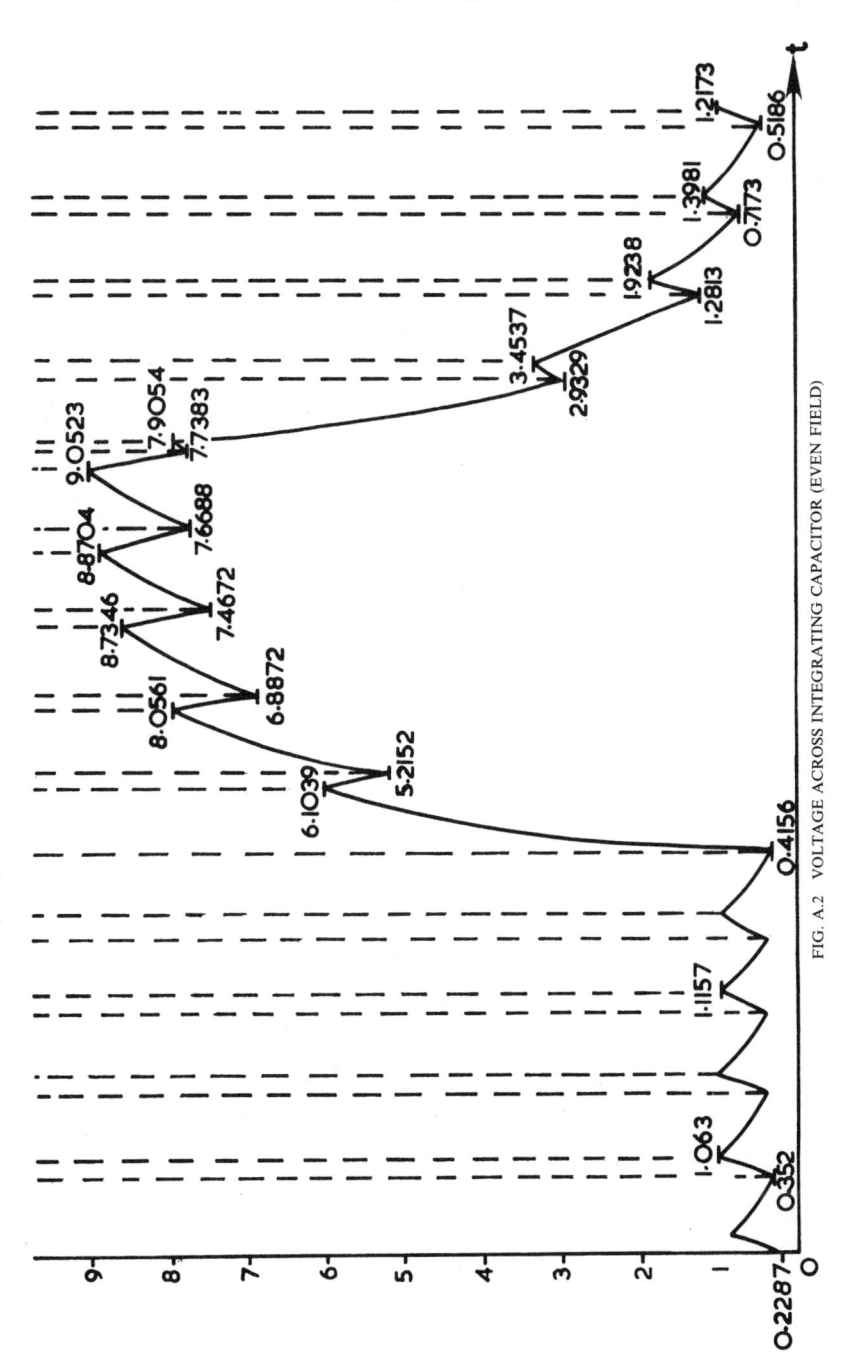

EVEN FIELD

Equalizing Pulses (a—e)

$$\left(1 - e^{\frac{-t_1}{CR}}\right) = 1 - \frac{1}{e^{\frac{2\cdot 3}{30}}} = 1 - 0\cdot 9263 = \underline{0\cdot 0737}$$

$$\left(e^{\frac{-t_2}{CR}}\right) = \frac{1}{e^{\frac{29\cdot 7}{30}}} = \underline{0\cdot 371}$$

Pulse a
Charge $V_C = 0\cdot 2287 + 9\cdot 7713\,(0\cdot 0737)$
              $V_C = \underline{0\cdot 9488\text{ V}}$
Discharge $V_C = 0\cdot 9488\,(0\cdot 371)$
              $V_C = \underline{0\cdot 3520\text{ V}}$

Pulse b
Charge $V_C = 0\cdot 3520 + 9\cdot 648\,(0\cdot 0737)$
              $V_C = \underline{1\cdot 063\text{ V}}$
Discharge $V_C = 1\cdot 063\,(0\cdot 371)$
              $V_C = \underline{0\cdot 3943\text{ V}}$

Pulse c
Charge $V_C = 0\cdot 3943 + 9\cdot 6056\,(0\cdot 0737)$
              $V_C = \underline{1\cdot 1022\text{ V}}$
Discharge $V_C = 1\cdot 1022\,(0\cdot 371)$
              $V_C = \underline{0\cdot 4089\text{ V}}$

Pulse d
Charge $V_C = 0\cdot 4089 + 9\cdot 5911\,(0\cdot 0737)$
              $V_C = \underline{1\cdot 1157\text{ V}}$
Discharge $V_C = 1\cdot 1157\,(0\cdot 371)$
              $V_C = \underline{0\cdot 4139\text{ V}}$

Pulse e
Charge $V_C = 0\cdot 4139 + 9\cdot 5861\,(0\cdot 0737)$
              $V_C = \underline{1\cdot 1203\text{ V}}$
Discharge $V_C = 1\cdot 1203\,(0\cdot 371)$
              $V_C = \underline{0\cdot 4156\text{ V}}$

Vertical Pulses (f—j)

$$\left(1 - e^{\frac{-t_1}{CR}}\right) = 1 - \frac{1}{e^{\frac{27}{30}}} = \underline{0\cdot 5935}$$

$$\left(e^{\frac{-t_2}{CR}}\right) = \frac{1}{e^{\frac{4\cdot 7}{30}}} = \underline{0\cdot 8549}$$

APPENDICES

Pulse f

Charge $V_C = 0.4156 + 9.5844 \, (0.5935)$
               $V_C = \underline{6.1039 \text{ V}}$

Discharge $V_C = 6.1039 \, (0.8549)$
               $V_C = \underline{5.2182 \text{ V}}$

Pulse g

Charge $V_C = 5.2182 + 4.7818 \, (0.5935)$
               $V_C = \underline{8.0561 \text{ V}}$

Discharge $V_C = 8.0561 \, (0.8549)$
               $V_C = \underline{6.8872 \text{ V}}$

Pulse h

Charge $V_C = 6.8872 + 3.1128 \, (0.5935)$
               $V_C = \underline{8.7346 \text{ V}}$

Discharge $V_C = 8.7346 \, (0.8549)$
               $V_C = \underline{7.4672 \text{ V}}$

Pulse i

Charge $V_C = 7.4672 + 2.5328 \, (0.5935)$
               $V_C = \underline{8.9704 \text{ V}}$

Discharge $V_C = 8.9704 \, (0.8549)$
               $V_C = \underline{7.6688 \text{ V}}$

Pulse j

Charge $V_C = 7.6688 + 2.3312 \, (0.5935)$
               $V_C = \underline{9.0523 \text{ V}}$

Discharge $V_C = 9.0523 \, (0.8549)$
               $V_C = \underline{7.7388 \text{ V}}$

Equalizing Pulses (k—o)

$$\left(1 - e^{\frac{-t_1}{CR}} \right) = \underline{0.0737}$$

$$\left(e^{\frac{-t_2}{CR}} \right) = \underline{0.371}$$

Pulse k

Charge $V_C = 7.7388 + 2.2612 \, (0.0737)$
               $V_C = \underline{7.9054 \text{ V}}$

Discharge $V_C = 7.9054 \, (0.371)$
               $V_C = \underline{2.9329 \text{ V}}$

Pulse l

Charge $V_C = 2.9329 + 7.0671 \, (0.0737)$
               $V_C = \underline{3.4537 \text{ V}}$

Discharge $V_C = 3.4537 \, (0.371)$
               $V_C = \underline{1.2813 \text{ V}}$

Pulse m

Charge $V_C = 1{\cdot}2813 + 8{\cdot}7187\,(0{\cdot}0737)$
$V_C = \underline{1{\cdot}9238\text{ V}}$

Discharge $V_C = 1{\cdot}9238\,(0{\cdot}371)$
$V_C = \underline{0{\cdot}7137\text{ V}}$

Pulse n

Charge $V_C = 0{\cdot}7137 + 9{\cdot}2863\,(0{\cdot}0737)$
$V_C = \underline{1{\cdot}3981\text{ V}}$

Discharge $V_C = 1{\cdot}3981\,(0{\cdot}371)$
$V_C = \underline{0{\cdot}5186\text{ V}}$

Pulse o

Charge $V_C = 0{\cdot}5186 + 9{\cdot}4814\,(0{\cdot}0737)$
$V_C = \underline{1{\cdot}21737\text{ V}}$

The results for an even field are as shown in figure A.2. The voltage across the integrating capacitor on an odd field may be determined in the same way but using a starting level of 0·662 V (see page 227).

APPENDICES

Figure A.3 overleaf shows the voltage waveform for an odd field. Comparison of the two diagrams indicates that the magnitude of the voltage across the integrating capacitor is very closely matched for the two fields just prior to the start of the first field pulse. In consequence, for engineering purposes, the two field locking pulses are identical.

234 INDUSTRIAL AND COMMERCIAL CCTV

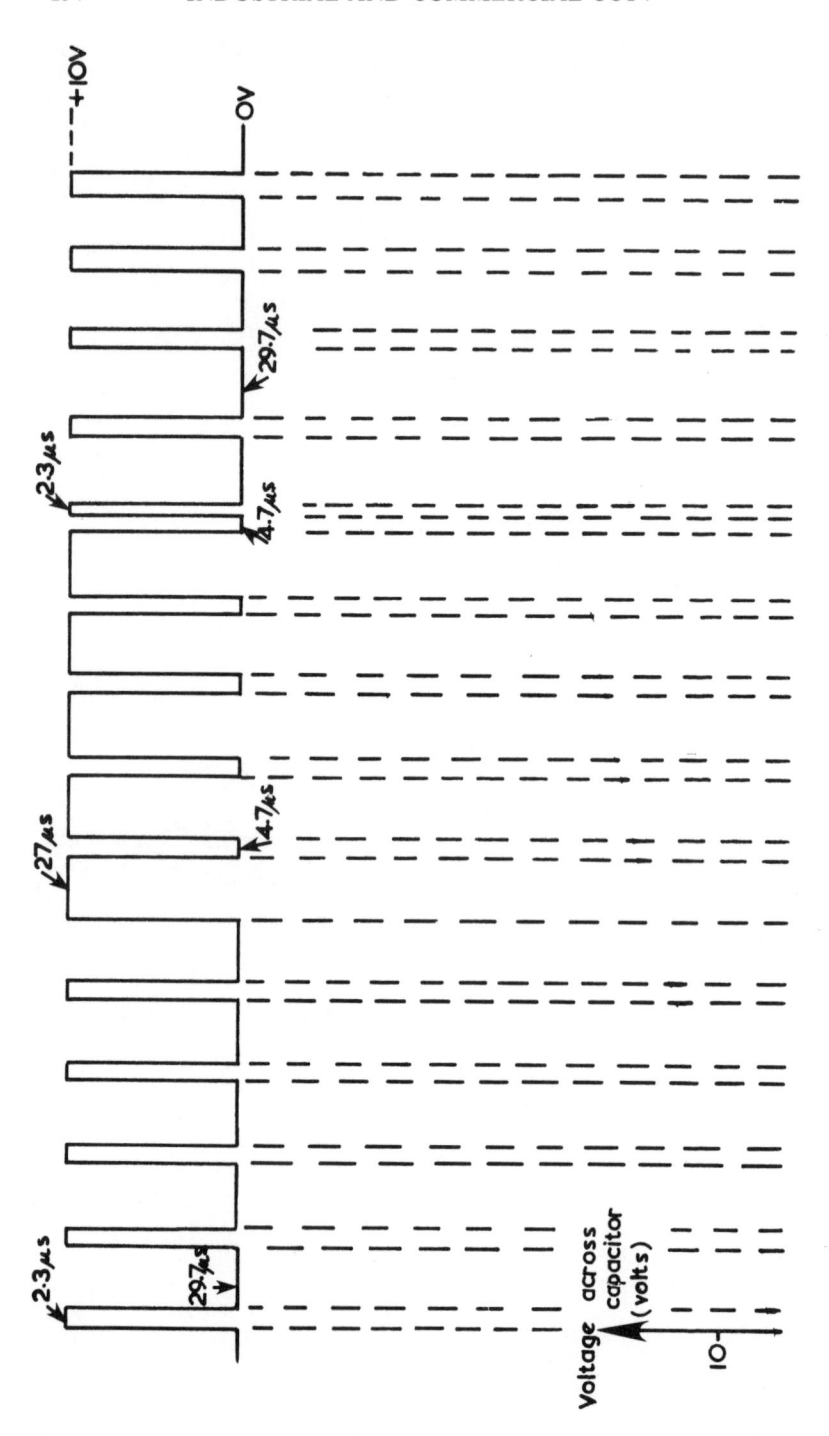

APPENDICES 235

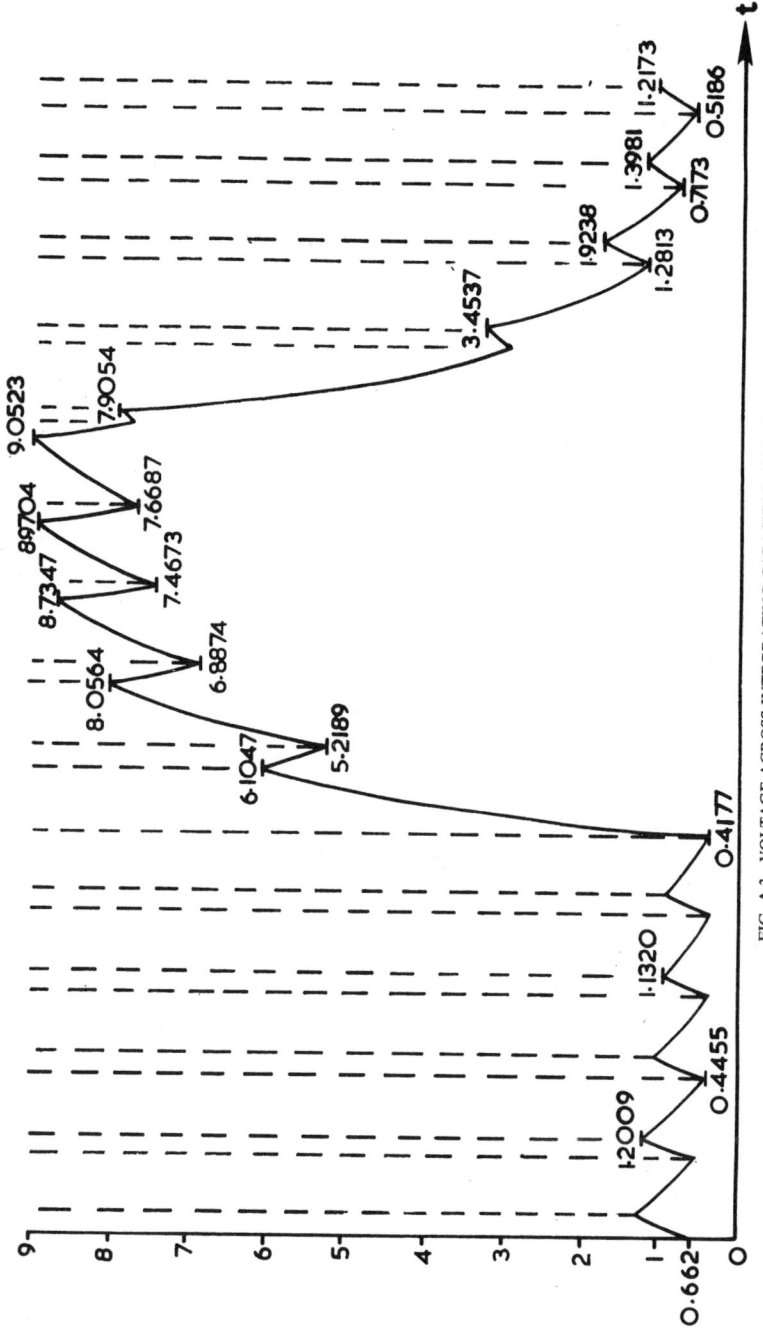

FIG. A.3 VOLTAGE ACROSS INTEGRATING CAPACITOR (ODD FIELD)

APPENDIX B

REACTANCE TRANSISTOR STAGE

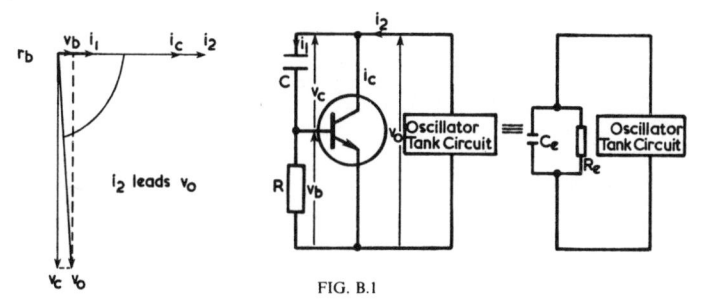

FIG. B.1

Considering figure B.1 where v_o is the oscillator tank circuit voltage which is applied across the RC network and transistor. If

$$\frac{1}{wC}$$

is large compared with R the current i_1 flowing in the network will lead v_o by almost 90°. The voltage v_b developed across R will be in phase with i_1 and this is the input voltage to the transistor. The voltage across R will be give rise to a collector current i_c in phase with v_b and of magnitude approximately gm v_b. Thus the resultant current i_2 leads v_o by almost 90°, *i.e.* the transistor and its RC network is equivalent to a capacitor C_e and a parallel resistance R_e connected in shunt with the oscillator tank circuit. By varying the gm of the transistor, *i.e.* by altering the d.c. base bias, the magnitude of i_c will be altered which thus varies the effective C_e and frequency of the oscillator tank circuit.

An analysis of the circuit using a simplified transistor equivalent circuit is as follows:

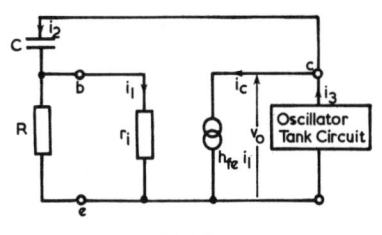

FIG. B.2

From figure B.2 $\quad i_2 = \dfrac{v_o}{x - \dfrac{j}{wC}} \quad$ where $x = \dfrac{Rr_i}{R + r_i}$

Therefore $i_1 = \dfrac{yv_o}{x - \dfrac{j}{wC}} \quad$ where $y = \dfrac{R}{R + r_i}$

Now $i_c = h_{fe} \, i_1 = \dfrac{yv_o \, h_{fe}}{x - \dfrac{j}{wC}}$

Thus $i_3 = \dfrac{yv_o\, h_{fe}}{x - \dfrac{j}{wC}} + \dfrac{v_o}{x - \dfrac{j}{wC}}$

or $\dfrac{i_3}{v_o} = \dfrac{yh_{fe} + 1}{x - \dfrac{j}{wC}} = Y_e$

Rationalising $Y_e = \dfrac{xyh_{fe} + x + \dfrac{jyh_{fe}}{wC} + \dfrac{j}{wC}}{x^2 + \dfrac{1}{w^2C^2}}$

Separating into real and unreal parts

$Y_e = \dfrac{x(yh_{fe} + 1)}{x^2 + \dfrac{1}{w^2C^2}} + \dfrac{\dfrac{j(yh_{fe} + 1)}{wC}}{x^2 + \dfrac{1}{w^2C^2}}$

Multiplying numerators and denominators by w^2C^2

$Y_e = \dfrac{w^2C^2 x(yh_{fe} + 1)}{w^2x^2C^2 + 1} + \dfrac{jwC(yh_{fe} + 1)}{w^2x^2C^2 + 1}$

Therefore $R_e = \dfrac{w^2x^2C^2 + 1}{w^2C^2 x(yh_{fe} + 1)}$ or when $w^2x^2C^2 \ll 1$

$R_e = \dfrac{1}{w^2C^2 x(yh_{fe} + 1)}$ or $\dfrac{1}{w^2C^2 \dfrac{Rr_i}{R + r_i}\left(\dfrac{Rh_{fe}}{R + r_i} + 1\right)}$

and $wC_e = \dfrac{wC(yh_{fe} + 1)}{w^2x^2C^2 + 1}$

or $C_e = \dfrac{C(yh_{fe} + 1)}{1}$ when $w^2x^2C^2 \ll 1$

or $C_e = C\left(\dfrac{Rh_{fe}}{R + r_i} + 1\right)$

APPENDICES

Thus the transistor and its network is equivalent to a high resistance R_e in parallel with a capacitance C_e, the magnitude of which may be adjusted by altering the gm of the transistor. R_e does not affect the oscillator tuning but causes some slight damping.

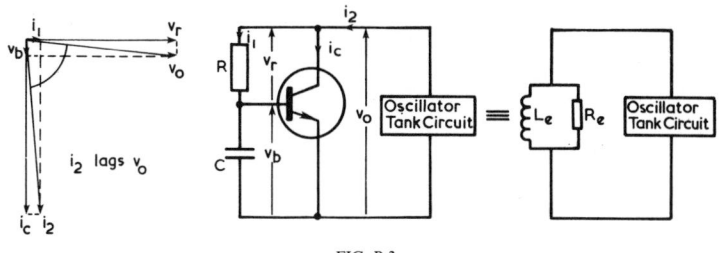

FIG. B.3

If R and C are interchanged as in figure B.3, the circuit is equivalent to an inductance *i.e.* in parallel with a high resistance R_e providing

$$R \gg \frac{1}{wC}$$

In this arrangement a suitable blocking capacitor must be inserted between R and the collector of the transistor.

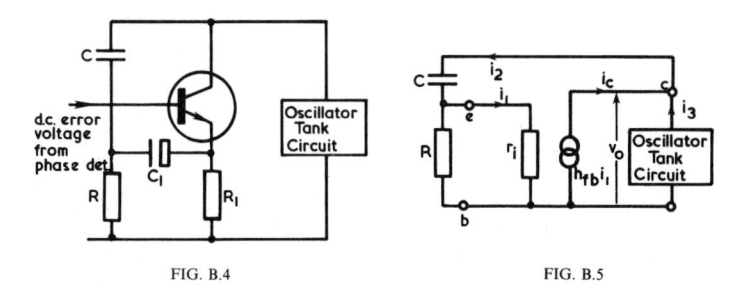

FIG. B.4   FIG. B.5

Instead of connecting the feedback network to the base of the transistor it may be connected to the emitter as in figure B.4. In this arrangement, the transistor and RC network presents a capacitive reactance to the oscillator tank circuit as for the configuration in figure B.1. However, the transistor *itself* viewed from the collector presents an inductive reactance to the oscillatory voltage. This may be explained with reference to figure B.5 which shows a simplified equivalent circuit but with the emitter resistor and the d.c. blocking capacitor C_1 omitted.

Now $i_2 = \dfrac{v_o}{x - \dfrac{j}{wC}}$ and $i_1 = \dfrac{yv_o}{x - \dfrac{j}{wC}}$ where $x = \dfrac{Rr_i}{R + r_i}$ and $y = \dfrac{R}{R + r_i}$

$\therefore\ i_c = -h_{fb}\, i_1$ (common base)

or $i_c = \dfrac{-h_{fb}\, y\, v_o}{x - \dfrac{j}{wC}}$

Thus $\dfrac{i_c}{v_o} = \dfrac{-y\ h_{fb}}{x - \dfrac{j}{wC}} = Y_t$ (where Y_t is the admittance of the transistor)

Rationalising we have $Y_t = \dfrac{-xy\ h_{fb} - jy\ h_{fb}}{x^2 + \dfrac{1}{w^2C^2}}$

Separating into real and unreal parts $\quad Y_t = \dfrac{-xy\ h_{fb}}{x^2 + \dfrac{1}{w^2C^2}} \quad \dfrac{\dfrac{jy\ h_{fb}}{wC}}{x^2 + \dfrac{1}{w^2C^2}}$

Multiplying numerators and denominators by w^2C^2 and when $x^2w^2C^2 \ll 1$ $\qquad Y_t = -xyw^2C^2h_{fb} - jy\ h_{fb}wC$

Thus $\quad R_t = -\dfrac{1}{xyw^2C^2h_{fb}}$

and $\quad L_t = \dfrac{1}{y\ h_{fb}w^2C}$

where R_t and L_t are the equivalent resistance and inductance of the transistor itself viewed from the collector.

APPENDIX C

MAXIMUM VIEWING DISTANCE

With average viewing conditions the eye can perceive an object distinctly with a viewing angle of about 3 minutes of arc, figure C.1. The distance of the eye from a particular displayed scene will thus determine the minimum height of any object in the scene that can be clearly resolved. When a person moves away from the monitor screen the minimum object height increases. One of the most important factors affecting the acuity of the eye is the contrast that an object makes with its background. Under the best viewing conditions the eye has an acuity of about 1 minute of arc.

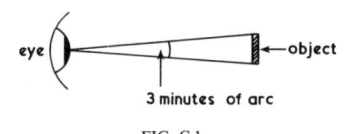

FIG. C.1

It is required to determine the maximum viewing distance from a reproduced object of particular dimensions in order to clearly discern the object, for different screen sizes. To establish a suitable expression the following will be used:

Let x = object height, h = height of picture, d = distance of observer from monitor screen and D = screen size (diagonal measurement)
All dimensions in the same units.

Let the ratio $\dfrac{x}{h} = A$

Now since the useful picture height is 0·63 times the diagonal measurement for the average tube,

$$\frac{x}{0\cdot 63\, D} = A$$

or $x = 0\cdot 63\ AD$ .... **(i)**

Considering figure C.2, the maximum viewing distance d may now be found.

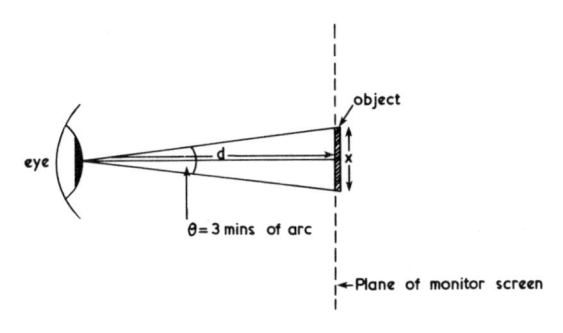

FIG. C.2

Now $\tan \dfrac{\theta}{2} = \dfrac{x}{2d}$

or $\tan 1\cdot 5 \text{ mins} = \dfrac{x}{2d}$

or $d = \dfrac{x}{2 \tan 1\cdot 5 \text{ mins}}$ .... (ii)

Substituting equation (i) for x in equation (ii)

we have $d = \dfrac{0\cdot 63\ AD}{2 \tan 1\cdot 5 \text{ mins}}$

$d = 721\cdot 93\ AD$

or $d \simeq \underline{722\ AD}$

A table of viewing distances for various screen sizes based on this expression is given on page 150.

APPENDIX D

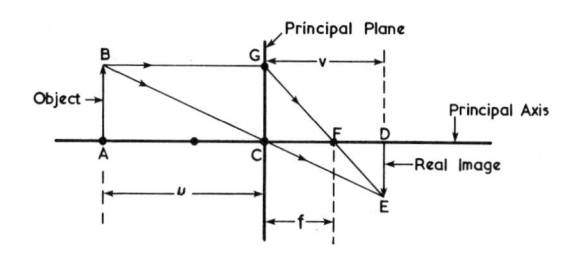

THE LENS EQUATION $\left(\text{Proof of } \dfrac{1}{f} = \dfrac{1}{u} + \dfrac{1}{v}\right).$

FIG. D.1

Considering figure D.1, the triangles CED and CAB are similar.

$$\text{Therefore } \quad \frac{DE}{AB} = \frac{DC}{AC} \qquad \text{(i)}$$

Also triangles DEF and CGF are similar.

$$\text{Thus } \quad \frac{DE}{CG} = \frac{DF}{CF} \qquad \text{(ii)}$$

But $CG = AB$, therefore the right-hand sides of equations (i) and (ii) are equal, i.e.

$$\frac{DC}{AC} = \frac{DF}{CF}$$

Now $DC = v$, $AC = u$, $CF = f$ and $DF = v-f$

$$\text{Thus } \quad \frac{v}{u} = \frac{v-f}{f}$$

$$\text{or } \quad vf = uv - uf$$

Dividing both sides by uvf and simplifying we have

$$\frac{1}{u} = \frac{1}{f} - \frac{1}{v}$$

$$\text{or } \quad \frac{1}{f} = \frac{1}{u} + \frac{1}{v}$$

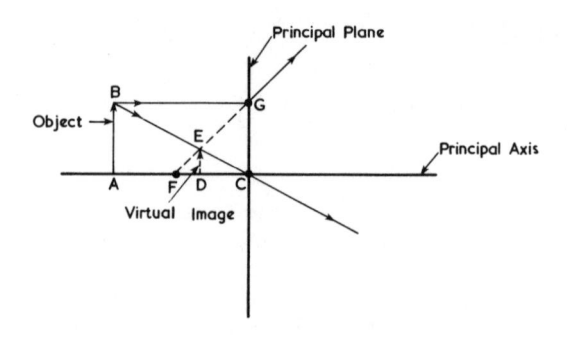

FIG. D.2

For the diverging lens, figure D.2, the triangles *CED* and *CAB* are similar also triangles *DEF* and *CGF*.

Thus $\dfrac{DC}{AC} = \dfrac{DF}{CF}$ (Since $CG = AB$)

Now for the diverging lens $DC = -v$, $AC = u$, $CF = -f$ and $DF = -f - (-v)$

Therefore $\dfrac{-v}{u} = \dfrac{-f+v}{-f}$

or $vf = uv - fu$

Again, by dividing both sides by uvf and simplifying we have

$$\dfrac{1}{f} = \dfrac{1}{u} + \dfrac{1}{v}$$ as for the converging lens.

BIBLIOGRAPHY

(1) Nelkon, M. "Light and Sound", Heinemann

(2) Daish, C. B. "Light", English Universities Press

(3) Amos, S. W. and Birkinshaw, D. C. "Television Engineering", Iliffe

(4) Hutson, G. H. "Colour Television Theory", McGraw Hill

(5) Millerson, G. "TV Camera Operation", Focal Press

(6) "Television Technology", British Kinematograph Sound and Television Society Education and Training

INDEX

Active Lines 72
Acuity of eye 74, 241–242
Additive colour mixing 14
Afterglow 141
Alignment coil, camera 77
Aluminium coating, c.r.t. 144
Aperture correction 100, 102
Aperture distortion 88
Aquadag, c.r.t. 144
Aspect ratio 73, 195
Astable multivibrator 182, 183
Astigmatism, c.r.t. 147
Automatic sensitivity control 134, 135

Back Porch 67
Barrel distortion, c.r.t. 152–154
Beam control 77
Beam current limiter, monitor 167
Bistable multivibrator 105–107
Black body radiator 18
Black level 66, 75, 99
Blanking level 66, 75
Blanking pulse
  generation 122–124, 128
Blocking oscillator 63
Brightness control 155, 156

Camera Adjustments 196–199
Camera, block diagram 58
Camera, test-bed 193
Camera tube, chalnicon 91
Camera tube, high sensitivity 91
Camera tube, image burns 80
Camera tube, plumbicon 84–86
Camera tube, silicon diode
  vidicon 89, 90
Camera tube, vidicon 76–84
Candela 19
Capture range 173, 174
Cathode modulation, c.r.t. 154–157
C.C.I.R. 131
Characteristic resistance 205, 206
Coaxial cable 204–210
Colour C.C.T.V. 14
Colour filters 16
Colour mixing 14, 15, 16
Colours, primary 14
Colour temperature 18, 19
Contrast control 155, 156

Cross-hatch pattern 190
C.R.T. 141–145, 150–158
Crystal oscillator 63

Dark Current 78
D.C. clamp 99, 188
D.C. restorer 99
Definition
  of picture 69, 71, 72, 73, 74, 75, 195
Deflection centre, c.r.t. 145
Deflection of beam, c.r.t. 145–147
Diascope 195
Diffraction of light 9, 10
Dioptre 33
Direct sync. 173
Divider–625 110–114
Divide-by-five stage 111–114
Divide-by-two stage 105

E.H.T. 144, 179
E.H.T. regulation 179–181
Electrical focus of camera 77, 78, 84
Electromagnetic spectrum 2, 3
Electromagnetic waves 2
Engraved disc 197, 198
Equalizing pulses 69, 70, 170
External drive, camera 128, 130, 131
Eye, human 16

Fibre Optics 94
Field blanking 132, 158
Field frequency 61
Field output stage 118–122, 183, 184
Field scan 56
Field sync. pulse 67–70
Field timebase 57, 118–122, 181–183
Flash-over, c.r.t. 157
Flat picture 83
Floodlamps 200–203
Fluorescence 141
Flyback 56, 59, 60
Flyback blanking, monitor 185
Flywheel sync. 172–179
Focus regulation 138, 139
Focusing lens, c.r.t. 143
Fresnel lens 201
Front porch 66

GAMMA CORRECTION	86–88, 135, 138
Ghosting	209
Grey-scale	66, 196
Grid modulation, c.r.t.	154–157
HORIZONTAL HOLD, MONITOR	176–177
Hue	11
Human eye	16
Human eye response	12, 13, 17
ILLUMINATION	20, 203
Illumination, inverse square law	20
Illuminant D	18
Image burns	80
Image intensifier	92, 93, 94
Image position	28, 29
Infra-red t.v.	90
Integrating network, monitor	169–172, 225–235
Interlace errors	65, 101
Interlace ratio	62
Interlaced scanning	59, 60, 61
Ion burn, c.r.t.	144, 146
KELL FACTOR	73
Kelvin scale	19
Kinescope	141
LATERAL DISPLACEMENT OF LIGHT	6, 7
Lens, aberrations	42, 43, 44
Lens, angle	39, 40, 41
Lens, aperture	25, 36
Lens, care of	52
Lens, combination	33
Lens, concave	23, 24
Lens, convex	23, 24
Lens, depth of field	34, 45, 46
Lens, f-number (stop)	36
Lens, focal length	25, 27
Lens, focal plane	26
Lens, focusing	38, 39
Lens, hyperfocal distance	35
Lens, image position	28
Lens, magnification	29
Lens, narrow angle	45, 46
Lens, optical centre	25
Lens, power of	33
Lens, principal axis	25
Lens, principal focus	25
Lens, principal plane	25
Lens, remote control	48, 49, 52
Lens, secondary axis	25
Lens, telephoto	45, 46
Lens, vignetting	52
Lens, wide-angle	46
Lens, zoom	47, 48, 49
Light beams	5
Light units	19, 20, 21
Lighting	200–203
Line blanking	66, 131, 158
Line frequency	61
Line pairing	65
Line scan	55
Line scan output, camera	107–110
Line scan output, monitor	179
Line sync. pulse	67
Line timebase	56
Lines per active picture height	71–75
Lock range	173, 174
Long focus coil	81
Looping-through	162, 163
Lumen	19
Luminaires	201, 202
Luminance levels	13
Luminous flux	19
Luminous intensity	19
Lux	20
MAGNIFYING GLASS	27
Mains lock	62, 63, 115–118
Mask, scan defining	196–197
Mirror images	4, 5
Mixed blanking	128, 131
Mixed sync.	131
Modulation	75, 76
Modulated r.f. output	139
Monitor, block diagram	58
Monitor picture geometry	190–193
Monostable multivibrator	124–127
ORTHOGONAL SCANNING	81, 82
Oscillator, blocking	63, 105
Oscillator, crystal	63, 104
Oscillator, Colpitts	105
Oscillator, master	63, 104, 105
Overshoot	97, 102, 165

INDEX

PEAK WHITE	66, 99
Pedestal (set-up)	67, 131
Persistance of vision	18, 55, 141
Phase detector	174–176
Phosphorescence	141
Picture Elements	53, 54
Pin-cushion distortion, c.r.t	152–154
Plastic mask	192
Plumbicon tube	84–86
Power supplies, monitor	159–162
RANDOM INTERLACE	62, 64, 65
Raster	60, 145
Raster correction magnets	152–154
Reactance stage	177, 178, 237–240
Real image	27
Rectilinear propagation	3
Reduced scan	184, 185
Reflection factor	22
Reflection of light	3, 4
Reflections, cable	207–209
Refraction of light	6, 7, 8
Refractive index	6
Resolution of picture	69, 71, 72, 73, 74, 75, 195
Ringing	97
Rounding	97
'S' CORRECTION	151, 152, 179
Sag	97
Saturation	11
Scan coils, camera	82, 109, 110
Scan coils, monitor	145–149
Scanning, interlaced	59, 60, 61
Scanning, simple	55
Screen size, c.r.t.	150, 151
Set-up	67, 99
Sequential transmission	55
Shift magnets, c.r.t.	152
Spark-gaps, monitor	157
Spectral colours	10, 11
Spectral response, camera tubes	90, 91, 92
Spot-light	200, 201

Standard output, camera	67
Step counter	115, 116
Streaking	97, 196
Subtractive colour mixing	14, 15, 16
Switch-off burns, monitor	185, 187
Sync. mixer	127, 128
Sync. pulse amplifier, monitor	168, 169
Sync. pulse generator	104, 124–126
Sync. pulse separation, monitor	167–169
Sync. pulses	57
TARGET CONTROL	83
Target limit	83
Target ring	78
Test chart	194, 195
Test equipment	212
Total internal reflection	7
Transmission factor	22
Transmission line	204
UNDERSHOOT	97, 102
VERTICAL AMPLITUDE	121, 122
Vertical linearity	121, 122
Video amplifiers, camera	95–102
Video amplifiers, monitor	162–167
Video distribution amplifier	163
Video signal	59, 65, 66
Video switching box	211, 212
Vidicon tube	76–84
Vidicon tube, integral mesh	78
Vidicon tube, lag	82
Vidicon tube, separate mesh	78
Viewfinder	139, 140, 165
Virtual image	4, 27
WAVELENGTH	2
Wavefront	2
White crush	136
White light	3, 18
White stretch	136